2019年四川省重点出版专项资金资助项目
"一带一路"输电线路建设管理专业丛书
OBOR Transmission Line Construction and Management - Book Series

带电作业工器具的检查、使用及保管

Inspection, Use and Storage of Tools and Instruments for Live Working

丛书主编／汤晓青
本册主编／杨　力
副 主 编／郑和平　全昌前　赵世林
编　　者／杜印官　魏　欣　陈劲松　陈　立
　　　　　谢勇泉　巫春玲　郑　勇
译　　者／成都优译信息技术股份有限公司
　　　　　孙语婷

电子科技大学出版社
University of Electronic Science and Technology of China Press

·成都·

图书在版编目(CIP)数据

带电作业工器具的检查、使用及保管 / 杨力主编. —— 成都：电子科技大学出版社，2019.12
ISBN 978-7-5647-7546-9

Ⅰ.①带… Ⅱ.①杨… Ⅲ.①带电作业工具 – 检修②带电作业工具 – 使用方法 Ⅳ.①TS914.53

中国版本图书馆CIP数据核字（2019）第287051号

带电作业工器具的检查、使用及保管
DAIDIANZUOYE GONGQIJU DE JIANCHA SHIYONG JI BAOGUAN

杨　力　主编

策划编辑	陈松明　陈　亮　罗国良	
责任编辑	高小红	
出版发行	电子科技大学出版社	
	成都市一环路东一段159号电子信息产业大厦九楼　邮编　610051	
主　页	www.uestcp.com.cn	
服务电话	028-83203399	
邮购电话	028-83201495	
印　刷	成都市火炬印务有限公司	
成品尺寸	185mm×260mm	
印　张	28.5	
字　数	742千字	
版　次	2019年12月第一版	
印　次	2019年12月第一次印刷	
书　号	ISBN 978-7-5647-7546-9	
定　价	114.00元	

版权所有，侵权必究

前　言

"'一带一路'输电线路建设管理专业丛书"共分十册，该丛书全面系统地介绍了国家电网公司在输电线路上的建设、运行及维护知识。每册图书都由中文版和英译文版组成，可作为"一带一路"上的国家和地区的电力职业教育规划教材，职业教育电力技术类专业培训用书。该丛书的出版对于推动"一带一路"的输电线路管理具有重要的意义。

自1954年鞍山地区的带电作业方法和工器具试验成功以来，带电作业技术已在我国电网运检工作中得到了广泛应用。随着带电作业人员的持续探索和创新，带电作业方法及工器具也得到了不断更新。为了规范带电作业工器具的生产、试验和使用，国家、行业制定和颁布了四十余个带电作业工器具的标准和规范，显著提升了带电作业工器具的制造质量和使用水平。

目前，带电作业相关书籍中介绍带电作业工器具的内容很少、资料不全且缺乏系统性，广大带电作业人员一直希望有一本内容全面、资料翔实、图文并茂的带电作业工器具专用书籍来指导实际工作，同时国内十余个输配电线路带电作业培训基地的教师和广大培训学员也迫切需要一本带电作业工器具培训教材。因此，编者从2011年开始着手搜集和整理输配电线路带电作业工器具型号、规格和使用的相关资料，拟定编写提纲，并对编写内容多次进行讨论和征求从业人员意见，不断改进完善，最终完成了本书的编写工作。

全书共8章，系统介绍了带电作业工器具的发展历程、材料、基本类型和试验、金属工器具、绝缘工器具、带电作业特种车辆、安全防护用具、带电作业常用仪器仪表、安全及绝缘手工工具和带电作业工器具的保管和运输等内容。针对国内外输配电线路带电作业常用的工器具，本书按照结构、分类、功能、检查、使用、保管存放、试验周期的体例来编写。在编写中严格遵循国家、行业颁布的带电作业工器具的标准、规程和规范，每种工器具都提供了适量的图片，以增加全书的可读性和实用性，

并适当体现当前带电作业工器具发展的新技术和新工艺水平。全书内容详尽、通俗易懂、图文并茂，具有很强的针对性和实用性。

本书的英文翻译主要由成都优译信息技术股份有限公司完成，仅有第7章安全及绝缘手工工具由孙语婷翻译。在此，对以上翻译单位和个人表示感谢。

由于编者知识、技能水平有限，书中尚有不足之处，恳请广大读者批评指正。

编 者

2019年9月

Foreword

OBOR Transmission Line Construction and Management - Book Series totally include ten volumes. The books comprehensively and systematically introduce the knowledge of the State Grid in the construction, operation and maintenance of power transmission lines. Each volume of the books consists of Chinese version and English translation version, which can be used as the electric power vocational education planning textbooks for the countries and regions on the "Belt and Road", and the professional training books for vocational education in electric power technologies. The publication of the book series is of great significance for promoting the management of the power transmission lines on the "Belt and Road".

Since the successful test of methods, tools and instruments for live working in Anshan region in 1954, the live working technologies have been widely applied in the operation and maintenance of power grids in China. With the continuous discovery and innovation by the live working personnel, the methods, tools and instruments for live working have been updated continuously.To standardize the production, test and application of tools and instruments for live working, the country and the industry have formulated and issued over forty standards and specifications for live working tools and instruments, which have significantly improved the manufacturing quality and usage level of live working tools and instruments.

At present, the books related to live working involve fewer contents introducing the tools and instruments for live working, and the information therein is incomplete and short of systematicness. The live working personnel have expected a dedicated book for live working tools and instruments complete in content, detailed and accurate information, and illustrated to guide their practical operations. Meanwhile, the teachers and trainees are a dozen of training bases for live working on power transmission and distribution lines in China is also eager for a training material for live working tools and instruments.Therefore, from 2011, the editor set about collecting and sorting the related data of the type, specification and usage of live working tools and instruments for power transmission and distribution lines, prepared the outline, discussed the contents and sought for opinions from related personnel, continuously improved the contents, and finally completed the edit of this book.

The whole book consists of 8 chapters, which systematically introduce the development history, materials, basic types and test of live working tools and instruments, metal tools and instruments, insulated tools and instruments, special vehicles for live working, safety protection equipment, instruments and apparatus frequently used in live working, keeping and transport of safety and insulated hand tools and live working tools and instruments, etc.For tools and instruments frequently used in the live working on power transmission and distribution lines in and out of China, the compilation is based on the style of structure, classification, functions, examination, usage, keeping and storage, and test period.In the compilation, the national and industrial standards, procedures and specifications for live working tools and instruments are strictly followed. Pictures in appropriate quantity are provided for each type of tool or instrument, so as to improve the readability and practicability of the whole book, and properly embody the current new technologies and new technological level in the development of live working tools and instruments.The whole book is detailed, popular and easy to understand, illustrated, and is strong in pertinence and practicability.

This Book is mainly translated by Chengdu UE Information Technology Corp., except Chapter 7 "Safety and Insulating Hand Tools" translated by Sun Yuting. Thanks are hereby given to the above company and person.

Due to the limited knowledge and skill level of the editor, there are still many places that can be improved in the book, and the criticism and correction from readers are appreciated.

Editor
September 2019

目 录

第1章 带电作业工器具概述 ·· 1
1.1 带电作业工器具的发展历程 ·· 1
1.2 带电作业工器具的材料 ·· 11
1.3 带电作业工器具基本类型 ·· 12
1.4 带电作业工器具的试验 ·· 14

第2章 金属工器具 ·· 24
2.1 绝缘子卡具及卡线器 ··· 24
2.2 滑车及飞车 ··· 37

第3章 绝缘工器具 ·· 46
3.1 硬质绝缘工具 ··· 46
3.2 软质绝缘工具 ··· 59
3.3 其他绝缘工具 ··· 67

第4章 带电作业特种车辆 ·· 74
4.1 绝缘斗臂车 ··· 74
4.2 旁路作业车及负荷开关车 ·· 83
4.3 移动箱变车 ··· 89
4.4 带电作业工具车 ··· 95

第5章 安全防护用具 ·· 101
5.1 绝缘防护用具 ··· 101

  5.2 绝缘遮蔽用具 ·· 117
  5.3 屏蔽用具 ·· 127

第6章 带电作业常用仪器仪表 ·· 133

  6.1 火花间隙检测装置 ·· 133
  6.2 绝缘电阻测试仪 ·· 138
  6.3 钳形电流表 ·· 146
  6.4 高压核相仪 ·· 150
  6.5 风速仪、温湿度仪 ·· 153

第7章 安全及绝缘手工工具 ·· 158

  7.1 安全工具 ·· 158
  7.2 绝缘手工用具 ·· 171

第8章 带电作业工器具的保管和运输 ································ 184

  8.1 带电作业工器具库房基本要求 ································ 184
  8.2 带电作业工器具库房管理系统 ································ 189
  8.3 带电作业工器具的存放方法 ·································· 197
  8.4 带电作业工器具的保管和运输 ································ 200

参考文献 ·· 202

CONTENTS

Chapter 1 Overview of Tools and Instruments for Live Working ········203

    1.1 Development History of Tools and Instruments for Live Working ············203

    1.2 Materials of Tools and Instruments for Live Working ·····························216

    1.3 Basic Types of Tools and Instruments for Live Working ·························218

    1.4 Tests of Tools and Instruments for Live Working ································220

Chapter 2 Metal Tools and Instruments ································232

    2.1 Insulator Clamp and Wire Clamp ···232

    2.2 Tackle and Hanging Wheel ···248

Chapter 3 Insulating Tools and Instruments ························259

    3.1 Hard Insulating Tools ···259

    3.2 Soft Insulating Tools ··274

    3.3 Other Insulating Tools ···283

Chapter 4 Special Vehicles for Live Working ·······················291

    4.1 Aerial Device with Insulating Booms ···291

    4.2 Bypass Work Vehicle and Load Switch Vehicle ·····································302

    4.3 Mobile Box-type Substation Vehicle ··309

    4.4 Live Working Tool Car ··317

Chapter 5 Safety Protection Devices ···································324

    5.1 Insulation Protective Equipment ···324

    5.2 Insulated Shielding Appliance ···344

  5.3 Shield Appliance ··357

Chapter 6 Common Instruments and Apparatus for Live Working ······364

  6.1 Spark Gap Detector ··364
  6.2 Insulation Resistance Tester ··369
  6.3 Clamp Ammeter··378
  6.4 High Voltage Phase Detector ··383
  6.5 Anemometer and Thermohygrometer ··387

Chapter 7 Safety and Insulating Hand Tools ··392

  7.1 Security Tools···392
  7.2 Insulating Hand Tools ···408

Chapter 8 Storage and Transportation of Tools and Instruments for Live Working ··422

  8.1 Basic Requirements of Live Working Tools and Instruments Warehouse ·········422
  8.2 Management System of Live Working Tools and Instruments Warehouse ······430
  8.3 Storage Method of Tools and Instruments for Live Working ····················440

References ···446

第1章

带电作业工器具概述

带电作业,是指作业人员使用高绝缘性能及机械强度的专业工器具,在带电运行的输配电线路及电气设备上,安全开展的运行维护工作。性能可靠的带电作业工器具是保证带电作业安全的必要条件。随着带电作业方法的普及,广大的从业人员研制出了出种类繁多、规格各异的带电作业工器具。目前,随着多条特高压交流1000kV、特高压直流±800kV输电线路工程的建成投运,制造和使用在特高压作业环境下开展带电作业需要的工器具迫在眉睫,并对其材料、性能和预防性试验提出了新的要求。

1.1 带电作业工器具的发展历程

带电作业的发展历史,也是带电作业工器具的发展历史。带电作业工器具是随着带电作业的开展而不断研发、使用和完善的。

1.1.1 中国带电作业开展及工器具使用历史

1. 带电作业工器具发展的第一个阶段

我国的带电作业技术起步于20世纪50年代,鞍山钢铁厂——当时我国最大的钢铁基地,对电力能源的依赖程度非常高,甚至电网检修必需的停电都不能实现。为解决电力线路要检修而用户又不能停电的矛盾,当时称之为"不停电检修技术"应运而生。

1953年,原鞍山电业局的工人开始研究带电清扫、更换和拆装配电设备及引线的简单工具。

1954年5月12日,在群众性的技术革新运动中,鞍山电业局技工刘长庚提出带电拆绑配电瓷瓶绑线的办法,还展示了他研制出的绝缘钳子、绝缘铁剪、瓷瓶线卡等工具,并提出不停电更换3.3kV线路直线杆、瓷瓶和横担技术分析报告和操作方案。在现场采用类似桦木木棒制作的工具，成功地完成了3.3kV配电线路不停电更换横担、木杆和瓷瓶的项目操作。他的技术创新得到了专家的认可，并通过了可行性论证。1954年5月12日，作为鞍山电业局带电作业创始日载入该局局志，也作为中国带电作业发展的开端载入中国电力发展的史册。

带电作业得到了鞍山电业局领导的高度重视，为了巩固和发展带电作业的初期成果，1956年6月，鞍山电业局正式成立了带电作业专业组，张仁杰任组长，刘长庚任专责工程师。进一步发展到带电更换44～66kV的木质直线杆、横担和绝缘子。

第一代工器具具有以下特征：使用瓷瓶、瓷套作为主绝缘工具，使用普通钢材及可锻铸铁作为金属部件，工器具笨重，缺乏通用性。如图1-1所示为使用第一代"升降涨缩"工具不停电更换66kV线路直线木杆操作，图1-2所示为使用第一代"绝缘夹板"工具不停电更换22～33kV线路耐张绝缘子串操作。

图1-1 不停电更换66kV线路直线木杆操作

图1-2 不停电更换22～33kV线路耐张绝缘子串操作

2. 带电作业工器具发展的第二个阶段

1957年10月,原东北电业局设计了第一套220kV高压输电线路带电作业工具,并成功应用于220kV高压输电线路的带电作业。同时3.3~33kV木杆和铁塔线路的全套检修工具,也得到了改进和完善,这就为各级电压线路推行不停电检修奠定了物质和技术基础。

1958年,当时的沈阳中心试验所开始了人体直接接触导线检修的试验研究。在学习国外经验的基础上,他们解决了高压电场的屏蔽问题,并在试验场成功地进行了我国第一次人体直接接触220kV带电导线的等电位试验。试验者为刘德威,首次在220kV线路上完成了等电位作业和修补导线的任务。这次等电位带电作业的试验成功,开创了中国带电作业的新篇章。从此,等电位作业技术在中国带电作业中得到了广泛的应用。

如图1-3所示为1958年鞍山电业局进行的220kV等电位试验实物图,图1-4为1958年鞍山电业局进行的220kV拆除避雷器引线操作。

图1-3　220kV等电位试验

图1-4　220kV拆除避雷器引线操作

1959年前后，鞍山局又在3.3～220kV户外输配电装置上，研究出了一套不停电检修变电设备的工具和作业方法。至此，中国带电作业技术已发展成为3.3～220kV包括输电、变电、配电三方面的综合性检修技术。

1958年3月，原辽宁省电业局在鞍山召开不停电检修线路会议，带电作业技术开始在辽宁省和东北电力系统的供电部门中推广。1958年4月12日，《人民日报》以《电力工业的重大技术革新——不停电检修电力线路》为题，报道了鞍山电业局带电作业技术试验成功。同年5月20日至7月1日，来自北京、上海、湖北、四川、陕西等省市供电部门的46名技术代表在鞍山接受了带电作业技术培训。那一年，全国各兄弟单位到鞍山参观学习带电作业技术的代表有5700多人。

1957—1960年使用的第二代带电作业工器具具有以下特征：使用酚醛木层压板、酚醛胶纸管、酚醛纸板、蚕丝制作主绝缘工具，使用普通钢材、可锻铸铁制作金属部件。此阶段的工器具笨重，有一定通用性。

3. 带电作业工器具发展的第三个阶段

1960年，原辽吉电管局制定了《高压架空线路不停电检修安全工作规程》，成为我国第一部具有指导性的带电作业规程。它标志着我国带电作业已步入正轨。在此期间，全国范围内的不停电检修工作从单纯的技术推广转入结合本地区具体条件和生产任务创新发展阶段。检修方法除了间接作业和直接等电位作业外，又向水冲洗、爆炸压接等方向迈进。检修工具从最初的支、拉、吊杆等较笨重的工具转向轻便化、绳索化。具有东方特色的绝缘软梯和绝缘滑车组也得到了广泛的应用。作业项目向更换导线、架空地线、移动杆塔和改造塔头等复杂项目进军。

1964年11月，在天津举行了带电检修作业表演，对促进全国范围内推广这些新技术产生了积极影响。

1966年，原水利电力部生产司在鞍山召开了全国带电作业现场观摩表演大会，标志着全国带电作业发展到普及阶段，同时也推动带电作业向更新更深的领域发展。

1968年，原鞍山电业局成功试验沿绝缘子串进入220kV强电场的新方法。用这种方法在具备一定条件的双联耐张绝缘子串上更换单片绝缘子很方便，因而，很快被推广到全国。

1973年，原水利电力部又在北京召开了第二次全国带电作业经验交流会。这次会议的技术组提出了带电作业安全技术有关专题的讨论稿，为制定全国性带电作业规程奠定了技术基础。

图1-5　更换220kV线路耐张整串绝缘子操作

图1-6　220kV线路上等电位检查，处理耐张引线触点发热等缺陷

图1-5所示为使用第三代带电作业工器具（紧线拉杆、托瓶架）更换220kV线路耐张整串绝缘子操作，图1-6所示为使用绝缘软梯在220kV线路上等电位检查、处理耐张引线接点发热等缺陷。

第三代带电作业工器具具有以下特征：使用环氧玻璃布层压板、环氧玻璃布管、玻璃丝引拔棒、尼龙绳、蚕丝绳制作主绝缘工具；使用合金钢材、可锻铸铁制作金属部件。此阶段带电作业工器具重量有所减轻，通用性得到加强。

4. 带电作业工器具后期的发展

1977年，原水利电力部将带电作业纳入部颁安全工作规程，进一步肯定了带电作业技术的安全性。同年，中国带电作业人员开始了国际交流，参加了国际电工委员会带电作业工作组的活动，成立了IEC/TC78标准国内工作小组，从事带电作业有关标准的制定工作。1984年5月，中国带电作业标准化委员会成立。

1979年，东北地区兴建了元宝山—锦州—辽阳—海城500kV输电线路，为确保500kV线路投产运行后安全经济供电，原鞍山电业局开展了500kV带电作业新技术的研究。为了攻破这个难题，张仁杰、葛元昌等退休员工又被聘请回来，进行500kV直线塔和耐张塔的带电作业研究工作。他们充分发扬当年不怕吃苦的精神，亲自制作工具、现场试验、编写操作规程。1984年，完成了500kV部分带电作业常规项目及工具的研制配套工作。此后不久，500kV带电更换直线绝缘子串、更换耐张绝缘子串、修补导线等工作方法和工器具均如期研制成功，并进入实施阶段。

2000年，原华北电网有限公司首次研究实现了500kV紧凑型线路带电作业。如图1-7所示为20世纪80年代锦州电业局采用"紧线拉杆—绝缘小车"更换500kV线路耐张整串绝缘子操作，如图1-8所示为东北地区采用的"沿绝缘子串进入强电场方法"更换500kV单片绝缘子操作图。

图1-7　更换500kV耐张整串耐张绝子操作

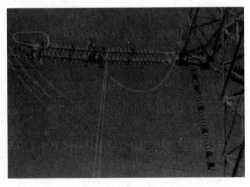

图1-8　更换500kV单片绝缘子操作

20世纪90年代，我国引进了绝缘斗臂车，解决了困扰配电线路带电作业的很多问题。目前，绝缘斗臂车在配电线路带电操作中已经得到广泛应用，且实现了工作电压10~500kV绝缘斗臂车国产化。

2007年，华北电网有限公司又研究实现500kV线路直升机带电作业，这一成果达到了国际领先水平。如图1-9所示为直升机等电位作业，图1-10所示为直升机带电水冲洗作业。

图1-9　直升机等电位带电作业

图1-10　直升机带电水冲洗作业

2006—2009年，国网上海市电力公司、国网北京电力公司研究并实施10kV配电线路完全不停电作业法。

2008—2009年，西北电网研究实施超高压交流750kV线路带电作业。华中、华北电网开始研究实施特高压1000kV带电作业，±500kV、±800kV输电线路带电作业工器具及项目实施也在研究实施中。

2009年4月14日，湖北超高压输变电公司，在国内首次进行的特高压带电消缺作业在1000kV南荆Ⅰ线上圆满完成，如图1-11所示。

2009年6月11日，全球首次±800kV特高压直流输电带电作业获得成功，这次作业任务由湖北超高压输变电公司的胡光等7名专业人员承担。这次带电作业操作采用了自行研制的带电作业工器具，如图1-12所示。

图1-11　1000kV特高压等电位消缺作业

图1-12　±800kV特高压直流输电带电作业

1.1.2　国外带电作业开展及工器具使用历史

1. 美国线路带电作业

世界上最早开展带电作业的国家是美国。1913年，在俄亥俄州最早实现线路带电作业，当时采用的是地电位方法，带电作业工具为木制。干燥的木质棒，由于其绝缘性能良好，完全能够耐受相对地电压，尽管当时制造的工具显得粗糙且笨重，但毕竟开创了带电作业的先河。随着电压等级的不断提高，需要绝缘性能更强的带电操作杆才能满足要求。塑料套木杆首先由美国的Chance公司引进并进行带电作业应用，随后出现了以玻璃纤维混合而成的增强型合成树脂管，这种树脂管具有耐压强度高、电场分布较为均匀等特点，随后广泛应用到输电线路带电作业中。目前，美国的各个电力公司均开展了输电线路带电作业检修和维护等项目，不少电力公司还开展了直升机带电巡线、带电检修等作业项目，如图1-13所示为美国的直升机等电位作业。

图1-13　美国开展的直升机等电位作业

2.日本线路带电作业

日本的输电线路带电作业相对较少，配电线路的带电作业比例较大，同时，日本对配电网的供电可靠性提出了更高要求，要求逐步实现完全不停电。与过去相比，日本的带电作业已逐步向自动化、智能化、机械化方向迈进，是世界上最早开展带电作业机器人研究和应用的国家。在20世纪90年代初，日本的四个电力公司开发了具有液压机械臂的带电作业机器人，在配电网系统检修中使用。为了提高供电可靠性，支持不停电作业的全面普及，日本研发了一系列机械化作业工具和设备，并不断推广应用，包括应急事故使用的高压发电机车、带电水冲洗变压器车、低压不停电切换装置，以及事故点检测车等直接带电作业法或间接带电作业法所需的工具及配套设施。这些机械化带电作业工器具的使用，在一定程度上降低了人员的劳动强度，提高了工作效率，如图1-14所示为日本开展的机器人带电作业。

图1-14　日本开展的机器人带电作业

3. 俄罗斯线路带电作业

俄罗斯（苏联）在20世纪30年代就开始进行输电线路的带电作业。在20世纪40年代，等电位新的作业方法已得到应用并大规模推广；20世纪50年代中期，绝大部分线路抢修工作已采用了带电作业。

俄罗斯330～750kV输电线路带电作业已发展出了一套完整、合理的操作方法。考虑到运行的可靠性，输电线路上配有适合于高电压、强电场作业的带电作业工器具。其带电作业主要包含绝缘测试、绝缘子串更换、导线补修、架空线路及带电设备的涂刷防腐漆等，在带电作业方面走到了世界前列。另外，俄罗斯还积极进行1150kV的带电作业技术研究，并在1150kV交流输电线路带电作业的安全防护方面有较深入的研究。

4. 英国、法国线路带电作业

英国从20世纪中叶开始使用操作杆进行带电作业，典型的有测量绝缘子串的电压分布等，直到1965年，才开始对输电线路带电作业进行研究。此时的法国，也成立了带电作业技术委员会和带电作业试验研究所，主要研究带电作业原理、带电作业安全规程、等电位作业法、使用工具的间接作业法等；其作业项目有带电更换绝缘子串、更换横担、更换防震装置。在超高压输电线路上80%的工作都是由等电位作业方法完成，在400kV以上电压等级的线路上，使用绝缘工具的间接作业方法和等电位作业方法联合作业。目前，英法两国带电作业主要应用在配电线路上，输电线路相对较少。这主要是因为输电线路有较多备用设备，并且英法两国因为地域面积和能源分布的原因，不需要建设跨区域的大型输电网络。

1.2 带电作业工器具的材料

1.2.1 常用绝缘材料

绝缘材料在一定电压作用下,只有极微的泄漏电流通过,绝缘电阻值较大。绝缘材料质量的好坏,直接关系到带电作业的安全,因此制作带电作业工具的绝缘材料必须是电气性能优良、机械强度高、重量轻、吸水性低、耐老化且易于加工的材料。

我国目前带电作业工器具使用的绝缘材料主要有以下几种。

（1）绝缘板材。其包括硬板和软板,其种类有层压制品,如3240环氧酚醛玻璃布板和工程塑料中的聚氯乙烯板、聚乙烯板等。

（2）绝缘管材。其包括硬管和软管,其种类有层压制品,如3640环氧酚醛玻璃布管、带或丝的卷制品。

（3）塑料薄膜。如聚丙烯、聚乙烯、聚氯乙烯、聚酯等塑料薄膜。

（4）橡胶。天然橡胶、人造橡胶、硅橡胶等。

（5）绝缘绳。天然蚕丝、人工化纤丝编织的绳索,如尼龙绳、绵纶绳和蚕丝绳（分生蚕丝绳和熟蚕丝绳两种）,其中包括绞制、编织圆形绳及带状编织绳。

（6）绝缘油、绝缘漆、绝缘黏合剂等。

按电气设备运行所允许的最高工作温度即耐热等级,国际电工委员会（简称IEC）把绝缘材料分为Y、A、Z、B、F、H、C七个等级,其允许工作温度分别为90℃、105℃、120℃、130℃、155℃、180℃及180℃以上。

1.2.2 常用金属材料

因直接关系到带电作业的安全,故制作带电作业工具的金属材料必须是机械强度高、重量轻的优质材料。目前带电作业使用的金属材料大致有以下几种。

（1）铝合金。航天使用的铝合金材料,一般为板材,用于加工各种卡具等。

（2）钛合金。航天使用的钛合金材料,强度比铝合金更高、重量更轻,但价格较高,适用于加工特高压各种卡具。

（3）高强合金钢。高强度合金钢,用于机械强度要求较高的工具部件。

1.3 带电作业工器具基本类型

我国开展带电作业多年，带电作业工器具早已经实现了标准化和生产专业化，为保证产品质量，国家和行业颁布了四十余种带电作业工器具的规程和标准。从目前带电作业项目使用的工器具来看，在大类上可以分为输电线路带电作业和配电线路带电作业工器具两大类，每一类工器具又可以分为多种形式。

1.3.1 输电线路带电作业工器具的基本类型

1. 硬质绝缘工具

硬质绝缘工具包括绝缘操作杆、绝缘托瓶架、绝缘横担、绝缘支拉吊杆、绝缘滑车等。

2. 软质绝缘工具

软质绝缘工具包括绝缘软梯、绝缘滑车组、绝缘绳索和后备保护绳等。

3. 金属工具

金属工具包括铝合金软梯头、飞车、卡线器、各类卡具、金属滑车、跟斗滑车等。

4. 检测工具

检测工具包括瓷质绝缘子零值检测器、风速仪、温湿度仪、绝缘电阻测试仪、钳形电流表和核相仪等。

5. 安全、防护用具

安全、防护用具包括绝缘服、屏蔽服、静电防护服、绝缘安全带、防坠器等。

1.3.2 配电线路带电作业工器具的基本类型

1. 遮蔽用具

遮蔽用具包括：导线遮蔽罩、耐张装置遮蔽罩、针式绝缘子遮蔽罩、棒式绝缘子遮蔽罩、横担遮蔽罩、电杆遮蔽罩、套管遮蔽罩、跌落式导管遮蔽罩、隔板和绝缘毯等。

2. 个人防护用具

个人防护用具包括绝缘袖套、绝缘服、绝缘鞋（靴）、绝缘安全帽、绝缘手套、防刺穿手套等。

3. 操作绝缘用具

操作绝缘用具包括绝绝缘支杆、拉杆、吊杆等。

4. 特种作业车辆

特种作业车辆包括绝缘斗臂车、旁路作业车、带电作业工具车、移动箱变车等。

5. 绝缘手工用具

绝缘手工用具包括绝缘螺钉旋具、绝缘扳手、拔销器、钢丝钳、剥皮器、齿轮断线钳等。

6. 安全用具

安全用具包括携带式接地线、安全带、脚扣、升降板等。

1.4 带电作业工器具的试验

带电作业工器具试验包括电气试验和机械试验两种。任何一种带电作业工器具都必须进行定期的电气和机械试验，以检验其是否达到规定的电气性能指标和机械强度。只有经过上述两种试验，才能做出合格与否的结论。因为工器具在制作、运输和保管各个环节中，都可能产生或遗留观察不到的缺陷，只有通过试验才能暴露出来，所以带电作业工器具的试验是检验工器具是否合格的唯一可靠手段。

1.4.1 带电作业工器具试验原则

1. 带电作业工器具试验类型与周期

带电作业工器具试验包括电气试验和机械试验，要定期进行，依据 DL/T 976-2005《带电作业工具、装置和设备预防性试验规程》的要求，带电作业工器具及设备试验项目和周期如表1-1所示。

表1-1 带电作业工器具及设备试验项目和周期

序号	器具	试验项目			
		电气试验	试验周期	机械试验	试验周期
1	绝缘支、拉、吊杆	工频耐压试验和冲击耐压试验	1年	静负荷试验、动负荷试验	2年
2	绝缘托瓶架	外观尺寸检查、工频耐压试验和冲击耐压试验	1年	抗弯静负荷试验、抗弯动负荷试验	2年
3	绝缘滑车	工频耐压试验	1年	拉力试验	1年
4	绝缘操作杆	外观尺寸检查、工频耐压试验和冲击耐压试验	1年	抗弯、抗扭静负荷试验、抗弯动负荷试验	2年
5	绝缘硬梯	工频耐压试验和冲击耐压试验	1年	抗弯静负荷试验、抗弯动负荷试验	2年
6	绝缘软梯	工频耐压试验和冲击耐压试验	1年	抗拉性能试验、软梯头静负荷试验、软梯头动负荷试验	2年

续表

序号	器具	试验项目			
		电气试验	试验周期	机械试验	试验周期
7	绝缘绳索类工具	工频耐压试验和操作冲击耐压试验	1年	静拉力试验	2年
8	绝缘手工工具	工频耐压试验	1年	—	—
9	绝缘（临时）横担、绝缘平台	工频耐压试验	1年	—	—
10	绝缘子卡具	—	—	静负荷试验、动负荷试验	1年
11	紧线卡线器	—	—	静态负荷试验、动态负荷试验	1年
12	屏蔽服装	成衣电阻试验、整套服装的屏蔽效率试验	半年	—	—
13	静电防护服装	整套防护服装的屏蔽效率试验	半年	—	—
14	绝缘服（披肩）	整衣层工频耐压试验	半年	—	—
15	绝缘袖套	标志检查、交流耐压和直流耐压试验	半年	—	—
16	绝缘手套	交流耐压和直流耐压试验	半年	—	—
17	防机械刺穿手套	交流耐压和直流耐压试验	半年	—	—
18	绝缘安全帽	交流耐压试验	半年	—	—
19	绝缘鞋	交流耐压试验	半年	—	—
20	绝缘毯	交流耐压试验	半年	—	—
21	绝缘垫	交流耐压试验	半年	—	—
22	导线软质遮蔽罩	交流耐压和直流耐压试验	半年	—	—
23	遮蔽罩	交流耐压试验	半年	—	—
24	绝缘斗臂车	交流耐压计泄漏电流试验	半年	额定荷载全工况试验	半年

续表

序号	器具	试验项目			
		电气试验	试验周期	机械试验	试验周期
25	核相仪	工频耐压及泄漏电流试验	半年	—	—
26	验电器	工频耐压及泄漏电流试验	半年	—	—
27	500kV四分裂导线飞车	—	—	静态负荷试验、动态负荷试验	1年
28	火花间隙检测装置	间隙调整与放电试验、工频耐压试验、操作冲击耐压试验	1年	—	—

2. 绝缘工具电气预防性试验项目及标准

绝缘工具电气预防性试验项目及标准如表1-2所示。

表1-2 绝缘工具电气预防性试验项目及标准

额定电压/kV	试验长度/m	1min高频耐压/kV		3min高频耐压/kV		15次操作冲击耐压/kV	
		出厂及型式试验	预防性试验	出厂及型式试验	预防性试验	出厂及型式试验	预防性试验
10	0.4	100	45	—	—	—	—
35	0.6	150	95	—	—	—	—
63（66）	0.7	175	175	—	—	—	—
110	1.0	250	250	—	—	—	—
220	1.8	450	440	—	—	—	—
330	2.8	—	—	420	380	900	800
500	3.7	—	—	640	580	1175	1050

注：试品应整根进行，不得分段。

3. 绝缘工具检查性试验标准

将绝缘工具分成若干段进行工频耐压试验，每300mm耐压为75kV，时间1min，以无闪络、击穿和过热为合格。

4. 带电作业工具的机械试验标准

（1）在工作负荷状态承担各类线夹和连接金具荷载时，应按有关金具标准进行试验。

（2）在工作负荷状态承担其他静荷载时，应根据设计荷载，按 DL/T 976—2005《带电作业工具、装置和设备预防性试验规程》的规定进行试验。

（3）带电作业工器具要定期进行荷载试验。

1.4.2 带电作业工具的机械试验

带电作业工具的机械试验包括在工作负荷状态下的静负荷试验和动负荷试验两种。有些带电作业工具，如绝缘拉板（杆）、吊线杆等，只做静负荷试验；而有些可能受到冲击荷载作用的工具，如操作杆、收紧器等除了做静负荷试验外还要做动负荷试验。

1. 静负荷试验

静负荷试验为了考核带电作业工具、装置和设备承受机械载荷（拉力、扭力、压力、弯曲力）的能力所进行的试验。静负荷一般指额定试验施加的荷载为被试品允许使用荷载的 2.5 倍，持续时间为 5min，卸载后试品各部件无永久变形即为合格负荷。

使用荷载可按以下原则确定。

（1）紧、拉、吊、支工具（包括牵引器、固定器），凡厂家生产的产品可把铭牌标注的允许工作荷载作为使用荷载；也可按实际使用情况来计算最大使用荷载。

（2）载人工具（包括各种单人使用的梯子、吊篮、飞车等）以人及人体随身携带工具的质量作为使用荷载。

（3）托、吊、钩绝缘子工具，以一串绝缘子的质量为使用荷载。

在进行静负荷试验时，加载方式为：将工具组装成工作状态，模拟现场受力情况施加试验荷载。图 1-15 为带电作业工器具的静负荷试验。

图 1-15 带电作业工器具的静负荷试验

2. 动负荷试验

动负荷试验是检验被试品在经受冲击时，机构操作是否灵活可靠的试验项目。因此，其所施负荷量不可太大。一般规定用1.5倍的使用荷载加在组装成工作状态的被试品上，操作被试品的可动部件（例如丝杠柄、液压收紧器的扳把及卸载阀等），操作三次，无受损、失灵及其他异常现象即为合格。

由于操作杆经常用来拔取开口销、弹簧销或拧动螺钉，因此也要做抗冲击和抗扭试验，冲击矩可取500N·cm，扭矩可取250N·cm。如图1-16所示为绝缘子卡具机械试验布置图。

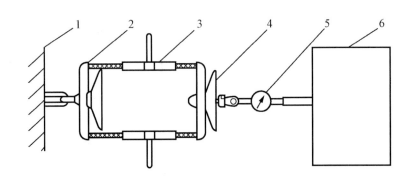

1—固定端；2—卡具；3—丝杠；4—绝缘子；5—拉力表（传感器）；6—拉力机

图1-16 绝缘子卡具机械试验布置图

1.4.3 带电作业工具的电气试验

带电作业用绝缘工器具出厂前应进行出厂试验，且试验项目和相关指标必须满足国标要求。由于产品长期积压、出厂运输以及有些厂家在出厂试验时只进行随机抽样试验等原因，产品到达用户手中时，还必须进行验收试验。试验标准应参照国家相应标准的规定。

除了以上两项试验外，带电作业工具经过一段时间的使用和储存后，无论是在电气性能方面还是机械性能方面，都可能会出现一定程度的损伤或劣化。所以，还应进行定期试验，也即预防性试验和检查性试验。下面就绝缘工器具的电气试验的内容、标准进行一些重点介绍。

绝缘工器具电气试验应定期进行，预防性试验每年一次，检查性试验每年一次，两种试验间隔半年，试验内容为工频耐压试验、操作冲击试验。而操作冲击耐压试验必须具备试验资质方可检测，故这里只简单介绍关于工频耐压试验的试验方法和试验条件。如图1-17为工频耐压及操作冲击耐压接线图。

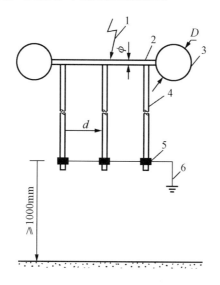

1—高压引线；2—模拟导线（$\phi>30mm$）；3—均匀球（$D=200\sim300mm$）；
4—试品（试品间距$d\geqslant500$）；5—下部试验电极；6—接地引线

图1-17 工频耐压及操作冲击耐压试验接线图

1. 试验方法

（1）绝缘杆的工频耐压试验。试验时绝缘操作杆，支、拉、吊杆等绝缘杆的金属头（或金属接头）部分应挂在施压的高压端（一般用长度不小于2.5m、直径为20mm的金属棒作高压端来模拟导线，并悬挂在空中），接地线接在握手部分与有效绝缘长度的分界线处（指操作杆）或原接地端（指支、拉、吊杆而言），然后进行整段施压试验。

（2）绝缘硬梯的工频耐压试验。绝缘硬梯包括直立梯、人字梯、水平梯和挂梯，试验时以一根直径为20mm、长2.5m的金属棒作施压的高压端，并水平悬挂来模拟导线。然后将绝缘硬梯的一端（一般指金属头的那端）挂在高压端上，接地线接在最短有效绝缘长度处（先用锡箔包绕其表面，然后再用裸铜线缠绕接地），进行施压试验。

（3）绝缘绳索、绝缘软梯的工频耐压试验。为了检验整副绝缘软梯或整根绝缘绳索的全部耐压水平，试验时，按图1-18所示的方法接线。在两根直径为20mm、长度适当的金属棒上缠绕悬挂绝缘软梯或绝缘绳索，其中一根金属棒为高压端，通过绝缘子水平悬挂于空中，另一根金属棒同时悬挂于空中并接地。两根金属棒之间的距离等于最高使用电压下的最短有效绝缘长度。

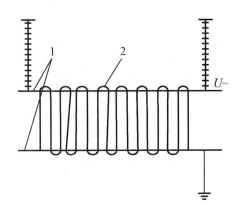

1—金属棒；2—绝缘绳

图 1-18　绝缘绳索的工频耐压试验

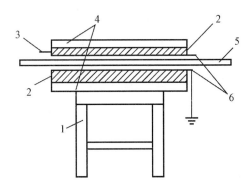

1—支承台；2—泡沫塑料；3—电源；4—连接片；5—被试物品；6—锡箔

图 1-19　绝缘遮蔽物的工频耐压试验

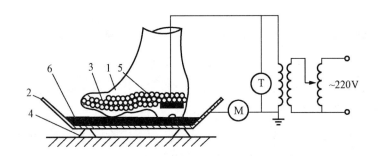

1—测试靴；2—金属盘；3—金属球；4—金属片；5—海绵加水；6—绝缘支架

图 1-20　绝缘靴的工频耐压试验

（4）绝缘遮蔽物的工频耐压试验。绝缘遮蔽物包括各种绝缘软板、硬板、薄膜等。在进行耐压试验时，如图1-19所示，将遮蔽物水平放置在绝缘支承台上，同时在被试物品的两侧用铝箔做电极,上下均用塑料连接片压紧,以保证其接触良好。然后将上面极板连接电源,下面极板接地。此外被试物品应按使用电压要求留边缘宽度。

(5) 绝缘靴工频耐压试验。将一个与试样鞋号一致的金属片上铺满直径不大于4mm的铜片,其高度不小于15mm,外接一片直径大于4mm的铜片,并埋入金属球内。外电极为置于金属器内的浸水海绵,试验电路如图1-20所示。以1kV/s的速度使电压从零上升到所规定电压的75%,然后再以100V/s的速度升到规定电压值,当电压升到规定电压时,保持1min,然后记录毫安表电流值。电流值小于10mA,则试验通过,绝缘靴合格。

(6) 绝缘手套的工频耐压试验。绝缘手套是配电线路带电作业中非常重要的人身防护用具,其种类非常多。其交流耐压试验如图1-21所示。10kV用绝缘手套露出水面部分的长度为75mm。

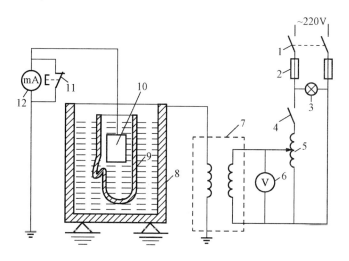

1—隔离开关;2—可断熔丝;3—电源指示灯;4—过负荷开关;
5—调压器;6—电压表;7—变压器;8—盛水金属器皿;
9—试样;10—电极;11—毫安表短路开关;12—毫安表
图1-21 绝缘手套交流耐压试验

绝缘手套的电气试验包括交流耐压试验和直流耐压试验。对绝缘手套做交流耐压试验时,加压1min,其交流耐压试验值见表1-3,无电晕、无闪络、无击穿、无明显发热为合格。

表1-3 绝缘手套的交流耐压值

型号	额定电压/V	交流耐压值（有效值）/V
1	3000	10 000
2	10 000	20 000
3	20 000	30 000

对绝缘手套做直流耐压试验时，加压时间保持1min，其直流耐压值见表1-4，无闪络、无击穿、无明显发热为合格。

表1-4 绝缘手套的直流耐压值

型号	额定电压/V	交流耐受电压（有效值）/V
1	3000	20000
2	10000	30000
3	20000	40000

（7）水冲洗工具的工频耐压试验。水冲洗工具的工频耐压试验组装时，应与地平面成30°~45°角倾斜放置。如图1-22所示，模拟导线为一根直径20mm、长2.5m的金属棒，且通过绝缘子串悬吊于空中，水枪喷嘴对模拟导线的距离以该水枪应用的电压等级而定，冲洗杆握手处和导水管距喷嘴1.8m处分别通过微安表接地，以测量泄漏电流。微安表上并联一接地开关，在加压和切换刻度旋钮时，接地开关应始终处于合闸状态，只有在读数时才可拉开接地开关，以防止高压作用在微安表上。此外模拟导线升压后，水枪对准模拟导线喷射（水电阻率10 000Ω·m）1min后再拉闸读微安表。

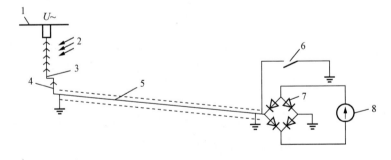

1—加压金属杆；2—淋水方向；3—雨天操作杆；4—流量防雨罩；
5—测量屏蔽线；6—开关；7—二极管；8—直流微安表
图1-22 水冲洗工具的工频耐压试验

考虑到水冲洗工具的组合绝缘（水＋绝缘杆＋导水管）问题，试验电压应按规程要求计算选取。

2. 试验条件

绝缘工具的绝缘强度包括外部绝缘和内部绝缘两部分，而影响外部绝缘的主要因素有气压、温度、湿度、雨水、污秽以及邻近物体的邻近效应等。因此，如果试验时的大气状态与标准大气状态不同，应将放电电压修正到标准大气状态下。相对湿度在80%以上时，会引起放电电压的变化，故在淋雨试验中，只进行相对空气密度的修正，不修正湿度。

第2章

金属工器具

输配电线路最常用的金属工器具有绝缘子卡具、铝合金卡线器、滑车和飞车等，它们主要用于带电更换绝缘子、导线修补和金具检修、更换等操作。良好的使用习惯和保管方法，能够防止金属工器具变形及损坏，延长金属工器具使用寿命，有效保证带电作业的安全。

2.1 绝缘子卡具及卡线器

绝缘子卡具从使用范围上可分为耐张串卡具、直线串卡具和单片绝缘子卡具，它们同铝合金卡线器一起，构成输配电线路带电更换单片或整串耐张或直线绝缘子的主要工具。

2.1.1 绝缘子卡具

带电更换绝缘子使用的主要工具有卡具、丝杠、绝缘拉板、吊线钩、绝缘操作杆等，其中绝缘子卡具是输电线路带电作业更换绝缘子最重要的工器具。更换时先将卡具安装在绝缘子串挂点的横担位置，通过丝杠和绝缘拉板收紧导线，使绝缘子串松弛，利用绝缘操作杆对绝缘子进行更换。

输电线路带电更换绝缘子针对不同的塔型、绝缘子串型、金具类型，最常用的绝缘子卡具有翼形卡、大刀卡、闭式卡三种，下面将详细介绍这三种绝缘子卡具。

1. 常用绝缘子卡具的基本结构、功能和使用方法

（1）翼形卡

①基本结构

组装在绝缘子串导线端的卡具叫前卡，组装在绝缘子串横担端的卡具叫后卡，如图2-1所示。

第2章　金属工器具

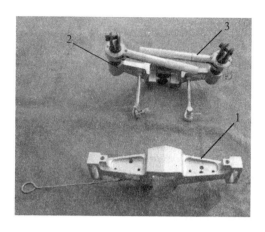

1—前卡；2—后卡；3—丝杆
图2-1　翼形卡结构图

②基本功能

主要用于带电更换单联耐张绝缘子串。作业时需要将卡具卡住绝缘子串两侧金具，通过丝杠和绝缘拉板收紧导线，使绝缘子串松弛到托瓶架上，然后对其更换（硅橡胶合成绝缘子不适用于托瓶架）。

③使用方法

电工上塔前，地面电工应先将绝缘子卡具、丝杠、绝缘拉板组装好。塔上作业人员到达指定位置后，将组装好的翼形卡具、托瓶架、操作杆相继传递到塔上。

1号电工持卡具，2号电工持绝缘操作杆，相互配合将翼形卡具前卡固定在螺栓型耐张线夹曲臂最低处，如图2-2所示。锁死防止线夹滑落的闭锁装置，如图2-3所示。将横担侧卡具固定在横担和绝缘子串连接金具上，确认金具嵌入卡槽正确到位，如图2-4所示。安装过程中，1号电工身体部位不得超过第一片绝缘子。

图2-2　搭放翼形卡

图2-3 安装前卡并锁死

图2-4 安装后卡

2号电工手持绝缘操作杆与1号电工配合将绝缘托瓶架的绝缘管插入导线侧支撑架,将绝缘托瓶架的三角固定架与横担侧卡具连接并上好螺栓。绝缘托瓶架安装前应调节好导线侧支撑架的高度,使托瓶架的绝缘管能与绝缘子串平行、紧贴。安装后应检查托瓶架两根绝缘管是否已插入导线支撑架,如图2-5所示。

图2-5 安装托瓶架

1号电工对绝缘子串及卡具做冲击试验，收紧丝杠后并对紧线工具做冲击试验检查。收紧丝杠时应保持两侧丝杠均衡受力且收紧长度大致相同，摇动丝杠时用力要均匀，尽量减小绝缘拉板的扭动幅度。

（2）大刀卡

①基本结构

装在绝缘子串导线端的大刀卡具叫前卡，组装在绝缘子串横担端的大刀卡具叫后卡，前卡顶端处有一个提升滑轮，如图2-6所示。使用大刀卡具同时需要配套拉板、丝杠等，如图2-7所示。

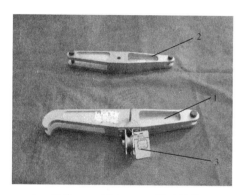

1—前卡；2—后卡；3—提升滑轮
图2-6 大刀卡

1—绝缘拉板；2—丝杠
图2-7 组装好的大刀卡

②基本功能

方法：主要用于地电位作业法更换耐张双联绝缘子串。

③使用

地面电工应先将大刀卡具、丝杠、绝缘拉板组装好，调整好丝杠、绝缘拉板的距离，各部分连接牢固可靠。塔上作业人员到达指定位置后，将组装好的大刀卡具、托瓶架、操作杆相继传递到塔上。

1号电工持绝缘拉板及卡具，2号电工持绝缘操作杆，相互配合将大刀卡具前卡固定在导线侧二联板，将大刀卡具后卡固定在横担二联板上。安装大刀卡具前卡时应检查卡具插入到位，如图2-8所示。安装大刀卡具后卡时应确认插销到位并锁紧，如图2-9所示。安装过程中1号电工身体不得超过第一片绝缘子。

图2-8　安装大刀卡前卡

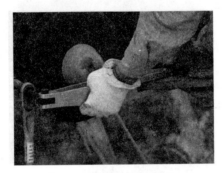

图2-9　安装大刀卡后卡

调整丝杠并使紧线装置轻微受力后，2号电工手持绝缘操作杆与1号电工配合将绝缘托瓶架的绝缘管插入导线侧支撑架，将绝缘托瓶架的三角固定架与大刀卡后卡连接并上好螺栓，如图2-10所示。绝缘托瓶架安装前应调节好导线侧支撑架的高度，使托瓶架的绝缘管能与绝缘子串平行、紧贴，并固定牢固前卡，如图2-11所示。安装后应检查托瓶架两根绝缘管是否已插入导线支撑架。1号电工安装托瓶架的三角固定架时，应注意手脚与横担下方的引流线保持1.8m及以上的安全距离。

工具组装完毕后，1号电工对绝缘子串及卡具做冲击试验。最后，1号电工匀速收紧丝杠，方可使绝缘子串松弛。

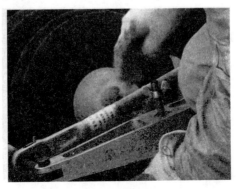

图2-10　安装托瓶架

图 2-11　固定前卡安装托瓶架

（3）闭式卡

①基本结构

闭式卡包括前卡活盖、前卡螺栓、后卡活盖、后卡螺栓等，如图 2-12 所示。使用闭式卡同时需要配套收紧丝杠，如图 2-13 所示。

②基本功能

适用于等电位作业法更换耐张串导线侧绝缘子。

③使用方法

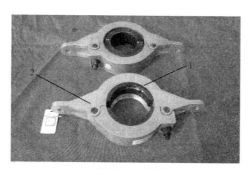

1—活盖；2—螺栓
图 2-12　闭式卡

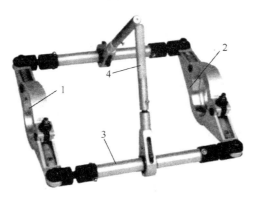

1—前卡；2—后卡；3—收紧丝杠；4—摇把
图 2-13　组装好的闭式卡

A. 若是瓷质绝缘子串,地面电工将火花间隙检测器传递给等电位电工,等电位电工检测绝缘子,判断是否满足作业所需良好绝缘子片数。

B. 等电位电工系好安全带及防坠落保护绳,携带绝缘传递绳,沿绝缘子串进入作业点,进入电位时双手抓扶作业串,双脚踩非作业串,采用"跨二短三"方式移动。

C. 地面电工将组装好的闭式卡传递至等电位电工作业位置。

D. 等电位电工将闭式卡安装在被更换绝缘子的相邻两片的钢帽上,如图2-14所示。

图2-14 安装闭式卡

图2-15 取出绝缘子

E. 等电位电工收紧丝杠,使之稍微受力,检查并确认各受力无异常后,取出被换绝缘子两侧的锁紧销。继续均衡收紧丝杠,方可取出绝缘子,如图2-15所示。

2. 绝缘子卡具的分类、技术参数和使用注意事项

(1) 分类

①卡具按功能分为以下系列:耐张串卡具、直线串卡具、单片绝缘子卡具。代号分别表示如下:

"N"——耐张串卡具；

"Z"——直线串卡具；

"D"——单片绝缘子卡具。

耐张串卡具用于更换耐张绝缘子及金具的卡具，按结构形状主要可分为翼形卡、大刀卡、翻板卡、弯板卡、斜卡等。直线串卡具用于更换直线绝缘子及金具的卡具，按结构形状主要可分为V形串卡、托板卡、花形卡、直线吊钩卡、钩板卡等。单片绝缘子卡具是用于更换单片绝缘子的卡具，其按结构形状主要可分为闭式卡、端部卡等。

②卡具的结构形式用卡具名称汉语拼音的第一个字母加"K"标示。如：翼形卡用YK表示，大刀卡用DK表示。

（2）技术参数

卡具额定荷重的取值一般为：

$$P=P_0 \times 25\% + 5$$

式中：P 表示卡具的额定荷载，kN；P_0 为适用的绝缘子或金具级别，kN。

①耐张串系列卡具性能参数

耐张串系列卡具规格及技术参数见表2-1。

表2-1 耐张串系列卡具的规格及技术参数

名称	型号	额定荷载/kN	动态试验荷载/kN	静态试验荷载/kN	破坏荷载/kN	适用绝缘子级别/kN
翼形卡	NYK80-300	80	120.0	200.0	240.0	300
	NYK60-210	60	90.0	150.0	180.0	210
	NYK45-160	45	67.5	112.5	135.0	160
	NYK35-120	35	52.5	87.5	105.0	120
	NYK30-100	30	45	75.0	90.0	≤100
大刀卡	NKD45-160	45	67.5	112.5	135.0	160
	NKD35-120	35	52.5	87.5	105.0	120
	NKD30-100	30	45.0	75.0	90.0	≤100
翻板卡	NFK60-210	60	90.0	150.0	180.0	210
	NFK45-160	45	67.5	112.5	135.0	160
	NFK35-120	35	52.5	87.5	105.0	≤120
弯板卡	NWK45-160	45	67.5	112.5	135.0	210、160
	NWK35-120	35	52.5	87.5	105.0	≤120
斜卡	NXK60-210	60	90.0	150.0	180.0	210、160
	NXK35-120	35	52.5	87.5	105.0	≤120

②直线串系列卡具技术参数

直线串系列卡具规格及技术参数见表2-2。

表2-2 直线串系列卡具的规格及技术参数

名称	型号	额定荷重/kN	动态试验荷重/kN	静态试验荷重/kN	破坏荷载/kN	适用绝缘子级别/kN
直线吊钩卡	ZDK60-210	60	90.0	150.0	180.0	210、160
	ZDK35-120	35	52.5	87.5	105.0	120
	ZDK30-100	30	45.0	75.0	90.0	≤100
V形串卡	ZVK60-210	60	90.0	150.0	180.0	210
	ZVK45-160	45	67.5	112.5	135.0	160
托板卡	ZTK60-210	60	90.0	150.0	180.0	210
	ZTK45-160	45	67.5	112.5	135.0	160
钩板卡	ZGK60-210	60	90.0	150.0	180.0	210
	ZGK45-160	45	67.5	112.5	135.0	160
花形卡	ZHK45-160	45	67.5	112.5	135.0	160
	ZHK35-120	35	52.5	87.5	105.0	120
	ZHK30-100	30	45.0	75.0	90.0	≤100

③单片绝缘子系列卡具技术参数

单片绝缘子系列卡具规格及技术参数见表2-3。

表2-3 单片绝缘子系列卡具的规格及技术参数

名称	型号	额定荷载/kN	动态试验荷载/kN	静态试验荷载/kN	破坏荷载/kN	适用绝缘子级别/kN
端部卡	DDK105-400	105	157.5	262.5	315.0	400
	DDK80-300	80	120.0	200.0	240.0	300
	DDK60-210	60	90.0	150.0	180.0	210
	DDK45-160	45	67.5	112.5	135.0	160
	DDK35-120	35	52.5	87.5	105.0	120
	DDK30-100	30	45.0	75.0	90.0	≤100
闭式卡	DBK105-400	105	157.5	262.5	315.0	400
	DBK80-300	80	120.0	200.0	240.0	300
	DBK60-210	60	90.0	150.0	180.0	210
	DBK45-160	45	67.5	112.5	135.0	160
	DBK35-120	35	52.5	87.5	105.0	120
	DBK30-100	30	45.0	75.0	90.0	100

（3）使用注意事项

①在选用卡具时应进行最大实际工作荷载的校核，若有特殊要求或不利气象环境条件致使工作荷载有可能超出卡具的额定荷载时，应选用大一级规格的卡具。

②使用前应对卡具外观进行检查，若发现裂纹应退出使用。运输中应避免碰撞，使用中不得用力敲打和摔落，避免影响其机械性能和工作性能。

③卡具与挂点的接触面应紧密可靠配合，非接触面应留有1～2mm间隙，以便于卡具安装或拆卸的方便、灵活。

④卡具各组成部分的零件应表面光滑、平整，无毛刺、尖棱、裂纹等缺陷。

3. 绝缘子卡具检查内容

绝缘子卡具在使用前必须进行检查，检查内容如下。

（1）卡具各组成部分零件表面应光滑、平整，无毛刺、尖棱、裂纹等缺陷。

（2）卡具各零件尺寸公差、形状公差、总体尺寸应符合设计图纸要求。

（3）卡具主体及其他主要受力零件所用的原材料，使用前需对其化学成分、力学性能进行复验，对铝合金材料还应进行低倍组织复验。

（4）卡具应与其挂点配装检验，确保其配合可靠，装卸方便灵活。

4. 绝缘子卡具的保管注意事项

（1）绝缘子卡具应存放在干燥、通风良好的工器具库房里面，可以放在专用工器具盒内或上架定点放置。如图2-16所示为存放绝缘子卡具的工具柜。

（2）卡具标志应标刻在易识别的部位，用压印法或其他方法，压痕深度不大于0.1mm。

（3）标示内容包括卡具型号规格、制造厂名简称或代号、商标、出厂编号等。

图2-16 绝缘子卡具存放柜

5. 绝缘子卡具的预防性试验

预防性试验需按要求对绝缘子卡具进行静态负荷试验、动态负荷试验，试验周期为1年。

2.1.2　铝合金紧线卡线器

1. 铝合金紧线卡线器的基本结构和功能

（1）基本结构

卡线器包括上夹板、下夹板、翼形拉板、拉板、压板，其结构与主要零件如图2-17所示。

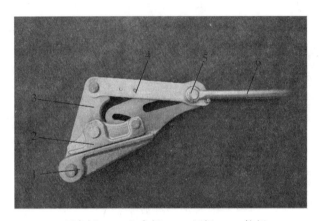

1—下夹板；2—上夹板；3—压板；4—拉板
5—翼形拉板；6—拉环
图2-17　铝合金紧线卡线器的结构与主要零件

（2）基本功能

在架空电力线路上进行松紧导线作业时，能自动夹牢导线，能方便拆装的用来连接导线和牵引机具的中间连接工具，适用于导地线弧垂调整和收紧导线。

2. 铝合金紧线卡线器的分类和技术参数

（1）分类

铝合金紧线卡线器按牵引方式分为单牵式（U形拉环式）和双牵式（翼形拉板式）两种。

图2-18所示为使用中的紧线卡线器，如图2-19所示为安装紧线卡线器。

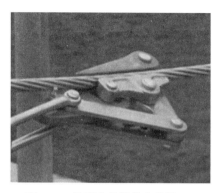
图2-18 使用中的紧线卡线器

图2-19 安装紧线卡线器

（2）技术参数

紧线卡线器型号、规格及尺寸见表2-4。

紧线卡线器技术性能指标见表2-5。

表2-4 紧线卡线器的型号、规格及尺寸

名称	a型	b型	c型	d型	e型	f型	g型	h型
型号	LJK_a25-70	LJK_b95-120	LJK_c150-240	LJK_d300	LJK_e400	LJK_f500	LJK_g630	LJK_h720
规格	ϕ12/14	ϕ16/20	ϕ22/24	ϕ26/28	ϕ30/32	ϕ31/33	ϕ35/37	ϕ37/39
钳口夹持长度mm	112	136	167	167	187	187	208	280
夹持弧面直径mm	12	16	22	26	30	31	34	37
最大开口mm	14	20	24	28	32	33	37	39

表2-5 紧线卡线器技术性能指标

型号	质量不大于/kg	额定负荷/kN	最大动试验负荷/kN	最大静试验负荷/kN	破坏负荷不小于/kN	在最大动试验负荷下允许导线相对滑移量不大于/mm
LJKa25-70	1.0	8.0	12.0	20.0	24.0	5
LJKb95-120	1.5	15.0	22.5	37.5	45.0	5
LJKc150-240	3.0	24.0	36.0	60.0	72.0	5
LJKd300	4.0	30.0	45.0	75.0	90.0	5
LJKe400	4.5	35.0	52.5	87.5	105.0	5
LJKf500	6.5	42.0	63.0	105.0	126.0	5
LJKg630	7.0	47.0	70.5	117.5	141.0	5
LJKh720	10.0	49.0	73.5	122.0	147.0	5

3. 铝合金紧线卡线器的检查内容

铝合金紧线卡线器在使用前必须进行检查，检查内容如下。

（1）铝合金紧线卡线器各零件尺寸公差、形位公差应符合设计要求。

（2）铝合金卡线器的上夹板、下夹板、翼形拉板、拉板、压板应采用超强度铝合金的模锻件，铝合金卡线器拉环应采用优质合金结构钢制成的模锻件。

（3）加工后的各零部件表面应光滑，无锐边、毛刺、裂纹及金属夹杂等缺陷。

4. 铝合金紧线卡线器的使用注意事项

铝合金紧线卡线器使用注意事项有以下几个。

（1）在额定负荷下，各种铝合金紧线卡线器与所夹持的导线不产生相对滑移，不允许夹伤导线表面。

（2）在最大试验负荷时，允许有一定的滑移量，但是最大滑移量不得大于5mm，卡线器各零件无变形，导线直径的平均值不得小于夹持前的97%，且导线表面应无明显压痕。

（3）在最大静试验负荷卸载后，各种铝合金紧线卡线器的各零件应不发生永久变形。

（4）各部件连接应紧可靠，开合夹口方便灵活，整体性好。

5. 铝合金紧线卡线器保管注意事项

（1）紧线卡线器应存放在干燥、通风良好的工器具库房里面，可以放在专用工器具盒内或上架定点放置。

（2）标示应标刻符号、制造厂商标、型号及出厂编号、额定负荷及出厂日期。

（3）在卡线器上应有一矩形标志，在矩形标志中标出检验周期和检测日期。

（4）标志采用压印法或铆贴标志牌。

6. 铝合金紧线卡线器预防性试验

预防性试验需按要求对铝合金紧线卡线器进行外观及主要尺寸检查和抗拉试验，试验周期为1年。

2.2 滑车及飞车

滑车和飞车是输电线路带电作业常用的金属工具,根据不同的作业情况滑车又分为铁滑车和铝滑车,飞车分为单导线飞车、220kV双分裂飞车和500kV四分裂飞车,它们都是输电线路等电位带电作业最常用的工具。

2.2.1 滑车

1. 滑车的基本结构和功能

(1)基本结构

滑车主要由吊钩(链环)、滑轮、轴、轴套(或轴承)以及夹板组成,如图2-20所示。

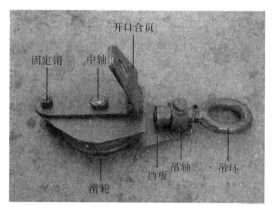

图2-20 滑车结构示例图

(2)基本功能

滑车是电力线路检修和施工安装的常用工具之一,它能借助起重绳索而产生旋转运动,从而改变作用力的方向。由于滑车使用方便而且便于携带,因此,在线路检修和施工安装中被广泛应用。

2. 滑车的分类和技术参数

(1)分类

①按照材质可分为铁滑车、铝滑车、不锈钢滑车和绝缘滑车等。

②按照用途可分为起重滑车、放线滑车、跟斗滑车。起重滑车分为吊环式起重滑车、吊钩式起重滑车、吊架式起重滑车、链环式起重滑车;放线滑车分为地缆滑轮、

电缆滑轮、朝天滑轮、尼龙滑轮、电缆导线放线滑轮、大直径放线滑车、起重滑轮。

③按照滑车的轮数可分为单轮滑车、双轮滑车和多轮滑车。

（2）技术参数

①滑车型号的命名方法

H系列滑车产品型号规格均用一组文字代号表示，代号由4部分组成。

②滑车型式代号

滑车型式代号如表2-6所示。

表2-6 滑车的型式代号

型式	开口	闭口	吊钩	链环	吊环	吊梁	挑式开口
代号	K	不加K	G	L	D	W	K_B

③滑车的选用

选用滑车时先根据起吊质量和需要的滑轮数，按表2-7查得滑车滑轮槽底的直径和配合使用的钢丝绳直径，核查所选用的钢丝绳是否符合规定。

表2-7 H型滑车选用表

轮槽底径 (mm)	起吊质量（t）												使用钢丝绳 ϕ (mm)	
	0.5	1	2	3	5	8	10	15	20	32	50	80	100	140	适用的	最大的
	滑轮数															
70	1	2													5.7	7.7
85		1	2	3											7.7	11
115			1	2	3	4									11	14
135				1	2	3	4								12.5	15.5
165					1	2	3	4	5						15.5	18.5
185						2	3	4	6						17	20
210									5						20	23.5
245							1	2		4	6				23.5	25
280									2	3	5	7			26.5	28
320								1			4	6	8		30.5	32.5
360									1	2	3	5	6	8	32.5	35

3. 滑车的检查内容

滑车在使用前必须进行检查内容如下。

（1）使用前必须进行外观检查，凡吊钩、吊环、护板、隔板等有严重变形裂纹，轮缘破损，轴承变形，轴瓦磨损以及转动不灵活，零部件不全，吨位不明等情况，均不得使用。

（2）检查吊环的扭转变形度、合页板错位变形度以及轮槽径向磨损和轮槽不均匀磨损是否影响使用。

（3）检查滑车外观是否存在裂纹、破碎等缺陷。

（4）检查滑车试验合格证是否在有效期内。

（5）主要检查滑车滑轮、吊环是否转动灵活，有无卡阻现象，开合合页、夹板转动是否灵活。

4. 滑车的使用方法

（1）按出厂铭牌安全起重使用，不得超载。

（2）使用开门式滑车必须将门扣锁好。

（3）滑车在装卸过程中，不得摔砸，以防部件变形。

（4）常用的多轮滑车均为不开口式，其连接部分有可旋转的吊环式和固定的吊环式，而固定吊环的滑车，悬挂时有方向性，当滑车受力后不得使滑车承受扭力，否则应加接U形环加以调整。

（5）在受力方向变化较大的场合或在高处使用单轮滑车时，应使用吊环式滑车，如果使用吊钩式滑车，必须对吊钩采取封口保险措施。因吊钩式滑车经反复受力后容易与绑扎绳脱钩。

（6）使用吊钩或滑车吊钩时封口板应装好，如果封口板损坏或丢失，应用10号铁线将弯钩封死，不得用铁线将绑扎绳绑扎在吊钩上，如图2-21所示。

（7）为防止单轮滑车的吊钩或吊环在受力过程中与滑车组相互转动而使牵引绳扭绞，应用8号铁线将吊钩或吊环与护板绑牢，如图2-22所示。

（8）滑车组是由单轮或多轮定滑车和动滑车组成的，使用时两个滑车的大小和型号应基本相同，不宜采用大小不同的滑车组成滑车组。

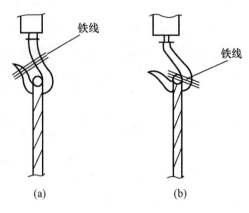

（a）正确；（b）不正确

图2-21　铁滑车使用规范（一）

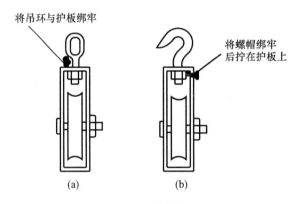

（a）正确；（b）不正确

图2-22　铁滑车使用规范（二）

（9）滑车组的穿绳方法有普通穿绕法和花穿绕法两种，一般多采用普通穿绕法。它是将两个滑车（单轮或多轮）相隔适当距离在地面上固定好，将钢丝绳端头从滑车的一侧轮槽内绕入，再绕向另一个相对应的滑车轮槽内，按同样的穿绕方向，依次绕到最后一个轮槽为止。钢丝绳穿好后，当拉紧时，绳股应相互平行，不得相互迭绞和串位。

5. 滑车保管注意事项

（1）滑车应存放保管在通风良好、干燥的环境内，并上架定置存放。如图2-23所示为铁滑车上架放置图。

（2）滑车长期不用时，应清洗并涂防锈油，避免锈蚀。每组滑车均应编号。

图2-23 铁滑车上架放置

6. 滑车的检验周期要求

滑车每季度做一次外观检查及静载试验，每半年在外观检查及静载试验的基础上需做拆解检查一次，新滑车使用半年后按照上述要求进行检验。

2.2.2 飞车

1. 飞车的基本结构和功能

（1）基本结构

飞车是用于架空导线上行驶的特殊工具，图2-24所示为双分裂导线人力飞车结构图。它由摆架、前行轮、辅轮、前行轮手柄、辅轮手柄、链条、链罩、坐垫、靠背、框架、后行轮手柄、后行轮、保险杠接头、保险杠、链条排档、脚踏板和刹车组成。如图2-25所示为单导线飞车。

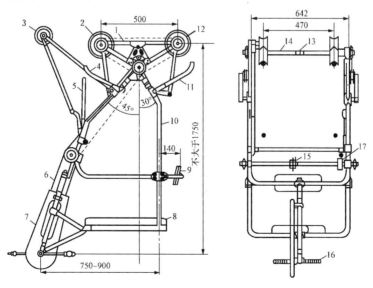

1—摆架；2—前行轮；3—辅轮；4—前行轮手柄；5—辅轮手柄；6—链条；7—链罩；8—坐垫；
9—靠背；10—框架；11—后行轮手柄；12—后行轮；13—保险杠接头；14—保险杠；
15—链条排档 16—脚踏板；17—刹车
图2-24 双分裂导线人力飞车结构图

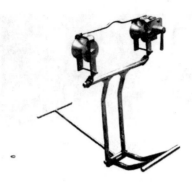

图 2-25 单导线飞车

（2）主要功能

飞车在输电线路的间隔棒、防震锤等附件安装、维修时或对线压接管进行检查、更换和维护等空中作业时，作为运载工具使用。

2. 飞车的基本类型和技术参数

（1）基本类型

①按轮子行驶的导线数目来分：有单导线、双分裂导线、三分裂导线和四分裂导线飞车等几种。单导线飞车一般采用两个轮子；其他多分裂导线飞车分别由 4~10 个轮子组成。

②按飞车行驶的驱动方式分：有人力飞车和机动力飞车两种。人力飞车大多数采用类似自行车脚蹬式驱动方式；机动飞车采用小型汽油机作为动力，通过液压传动或链条传动方式驱动。

③按飞车整体结构分：可以分为筐式飞车和架式飞车两种。筐式飞车的传动装置安装在易购金属筐内，操作人员作业时站在筐内进行。车架式飞车的外形同自行车或轻型摩托车相似，操作人员坐在坐垫上进行作业。

如图 2-26 所示为双分裂导线脚蹬式飞车，如图 2-27 所示为常见的几类飞车示意图。

图 2-26 及分裂导线脚蹬式飞车

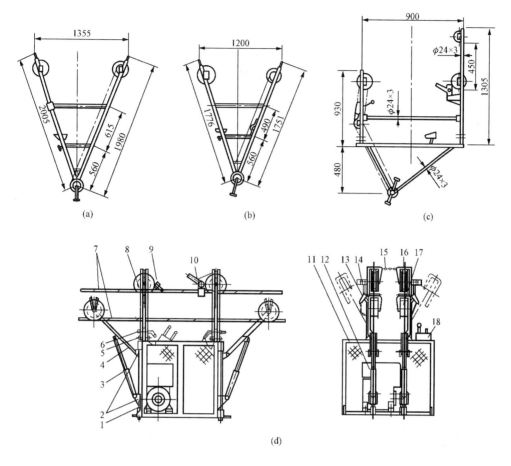

(a) SFS-1×150 单线飞车；(b) SFC-2×150 双分裂导线飞车；
(c) 人力脚蹬式四分裂导线飞车；(d) 四线八轮框式机动飞车
1—长销轴；2—绞支连接板；3—液压缸；4—从动轮臂；5—销轴；6—插销；7—导线；8—主动轮；9—距离测量装置；10—线夹式制动器；11—筐架；12—液压泵；13—防止脱落插销；14—液压马达；15—防止坠落链；16—主动轮臂；17—从动轮；18—操作盒

图 2-27　飞车示意图

（2）性能参数

①产品代号

产品代号由电压等级、飞车符号、特性、适用范围组成。如 F500-SB-ZA（B）。

其中：F 表示飞车；

500 表示 500kV；

SB 表示双驱动摆滚式；

Z 表示直线型；

A 表示最大坡度 14°；

B—最大坡度20°。

② 技术要求

a. 材料要求：飞车应使用符合国家标准规定的强度高、质量轻、非脆性材料。

b. 额定负荷：飞车试验额定负荷应不小于900N。

c. 保险：飞车应有双重保险装置。

③ 行驶性能要求

a. 飞车应能通过间隔棒、防震锤及悬垂绝缘子串。

b. 主动行轮轮槽应镶嵌导电耐磨橡胶。

c. 在带电线路上行驶不应发生充、放电现象；

d. 飞车轮距应能在450（-30～+50）mm的范围内变化。

e. 飞车外形尺寸应紧凑，不大于图2-2-8所示的数据。

f. A型飞车的爬坡能力应达到14°，B型飞车的爬坡能力应达到20°。

g. 飞车应设前进挡、倒退挡和空挡三挡，倒退时同样能通过间隔棒、防震锤及悬式绝缘子串，B型飞车应设快速、慢速两挡。

④ 结构要求：飞车结构应轻巧，操作方便，便于运输存放，A型飞车重量在40kg以下，B型飞车重量在50kg以下。

⑤ 工艺要求：飞车各部件外表不得有尖锐棱角，各接口应倒圆角处理，加工、热处理、装配按国标的规定进行，材料表面进行防腐处理。

3. 飞车使用前的检查内容

（1）飞车使用前应该进行外观检查，包括整车外形、材料、标志、工艺等项目的检查。

（2）飞车使用前应该检查各部分结构是否牢固可靠，各传动结构是否灵活，各个保险装置是否完好可靠。

（3）飞车使用前应检查其是否进行机械载荷试验，其试验周期是否在有效期内。

4. 飞车使用注意事项

（1）在连续档距的导线、地线上挂梯（或飞车）作业时，其导线、地线的截面面积不应满足钢芯铝绞线不小于120mm²，钢绞线不小于50mm²，铜绞线不小于95mm²。

（2）有下列情况之一的，应经过验算，并经单位总工程师批准后才能进行的作业：在弧立档内的导线、地线上作业；在有断股的导线、地线上作业；在锈蚀的地线上作业；在截面较小或其他型号的导线、地线上作业；二人以上在导线、地线上作业。

（3）在导线、地线上挂梯（或飞车）前，必须检查本档两端杆塔处导线、地线的紧固情况，挂梯（或飞车）载荷后，地线及人体对带电体的最小距离比人体与带电体的最小安全距离增大0.5m，导线及人体对被跨越的电力线路、通信线路和其他建筑物的最小距离应比人体与带电体的最小安全距离增大1.0m。

（4）在瓷横担线路上严禁挂梯（或飞车）作业，在转动横担的线路上挂梯（或飞车）前，应将横担固定。

5. 飞车的保管注意事项

（1）飞车应存放保管在通风良好、干燥的环境内，并上架定置存放。

（2）飞车长期不用时，应清洗并涂防锈油，避免锈蚀。

6. 飞车的预防性试验

飞车的预防性试验包括外观检查、踏行试验、静态负荷试验、动态负荷试验、保险杠试验和刹车试验，其内容如下。

（1）外观检查。主要检查整车外形、材料、工艺和标志，周期为每年1次。

（2）踏行试验。在900N额定负荷下，飞车在模拟线路上踏行应正常，并能顺利通过间隔棒、防震锤和悬垂绝缘子串的模拟线路，周期为每年1次。

（3）静态负荷试验。在飞车坐垫上施加2250N的负荷，在模拟线路上持续5min，飞车坐垫应完整、良好、无永久形变，试验周期为1年。

（4）动态负荷试验。在飞车坐垫上施加900N的负荷（包括操作人员负荷），在装有间隔棒、防震锤和悬垂绝缘子窜的模拟线路上能来回3次正常踏行，试验周期为1年。

（5）保险杠试验。飞车一侧前后主动行轮从导线上脱出，使两道保险杠的一侧直接搁在导线上，施加1350N的负荷应正常，无永久形变，试验周期为1年。

（6）刹车试验。飞车的刹车性能应良好，按下列方法试验满足要求。A型飞车在$\alpha=14°$坡上载人（总负荷900N），以3km/h的速度下滑，刹车位移应≤0.2m；B型飞车在$\alpha=20°$坡上载人（总负荷900N），以3km/h的速度下滑，刹车位移应≤0.3m，试验周期为1年。

第3章

绝缘工器具

带电作业用绝缘工具应具有良好的电气绝缘性能和较高的机械强度，同时还应具有吸湿性低、耐老化、重量轻、操作方便、不易损坏等特点。目前带电作业用绝缘工具大致可分为硬质绝缘工具、软质绝缘工具和其他绝缘工具三类。硬质绝缘工具主要指以绝缘管、棒、板为主体的工具，软质绝缘工具主要指以绝缘绳为主体的工具，其他绝缘工具是指绝缘滑车、绝缘紧线工具等兼具硬质和软质特点的绝缘工具。

3.1 硬质绝缘工具

在硬质绝缘工具中，使用最广泛的是绝缘杆，以及利用绝缘管材或板材制成的绝缘平台和绝缘硬梯等。绝缘杆根据用途和操作方法分为绝缘操作杆、绝缘支杆和拉（吊）杆三类。此外，绝缘横担和绝缘托瓶架也是经常使用的硬质绝缘工具。硬质绝缘工具基本上都采用玻璃纤维板或环氧玻璃钢为原材料。特别是环氧玻璃钢（简称为玻璃钢）是由玻璃纤维与环氧树脂复合而成，由于玻璃纤维和环氧树脂的电气绝缘性能都十分优良，因此由它们复合而成的玻璃钢具有优良的机械和电气性能。

3.1.1 绝缘操作杆

1.绝缘操作杆的基本结构和功能

（1）基本结构

①绝缘操作杆由接头、绝缘杆两部分组成，有的绝缘操作杆端部还有握柄部分。其接头可为固定式或拆卸式，固定在操作杆上的接头为高强度材料。如图3-1所示为绝缘操作杆结构。

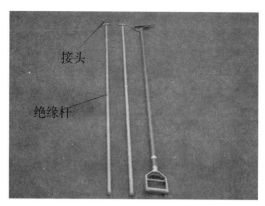

图3-1 绝缘操作杆结构

②绝缘操作杆各部分长度应满足表3-1的要求。

表3-1 绝缘操作杆各部分长度要求

额定电压（kV）	最短有效绝缘长度（m）	端部金属接头长度（m）	手持部分长度（m）
10	0.70	≤0.10	≥0.60
35	0.90	≤0.10	≥0.60
66	1.00	≤0.10	≥0.60
110	1.30	≤0.10	≥0.70
220	2.10	≤0.10	≥0.90
330	3.10	≤0.10	≥1.00
500	4.00	≤0.10	≥1.00
750	5.00	≤0.10	≥1.00
±500	3.50	≤0.10	≥1.00

（2）功能

在带电作业中，作业人员在绝缘操作杆上装上对应的绝缘操作杆附件，就可在与带电体保持安全距离的情况下，使用绝缘操作杆完成断、接引线等多项操作。

2. 绝缘操作杆的分类和技术规格

（1）分类

①绝缘操作杆按长度可分为：3m，4m，5m，6m，8m，10m。

②绝缘操作杆按电压等级可分为：10kV、35kV、110kV、220kV、330kV、500kV。

③绝缘操作杆按材质分类：绝缘操作杆一般采用玻璃钢环氧树脂杆手工卷制成型杆和机械拉挤成型杆两种。两种绝缘杆各有优势：手工卷制杆的优点是张性大，但是纵向强度相对机制拉挤成型杆小；机械拉挤成型杆的优点是强度大，但是横向张性相对手工卷制杆要小些。

④按照绝缘操作杆的结构可分为：接口式绝缘操作杆、伸缩式绝缘操作杆和游刃式绝缘操作杆。接口式绝缘操作杆是比较常用的一种绝缘杆，分节处采用螺旋接口，最长可做到10米，可分节装袋携带方便。伸缩式绝缘操作杆分3节伸缩设计，一般最长做到6米，重量轻，体积小，易携带，使用方便，可根据使用空间伸缩定位到任意长度，有效地克服了接口式绝缘操作杆因长度固定而使用不便的缺点。游刃式绝缘操作杆接口处采用游刃设计，旋紧后不会倒转。

（2）技术规格

①绝缘操作杆的绝缘部分长度不应小于0.7m。

②绝缘操作杆的材料要耐压强度高、耐腐蚀、耐潮湿、机械强大、质量轻、便于携带。

③三节之间的连接应牢固可靠，不得在操作中脱落。

3. 绝缘操作杆的检查内容

绝缘操作杆在使用前必须进行检查，检查内容如下。

（1）首先检查绝缘操作杆的规格是否同工作线路电压等级一致，严禁使用不符合规格的绝缘操作杆。

（2）检查绝缘操作杆的试验合格证是否在有效期内。

（3）检查绝缘操作杆外观上是否能有裂纹、划痕等外部损伤。

（4）绝缘操作杆节与节之间应连接牢固可靠。

4. 绝缘操作杆的使用方法

（1）雨雪天气必须在室外进行操作的要使用带防雨雪罩的特殊绝缘操作杆。

（2）连接绝缘操作杆时，连接节与节的丝扣要离开地面，不可将杆体置于地面上进行，以防杂草、土进入丝扣中或黏附在杆体的外表，丝扣要轻轻拧紧，丝扣未拧紧前不得使用。

（3）使用时压尽量减少对杆体的弯曲力，以防损坏杆体。绝缘操作杆使用如图3-2所示。

图3-2 绝缘操作杆使用

5. 绝缘操作杆的保管注意事项

（1）绝缘操作杆使用后要及时将杆体表面的污迹擦拭干净。

（2）绝缘操作杆应存放保管在通风良好、干燥的环境内，存放在屋内通风良好、清洁干燥的支架上或悬挂起来。

（3）绝缘操作杆要有专人保管。

如图3-3所示为存放绝缘操作杆的工具柜。

图3-3 存放绝缘操作杆的工具柜

6. 绝缘操作杆预防性试验

绝缘操作杆例行试验内容如下。

（1）电气试验。绝缘操作杆的电气试验要求是220kV及以下电压等级的试品应能通过短时工频耐受电压试验，330kV及以上电压等级的试品应能通过长时间工频耐受电压试验、操作冲击耐受电压试验，试验周期为1年。

（2）机械试验。绝缘操作杆的机械试验项目有抗弯静负荷试验和抗弯动负荷试验，试验周期为2年。

3.1.2 绝缘支、拉、吊杆

1. 基本结构和功能

（1）基本结构。绝缘支、拉、吊杆一般由金属配件与空心管、填充管、绝缘板连接组合而成。

（2）绝缘支拉杆的功能。它是以电杆杆身为依托的绝缘承力工具，它可以使导线同时做水平及垂直方向的位移。此种工具不仅能更换绝缘子，而且也适用于更换横担及电杆的工作。

（3）绝缘紧线拉杆（板）的功能。它是承担水平荷载（导线紧力）的绝缘部件。更换单串耐张绝缘子需用两根拉杆，更换双联串绝缘子多数情况只用一根拉杆单拉杆并只能配合固定卡具（如大刀卡）使用。

（4）绝缘吊线杆的功能。承受垂直荷载的绝缘部件一般两根为一组（也有只用单根的），使用于垂直荷载较大的场合。

2. 绝缘支、拉、吊杆的类型和性能参数

（1）绝缘吊线杆按结构可分为共用型、固定型、压杆型以及绝缘子托架型，其中前两种需配合固定器和牵引机具使用。

（2）性能参数：绝缘支、拉、吊杆的最短有效绝缘长度、总长度和各部分长度应符合表3-2的规定。需要注意的是，支杆的总长度由支杆的最短有效绝缘长度、固定部分长度和活动部分长度的总和决定，吊、拉杆的总长度由最短有效绝缘长度和固定部分长度的总和决定。

表3-2 绝缘吊、拉、支杆的长度要求

额定电压/kV	最短有效绝缘长度/m	固定部分长度/m		支杆活动部分长度/m
		支杆	吊、拉杆	
10	0.4	0.6	0.2	0.5
35	0.6	0.6	0.2	0.6
63	0.7	0.7	0.2	0.6
110	1.0	0.7	0.2	0.6
220	1.8	0.8	0.2	0.6

3. 绝缘支、拉、吊杆的使用

绝缘支、拉、吊杆主要用于转移导线、金具、绝缘子等操作项目。

如图3-4所示为绝缘吊杆的使用。

图3-4　绝缘吊线杆的使用

4. 绝缘支、拉、吊杆的检查内容

（1）首先检查绝缘支、拉、吊杆的规格是否同工作线路电压等级一致。

（2）检查绝缘支、拉、吊杆的试验合格证是否在有效期内，严禁使用不符合规格的绝缘支、拉、吊杆。

（3）检查绝缘支、拉、吊杆的外观是否光滑，不能有裂纹、划痕等外部损伤，杆段间连接牢固。

（4）支、拉、吊赶上的金属配件与空心管、泡沫填充管、实心棒、绝缘板的连接应牢固，使用时应灵活方便。

5. 绝缘支、拉、吊杆的保管注意事项

（1）绝缘支、拉、吊杆应存放保管在通风良好、干燥的环境内。

（2）绝缘支、拉、吊杆应竖直存放在专用的工具柜内，不得横搁在架上。

6. 绝缘支、拉、吊杆的预防性试验

绝缘支、拉、吊杆例行试验内容如下。

（1）电气试验。绝缘支、拉、吊杆的电气试验项目有工频耐受电压试验和操作冲击耐受电压试验，试验周期为1年。

（2）机械试验。绝缘支、拉、吊杆的机械试验项目有抗弯静负荷试验和抗弯动负荷试验，试验周期为2年。

3.1.3 绝缘托瓶架

1. 绝缘托瓶架的基本结构和功能

（1）基本结构

绝缘托瓶架是输电线路带电更换耐张、直线绝缘子串的重要工具。它两侧一般由椭圆形空心管或矩形绝缘板构成，主材之间用多根弧形板连接，再在弧形板上固定2~3根绝缘管，以便绝缘子串动。如图3-5所示为绝缘托瓶架。

图3-5　绝缘托瓶架

（2）基本功能

在更换耐张绝缘子串作业中托住绝缘子串，使其保持松弛状态，以便于对绝缘子串的更换。

2. 绝缘托瓶架的分类和技术规格

（1）分类

绝缘托瓶架根据其功能的不同，分为直线托瓶架和耐张托瓶架两种。如图3-5所示为绝缘直线和耐张托瓶架。

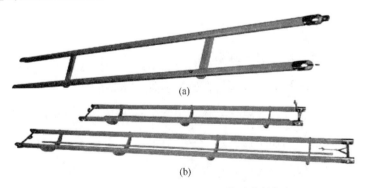

a—绝缘直线托瓶架；b—绝缘耐张托瓶架
图3-6　绝缘直线和耐张托瓶架

（2）技术规格

①绝缘托瓶架两端的金属附件，应做镀铬等表面防腐处理。绝缘层压类材料制件成件后，加工表面应进行绝缘处理，即各接口内孔接缝处，应采用高强度绝缘粘接胶

填实，表面再涂以绝缘漆。

②绝缘托瓶架的部件加工成形后，各加工表面应规则平整。

③绝缘托瓶架中的绝缘撑板间距：220kV及以下电压等级的托瓶架一般以500mm左右（3片绝缘子高度）为宜，330kV及以上电压等级的托瓶架一般以600~800mm左右（3片~4片绝缘子高度）为宜。

④绝缘托瓶架中的两根上连杆的宽度应与绝缘子的盘径相适应，上下连杆的距离（深度），应达到绝缘子盘径的一半以上。

3. 绝缘托瓶架的检查内容

（1）外观检查。绝缘托瓶架表面应光滑，无气泡、皱纹、开裂，玻璃纤维布与树脂间粘接完好，杆、段、板间连接牢固。

（2）绝缘托瓶架使用前应进行绝缘检查，如图3-7所示。

（3）绝缘托瓶架在使用前必须检查电气和机械试验合格证是否在有效期内，对超过有效试验期的绝缘托瓶架不能使用。

图3-7 检查绝缘托瓶架

4. 绝缘托瓶架的使用方法

（1）绝缘托瓶架在使用前要进行外观检查，检查其主材、弧形板和绝缘管是否变形或连接松动。

（2）在更换绝缘子串时必须注意同卡具的配合，保证其连接牢固可靠，并做冲击试验验证其连接情况。

（3）绝缘托瓶架在传递过程中要注意不要同杆塔和金具碰撞，以免导致其损伤。

如图3-8所示是使用绝缘托瓶架更换绝缘子串。

图3-8 使用托瓶架更换绝缘子串

5. 绝缘托瓶架的保管

绝缘托瓶架应置于通风良好的带电作业专用工具房间存放。贮存期超过生产日期12个月,要按相关标准进行电气性能检验。绝缘托瓶架在运输时,应装入专用工具袋、工具箱或专用工具车内,以免受潮和损伤。如图3-9所示为绝缘托瓶架的上架存放。

图3-9 绝缘托瓶架的上架存放

6. 绝缘托瓶架的预防性试验

绝缘托瓶架的预防性试验内容如下。

(1)电气试验。工频耐压试验和操作冲击耐压试验:220kV及以下电压等级的试品做短时工频耐压试验,330kV及以上电压等级的试品做长时工频耐压试验及操作冲击耐受电压试验,试验周期为1年。

(2)机械试验。抗弯静负荷试验和抗弯动负荷试验,试验周期为2年。

3.1.4 绝缘横担

1. 绝缘横担的基本结构和功能

（1）基本结构

绝缘横担一般由玻璃纤维板和金属配件构成，近年来又出现了玻璃钢和复合材料构成的绝缘横担，它们都具有重量轻、绝缘性能强、耐腐蚀性和体检小等特点。

复合绝缘横担由玻璃纤环氧树脂芯棒、硅橡胶伞裙、前端连接金具及连接法兰组成，如图3-10所示。芯棒与金具连接采用胶装工艺，芯棒与伞裙采用整体一次成型注压工艺。芯棒机械强度高、抗张抗弯强度比普通钢材高。

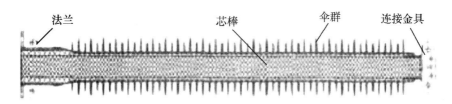

图3-10　复合绝缘横担

（2）基本功能

绝缘横担主要作业是支撑导线和保持导线与杆塔绝缘。它广泛应用于10～35kV配电线路杆塔带电更换直线横担作业中，以及输电线路110kV～220kV钢管杆更换钢制横担作业中。

2. 绝缘横担的分类和技术规格

（1）分类

①按组成材料可以分为玻璃纤维板绝缘横担、玻璃钢绝缘横担、复合材料绝缘横担等。

如图3-11所示为玻璃纤维板绝缘横担，如图3-12所示为使用玻璃钢横担，如图3-13所示为复合绝缘模担。

图3-11　玻璃纤维板绝缘横担

图3-12 玻璃钢横担

图3-13 复合绝缘横担

②按外形用途可分为绝缘支线横担、电杆临时横担、橡胶棒辅助横担、电杆用三相绝缘支线横担等。如图3-14所示。

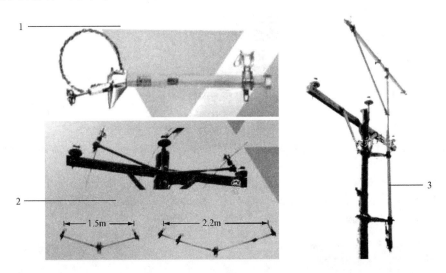

1—绝缘支线横担；2—电杆临时横担；3—电杆用三相支线横担
图3-14 配电线路用绝缘横担

（2）性能规格

带电更换直线杆横担最常用的玻璃钢绝缘横担技术规格见表3-3。

表3-3　玻璃钢绝缘横担和绝缘撑足的规格

角形玻璃钢规格（mm）	角边厚度（mm）	角形玻璃钢长度（mm）	适用范围
∠50×∠50（±2）	6~8（±0.5）	1200~2000（±10）	撑足
∠50×∠50（±2）	6~8（±0.5）	1200~2000（±10）	横担
∠50×∠50（±2）	6~8（±0.5）	1200~2000（±10）	横担
∠50×∠50（±2）	6~8（±0.5）	1200~2000（±10）	横担
∠50×∠50（±2）	6~8（±0.5）	1200~2000（±10）	横担

3. 绝缘横担的检查内容

（1）绝缘横担在使用前，必须进行外观检查，主要对绝缘主材（或绝缘杆）和金属部件进行检查；检查绝缘部分是否可靠（可以借助相关仪器），金属部件是否变形、损坏；绝缘部分同金属部件的连接是否牢固可靠。

（2）对玻璃钢绝缘横担除检查其外观是否破裂和脏污外，在使用前还必须进行受力检查，无问题才能使用。严防短接玻璃钢横担。

（3）复合绝缘横担在使用前必须检查其伞裙、芯棒和前后端连接金具部分是否破裂和损坏，必须进行载荷试验和工频耐压试验。

4. 绝缘横担的使用方法

（1）配电线路使用的绝缘支线横担主要用于重新架线或维护单相及三相的导线支撑架，电杆临时横担主要用于更换电杆横担时作临时横杆架接带电电缆，电杆用三相支线横担主要用于在替换电杆、支架和绝缘瓷瓶等作业时使用。

（2）玻璃钢绝缘横担应用于10~35kV带电更换直线杆横担，具有不需要移动导线，保持导线原来位置，操作更为简单和可靠的特点。

（3）复合绝缘横担是一种新型材料的横担，它体积小、质量轻，电气和机械性能优良，多用于110kV及220kV双回路钢管杆上横担，可以进一步缩小线路走廊。

此外，绝缘横担还可以用于带电状态下直线杆耐张杆作业中。

5. 绝缘横担的保管事项

绝缘横担应置于干燥、温度合适、通风良好的带电作业专用工具房间定点上架存放。如图3-15所示为绝缘横担存放。

图3-15 绝缘横担保管存放

6. 绝缘横担的预防性试验

绝缘横担的预防性试验，内容如下。

（1）电气试验。在10kV及35kV电压等级应通过对应45kV或95kV短时工频耐受电压试验（以无击穿、无闪络及无明显发热为合格），周期为1年。

（2）机械试验。包括静负荷和动负荷试验，周期为2年。

3.2 软质绝缘工具

在软质绝缘工具中,使用最广泛的是绝缘绳。绝缘绳是广泛应用于带电作业的绝缘材料之一,可用作运载工具、攀登工具、吊拉绳、连接套及保安绳等。以绝缘绳为主绝缘部件制成的工具,具有灵活、轻便、便于携带、适于现场作业等特点。此外,利用绝缘绳或绝缘带又可以制成绝缘软梯、腰带等。软质绝缘工具主要采用蚕丝或合成纤维为原料,其中以蚕丝绳应用得最为普遍。

3.2.1 绝缘绳

1. 绝缘绳的基本结构和功能

(1) 基本结构

绝缘绳主要采用蚕丝或合成纤维为原料,其中以蚕丝绳应用得最为普遍。蚕丝在干燥状态时是良好的电气绝缘材料,但由于蚕丝的丝胶具有亲水性及纤维具有多孔性,因此蚕丝具有很强的吸湿性,当蚕丝作为绝缘材料使用时,应特别注意避免受潮。以绝缘绳为主绝缘部件制成的工具,具有灵活、轻便、便于携带、适于现场作业等特点。如图3-16所示为绝缘绳。

图3-16 绝缘绳

(2) 基本功能

绝缘绳广泛应用于带电作业工作中,是带电作业中最重要的绝缘材料之一,可用作运载工具、攀登工具、吊拉绳、连接套及保安绳等。此外,利用绝缘绳或绝缘带又可以制成绝缘软梯、腰带等。

带电作业工器具的检查、使用及保管

带电作业常用的绝缘绳有蚕丝绳（分生蚕丝绳和熟蚕丝绳）和尼龙绳（分尼龙丝绳和尼龙线绳）等。在带电作业中绝缘绳用作牵引、提升物体、临时拉线、制作软梯和滑车绳等。绝缘绳不但应有较高的机械强度，而且还应具有耐磨性。

2. 绝缘绳的分类和型号规格

（1）分类

①根据材料，绝缘绳可分为天然纤维绝缘绳和合成绝缘绳。

②根据在潮湿状态下的电气性能，绝缘绳可分为常规型绝缘绳和防潮型绝缘绳索。

③根据机械强度，绝缘绳可分为常规强度绝缘绳和高强度绝缘绳。

④根据编织工艺，绝缘绳可分为编织绝缘绳、绞制绝缘绳和套织绝缘绳。

⑤根据使用用途，绝缘绳可分为消弧绳、绝缘绳套、绝缘保险绳、绝缘测距绳和吊绳。

（2）型号规格

①消弧绳

消弧绳的型号由材料、代号、规格几部分组成。

示例：SCXHS—10×20m

其中：SC—桑蚕；XHS—消弧绳；10—绳径10 mm；20m—长度20m。消弧绳的绳径12mm为宜，长度可分为15m、20m、30m三种。

②绝缘绳套

a. 无极绝缘绳套。无极绝缘绳套的型号由材料、代号、规格几部分组成。无极绝缘绳套的规格可以根据需要做成各种绳径和长度。

示例：JCSTW—16×400m

其中：JC—锦纶长丝；ST—绳套；W—无极绳套；16—绳径16mm；400—长度400m。

b. 两眼绝缘绳套。两眼绝缘绳套的型号由材料、代号、规格几部分组成。两眼绝缘绳套的规格可以根据需要做成各种绳径和长度。

示例：JCSTL—16×400m

其中：JC—锦纶长丝；ST—绳套；L—两眼绳套；16—绳径16mm；400—长度400m。

③绝缘保险绳

a. 人身绝缘保险绳。人身绝缘保险绳的型号由材料、代号、规格几部分组成。人身绝缘保险绳的绳径不得小于14mm，长度可以做成2～7m各种规格。

示例：JCSTL—14×4m

其中：SC—桑蚕；RBS—人身保险绳；14—绳径14mm；4—长度4m。

b. 导线绝缘保险绳。导线绝缘保险绳的型号由材料、代号、规格几部分组成。

示例：SCDBS—30×4.5m

其中：SC—桑蚕；DBS—导线保险绳；30—绳径30mm；4.5—长度4.5m。

导线绝缘保险规格分类见表3-4。

表3-4　导线绝缘保险绳规格分类

型号	绳径（mm）	绳长（m）	额定负荷（kN）	电压等级（kV）	备注
SCDBS—18×2.5m	18	2.5	8	35～110	
SCDBS—22×3.5m	22	3.5	12	220	
SCDBS—34×3.5m	34	3.5	24	220	适用于2分裂导线
SCDBS—34×4.5m	34	4.5	24	330	适用于2分裂导线
SCDBS—34×5.5m	2×34	5.5	2×24	500	适用于4分裂导线

注：500 kV四分裂导线的绝缘保险绳应用2根SCDBS—34×5.5m

④ 绝缘测距绳

绝缘测距绳的型号由材料、代号、规格几部分组成。

示例：SCCS—4×50m

其中：SC—桑蚕；CS—测距绳；4—绳径4 mm；50m—长度50m。绝缘测距绳的直径一般为4～5mm，长度为50m。

3. 绝缘绳的检查内容

（1）绝缘绳在使用前，必须做外观检查，严禁有金属丝物夹带缠绕。所有由绝缘绳类工具捻合成的绳索合绳股应紧密绞合，不得有松散、分股的现象。绳索各股及各股中丝线不应有叠痕、凸起、背股、抽筋等缺陷。

（2）绝缘绳在使用前必须检查其试验合格证是否在有效期内，对超过有效试验期的绝缘绳不能使用。

（3）绝缘绳使用前根据工作荷载选用绝缘绳种类和直径；并使用2500V及以上兆欧表或绝缘检测仪进行分段绝缘检测（电极宽2 cm，极间宽2 cm），阻值应不低于700 MΩ。如图3-17所示为绝缘绳绝缘性能检查。

图3-17 绝缘绳绝缘性能检查

4. 绝缘绳的使用方法

（1）绝缘绳必须经试验合格后方可使用，严禁使用不合格的绝缘绳；须根据环境湿度选择防潮绝缘绳索；脏污严重的绝缘绳严禁在带电作业工作中使用。

（2）绝缘绳使用人员应戴清洁、干燥的手套。

（3）绝缘绳使用过程中必须放在防潮垫布上，防止受潮和脏污。

（4）绝缘绳在使用过程中应满足相对应电压等级的安全距离。

如图3-18所示为绝缘绳的使用。

图3-18 绝缘绳的使用

5. 绝缘绳的保管注意事项

绝缘绳在运输过程中应装专用工具袋、工具车或专用工具车内，以防受潮和损伤。绝缘滑车应置于通风良好、有红外线灯泡或除湿设备的清洁干燥的专用房间存放。贮存期超过生产日期12个月，要按相关标准进行电气性能检验。如图3-19所示为绝缘绳的存放。

图3-19 绝缘绳的存放

6. 绝缘绳的预防性试验

绝缘绳的预防性实验内容如下。

（1）电气试验。主要是工频耐压试验和操作冲击耐压试验，周期为1年。

（2）机械试验。主要是静拉力试验，周期为2年。

3.2.2 绝缘软梯

1. 绝缘软梯的基本结构和功能

（1）基本结构

绝缘软梯的边绳和环形绳应采用桑蚕丝或不低于桑蚕丝性能的阻燃绝缘纤维原材料制作。横蹬应采用环氧酚醛层压玻璃布管原材料制作。

如图3-20所示为绝缘软梯。

图 3-20　绝缘软梯

（2）基本功能

带电作业过程中，绝缘软梯用于高处攀登、高处作业及等电位作业中。

2. 绝缘软梯的分类和技术规格

（1）分类

① 根据绝缘软梯步距，可分为300mm、400mm、500mm三种梯步距离的绝缘软梯。

② 根据在潮湿状态下的电气性能，绝缘软梯分为常规型绝缘软梯和防潮型绝缘软梯。

（2）技术规格

① 用作横蹬的环氧层压玻璃布管，其外径为22mm，壁厚为3mm，长度为300mm，两端管口呈$R1.5$的圆弧状，且应平整、光滑，外表面涂有绝缘漆。

② 环形绳与边绳的绳径为10mm，绳股的捻距为32±0.3mm。

③ 软梯头的所有部件表面均应做防腐蚀处理。

3. 绝缘软梯的检查内容

（1）各部件连接应紧密牢固，整体结构性好。

（2）软梯头的主要部件应表面光滑，无尖边、毛刷、缺口、裂纹等缺陷。

（3）环形绳与边绳的连接应牢固、平服，捻合成的绳索合绳股应紧密绞合，不得有松散、分股的现象。绳索各股及各股中丝线不应有叠痕、凸起、压伤、背股、抽筋等缺陷，不得有错乱、交叉的丝、线、股。

（4）绝缘软梯使用前对绝缘绳部分使用2500V及以上兆欧表或绝缘检测仪进行分段绝缘检测（电极宽2cm，极间宽2cm），阻值应不低于700MΩ。

（5）绝缘软梯在安装完毕后，必须首先进行受力的冲击试验，以检验绝缘软梯的安装及承力情况。如图3-21所示为绝缘软梯使用前的受力检查。

图3-21 绝缘软梯使用前的受力检查

（6）绝缘软梯在使用前必须检查其试验合格证是否在有效期内，不能使用超过有效试验期的绝缘软梯。

4. 绝缘软梯的使用注意事项

（1）根据作业点的高度选择绝缘软梯的长度。

（2）进等电位进电场时，先挂跟斗滑车及吊绳；通过吊绳吊起绝缘软梯及后备保护绳。

（3）攀登软梯必须使用后备保护绳。攀登有走侧边攀登和正面攀登两种方式。

（4）在攀登软梯与梯头之间转移电位时，要注意人体裸露部分与带电体之间的距离。

（5）在架空线路上使用软梯作业或用梯头进行移动作业时，软梯或梯头上只准一人工作。工作人员到达梯头上进行工作和梯头开始移动前应将梯头的封口可靠封闭，否则应使用保护绳防止梯头脱钩。如图3-22所示为绝缘梯在等电位作业中的使用。

图3-22 绝缘软梯在等电位作业中的使用

5. 绝缘软梯的保管注意事项

绝缘软梯在运输过程中应装在专用工具袋、工具车或专用工具车内，以防受潮和损伤。绝缘软梯应置于通风良好、备有红外线灯泡或除湿设备的清洁干燥的专用房间存放。贮存期超过生产日期12个月时,要按相关标准进行电气性能检验。如图3-23所示为绝缘梯的存放。

图3-23　绝缘软梯存放示例图

6. 绝缘软梯的预防性试验

绝缘软梯的预防性试验内容如下。

(1) 电气试验。其主要包括工频耐压试验和操作冲击耐压试验,试验周期为1年。

(2) 机械试验。其主要包括绝缘软梯拉力试验和软剃头动、静负荷试验,周期均为2年。

3.3 其他绝缘工具

其他绝缘工具包括绝缘滑车、绝缘紧线工具等,它们在输配电线路带电作业操作中应用也非常广泛。

3.3.1 绝缘滑车

1. 绝缘滑车的基本结构和功能

(1)基本结构

绝缘滑车由挡板、滑轮、轴承、中轴、拉板、吊轴和吊钩组成,如图3-24所示。滑轮一般用尼龙、有机玻璃或工程塑料等材料,采用车制或压塑方法成形,内径装有轴承,其个数根据荷载的大小确定;隔板及加强板均用3240环氧玻璃布板制成;吊钩和一般滑车相同,由优质结构钢锻造而成,少数关键绝缘部位的吊钩也可用3240环氧玻璃布板制作。

(2)基本功能

在带电作业吊装过程中,用于绳索导向或承担负载的工具。在线路作业中,起吊绝缘子串、带电作业工器具和等电位作业人员进出电场,作业力转向、换向均需要使用绝缘滑车。根据作业位置的不同,可以选择不同形状的挂钩;根据承受力的大小也可以选择承受对应拉力滑车组,使工作省力。

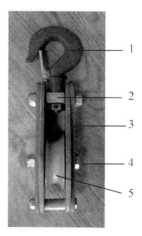

1—吊钩;2—吊轴;3—拉板;4—滑车;5—中轴
图3-24 绝缘滑车结构

2. 绝缘滑车的分类和技术要求

（1）分类

绝缘滑车共分为15种型号，其型号规格见表3-5。型号编制采用汉语拼音第一个字母及阿拉伯数字表示的方法：JH表示绝缘滑车，JH之后的数字表示额定负荷，短横线后的数字表示滑轮个数，最后一个字母表示结构特点。如B—侧板闭口型，K—侧板开口型，D—短钩型，C—长钩型，J—绝缘钩型，X—导线钩型。

表3-5 绝缘滑车型号规格

型号	名称	额定负荷	滑轮个数	备注
JH5-1B	单轮闭口型绝缘滑车	5	1	
JH5-1K	单轮开口型绝缘滑车	5	1	
JH5-1DY	单轮多用钩型绝缘滑车	5	1	
JH5-2D	双轮短钩型绝缘滑车	5	2	
JH5-2X	双轮导线钩型绝缘滑车	5	2	
JH5-2J	双轮绝缘钩型绝缘滑车	5	2	
JH5-3D	三轮短钩型绝缘滑车	5	3	
JH5-3X	三轮导线钩型绝缘滑车	5	3	
JH10-2D	双轮短钩型绝缘滑车	10	2	
JH10-2C	双轮长钩型绝缘滑车	10	2	
JH10-3D	三轮短钩型绝缘滑车	10	3	
JH10-3C	三轮长钩型绝缘滑车	10	3	
JH15-4D	四轮短钩型绝缘滑车	15	4	
JH15-4C	四轮长钩型绝缘滑车	15	4	
JH20-4D	四轮短钩型绝缘滑车	20	4	
JH20-4C	四轮长钩型绝缘滑车	20	4	

（2）技术要求

①零件及组合件按图纸检查合格后才能使用装配。

②装配后滑轮在中轴上应转动灵活，无卡阻和碰擦轮缘现象。

③吊钩、吊环在吊梁上应转动灵活。

④各开口销不得向外弯，并切除多余部分。

⑤侧向螺栓高出螺母部分不大于2mm。

⑥侧板开口在开合90°范围内无卡阻现象。

⑦各种型号绝缘滑车均应按相关规定进行电气性能试验，交流工频耐压试验1min后应不发热、不击穿。

⑧机械性能试验按1.6倍额定负荷，持续5 min无永久变形或开裂，滑车的破坏拉力不得小于3倍额定负荷。

3. 绝缘滑车的检查内容

（1）绝缘滑车的绝缘部分应光滑，无气泡、皱纹、开裂等现象。

（2）滑轮在中轴上应转动灵活，无卡阻和碰擦轮缘现象。

（3）侧板开口在开合90°范围内无卡阻现象。

（4）绝缘滑车在使用前必须检查其试验合格证是否在有效期内，对超过有效试验期的绝缘滑车不能使用。

（5）检查绝缘滑车受力是否满足要求。

4. 绝缘滑车的使用方法

（1）根据使用需要选择绝缘滑车承重吨位。

（2）根据使用地点特点不一样可以选择短钩型或长钩型。

（3）更换直线绝缘子时采用滑轮组省力，距离控制更好。

如图3-25所示为绝缘滑起车吊工具。

图3-25　绝缘滑车起吊工具

5. 绝缘滑车的保管注意事项

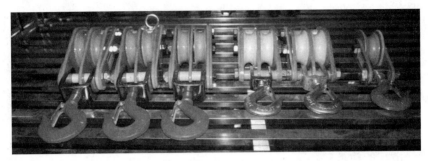

图 3-26　绝缘滑车存放示例图

绝缘滑车应置于通风良好、清洁干燥的专用房间存放贮存期超过生产日期 12 个月，要按相关标准进行电气性能检验。如图 3-26 所示为绝缘滑车的存放。

6. 绝缘滑车的预防性试验

绝缘滑车的预防性试验内容如下。

（1）电气试验。主要是工频耐压试验，试验周期为 1 年。

（2）机械试验。主要是拉力试验，试验周期为 1 年。

3.3.2　绝缘紧线工具

1. 绝缘紧线工具的基本结构和功能

（1）基本结构

绝缘棘轮紧线器由操作杆、棘轮机构、绝缘带和紧线挂钩等部分组成，如图 3-27 所示。

图 3-27　绝缘棘轮紧线器

（2）基本功能

常常用于收紧钢芯铝绞线、钢绞线和线路。

2. 绝缘紧线工具的分类和技术规格

（1）分类

绝缘紧线工具常用的有两大类：绝缘双沟紧线器、绝缘棘轮紧线器。绝缘双钩紧线器如图3-28所示。

图3-28 绝缘双钩紧线器

（2）技术规格

①绝缘棘轮紧线器绝缘杆由玻璃纤维材料构成，操作安全可靠，挂钩可以360°旋转，配有安全锁扣装置，不锈钢弹簧可以防腐蚀，铜制的装接器可以防潮湿。

②绝缘双钩紧线器由操作杆和外包橡胶组成。用操作杆可以操作绝缘双钩。导线钩上有连接操作杆的小孔。操作杆上装有弹簧锁，可以确保安全。金属部分由热处理过的铝合金和铜合金组成，牢固可靠。

美国只生产绝缘棘轮紧线器，其技术参数见表3-6，日本生产的双沟绝缘紧线器和绝缘棘轮紧线器参数见表3-7和3-8所示。

表3-6 美制绝缘棘轮紧线器技术参数

型号	单带			双带			手柄长 (mm)	质量 (kg)
	负荷 (kg)	提线长度 (m)	最小头距 (mm)	负荷 (kg)	提线长度 (m)	最小头距 (mm)		
E153-10	1500	10	22	3000	5	27	20	12
E153-10	1500	10	22	3000	5	27	20	12
EH24-12	2000	15	17	4000	7.5	30	30	12

表3-7 日制绝缘双钩紧线器技术参数

型号	使用电压 (kV)	最大负荷 (kg)	最大紧线距离 (mm)	缩长 (mm)	伸长 (mm)	绝缘长度 (mm)	质量 (kg)	备注
3464	5/35	1814	305	1499	1804	660~914	5.4	双带式
3465	46/69	1814	305	1804	2108	965~1219	5.6	双带式

表3-8　日制绝缘棘轮紧线器技术参数

型号	负荷(t)	织带尺寸(mm)	缩长(mm)	扬程(mm)	手柄长度(mm)	质量(kg)	备注
N-1000	1	2	420	850	400	3.2	双带式
N-1500R	1.5	2	450	850	460	4.5	双带式

3. 绝缘紧线工具的检查内容

（1）检查机械部分转动灵活、无卡瑟。

（2）使用时，请尽量使绝缘织带远离锋利棱角，避免受磨损或切割。

（3）不要把紧线工具当作负载起重调整使用。

（4）绝缘编织带或绝缘管不得有脏污、烧伤等。

（5）检查绝缘紧线工具铭牌，不可以超载使用。

（6）绝缘紧线工具在使用前必须检查其试验合格证是否在有效期内，对超过有效试验期的绝缘紧线工具不能使用。

如图3-29所示为日制绝缘棘轮紧线器的绝缘检查。

图3-29　日制绝缘棘轮紧线器检查

4. 绝缘紧线工具的使用方法

绝缘紧线工具是在配电架空线路带电检修中作拉紧导线用的。根据线路的荷载选择相应的绝缘紧线工具。以绝缘棘轮紧线器为例叙述使用方法。使用时先把紧线器上的绝缘织带松开，将操作手柄端固定在横担上，另一端用卡线器夹住导线，其中绝缘织带经过绝缘子部分时扳动专用扳手，由于棘爪的防逆转作用，逐渐把绝缘织带绕在棘轮滚筒上，使导线收紧，把收紧的导线固定在绝缘子上。然后先松开棘爪，使绝缘织带松开，再松开卡线器，最后把绝缘织带绕在棘轮的滚筒上。绝缘紧线器的使用如图3-30所示。

图3-30 绝缘紧线器的使用

5. 绝缘紧线工具的保管注意事项

绝缘紧线工具应置于通风良好、清洁干燥的专用房间上架存放。贮存期超过生产日期1年时要按相关标准进行电气性能检验。图3-31所示为绝缘棘轮、绝缘双钩紧线器的存放。

1—绝缘棘轮紧线器；2—绝缘双钩紧线器

图3-31 绝缘棘轮、绝缘双钩紧线器的存放

6. 绝缘紧线工具的试验

绝缘紧线工具的预防性试验内容如下。

（1）电气试验。工频耐压试验，试验周期为1年。

（2）机械试验。机械拉力试验，试验周期为1年。

第4章

带电作业特种车辆

输配电线路带电作业特别是配电线路带电作业中，随着国外带电作业设备的引进和国内带电作业技术的进步，带电作业工器具制造厂家逐步发展出将带电作业工器具与汽车相结合的复合型工具，比如绝缘斗臂车、带电作业工具车、旁路电缆绝缘斗臂车、移动箱变车等一系列带电作业特种车辆，初步实现了带电作业机械化，有效地降低了带电作业的劳动强度，提升了安全作业水平，提高了作业效率。本章将重点介绍绝缘斗臂车、旁路作业车、负荷开关车、移动箱变车、带电作业工具车等。

4.1 绝缘斗臂车

绝缘斗臂车，是一种在交通方便且布线复杂的场合进行等电位作业的特种车辆，通过其绝缘臂、工作斗等能够实现带电作业必需的主绝缘，能在10kV及以上电力线路上进行带电高处作业的特种车辆。只采用工作斗绝缘的高空绝缘斗臂车一般不列入绝缘斗臂车范围。在实际工作中，受线路杆塔位交通条件、配套底盘、使用效率、产品价格等多种因素的限制，输电线路带电作业极少使用绝缘斗臂车，绝缘斗臂车通常在10kV、35kV和66kV的城区配电线路检修工作中使用，因此绝缘斗臂车在配电线路带电作业中起到非常重要的作用。

4.1.1 绝缘斗臂车的基本结构和功能

绝缘斗臂车一般采用汽车发动机和底盘改装而成，车的后部安装动转盘和液压斗臂装置，臂采用折叠或伸缩结构，前端带方形或圆形绝缘斗。整个斗臂装置安装在一个转盘上，可以360°旋转。为了增加车子的稳定性，从而最大限度地加大斗臂的长度，一般都装有液压支腿，其结构如图4-1所示。

绝缘斗臂车的工作斗、工作臂、控制油路、斗臂结合部都能满足一定的绝缘性能指标，并带有接地装置。绝缘臂一般采用玻璃纤维增强型环氧树脂材料制成，其一般绕制成圆柱形或矩形截面结构，具有质量轻、机械强度高、电气绝缘性能好、憎水性强等优点，在带电作业时为人体提供相对地的绝缘防护。绝缘斗有单层斗和双层斗，外层斗一般采用环氧玻璃钢制作，内层斗采用聚四氟乙烯材料制作。

1—绝缘工作斗；2—动转盘；3—液压支腿；4—绝缘臂；5—接地线
图4-1 绝缘斗臂车结构

4.1.2 绝缘斗臂车的分类

（1）根据工作臂的结构形式，可将绝缘斗臂车分为折叠臂式、直伸臂式、多关节臂式、垂直升降式和混合式。最常见的是直伸臂式和折叠臂式，如图4-2所示。

（2）根据升降高度，可将绝缘斗臂车可分为6m、8m、10m、12m、16m、20m、25m、30m、35m、40m、50m、60m、70m等。

（3）根据作业线路电压等级，可将绝缘斗臂车分为10kV、35kV、46kV、63（66）kV、110kV、220kV、330kV、345kV、500kV、765kV等。我国的绝缘斗臂车通常在10kV、35kV和66kV的线路上使用。

1—直伸臂式绝缘斗臂车；2—折叠臂式绝缘斗臂车
图4-2 绝缘斗臂车

4.1.3 绝缘斗臂车的技术要求

1. 工作条件

（1）风速不超过 10.8m/s。

（2）环境温度为 -25~40℃。

（3）相对湿度不超过 90%。

（4）对海拔 1000m 及以上地区的要求：选用的底盘动力应适应高原行驶和作业，海拔每增加 100m，绝缘体的绝缘水平应相应增加 1%。

（5）地面坚实、平整，作业过程中支腿不下陷。

（6）转台平面处于水平状态。

（7）绝缘工作斗电气绝缘性能必须满足相应要求。

（8）绝缘工作斗的表面平整、光洁，无凹坑、麻面现象，憎水性强。

（9）绝缘工作斗的高度宜在 0.9~1.2m。

（10）绝缘工作斗上应醒目注明工作斗的额定载荷量。

2. 绝缘臂的要求

（1）绝缘臂的电气性能试验需满足相应要求。

（2）绝缘臂的表面平整、光洁，无凹坑、麻面现象，憎水性强。

（3）各电压等级绝缘斗臂车绝缘臂的最小绝缘长度不宜小于相应规定要求。

3. 整车的要求

（1）接地部分与工作斗之间仅绝缘臂绝缘的绝缘斗臂车，其整车电气绝缘性能应符合相应要求。

（2）具有上下操作功能及自动平衡功能的绝缘斗臂车，其整车电气绝缘性能还需要满足部分单独试验泄漏电流小于 $200\mu A$ 要求。

4. 绝缘液压油的要求

用于承受带电作业线路相应电压的液压油，应对其进行击穿强度试验。

5. 绝缘性能要求

绝缘斗臂车应在说明书和铭牌上清楚标明额定电压，每台车出厂前应进行例行绝缘性能试验。

4.1.4　绝缘斗臂车的使用方法

正确使用和操作绝缘斗臂车是保障作业人员的人身安全和车辆安全的基础。

1. 发动机的启动和取力器及支腿的操作

（1）挂好手刹车，垫好三角块。

（2）确认变速器杆处于正确位置。

（3）将离合器踏板踩到底，启动发动机。

（4）踩住离合器踏板，将取力器开关扳至"开"的位置。

（5）缓慢地松开离合器踏板。

（6）油门高低速的操作。

（7）水平支腿操作。在转换杆中，选出欲操作的水平支腿的转换杆，切换至"水平"位置，当将"伸缩"操作杆扳至"伸出"位置时，水平支腿就会伸出，如图4-3所示。

图4-3　绝缘斗臂车支腿操作杆

（8）垂直支腿操作。正确操作支腿，以免支腿跑出或收缩，造成车辆损坏。绝缘斗臂车工作时的支腿如图4-4所示。

图4-4　绝缘斗臂车工作时的支腿

(9）收回操作方法，将各支腿收回到原始状态，根据"垂直支腿—水平支腿"的顺序，按（8）项和（9）项相反的顺序进行收回操作，收回后，各操作杆定要返回到中间位置。

2. 安装接地线

回出线盘上的接地线，将绝缘斗臂车的接地线与杆塔接地引线可靠连接。绝缘斗臂车安装接地线如图4-5所示。

图4-5　安装接地线

3. 上部操作（工作斗操作）

（1）工作臂的操作

①下臂操作（臂的升降操作），如图4-6所示为工作斗内操作盘。

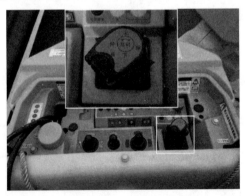

图4-6　工作斗内操作盘

②回转操作。

③上臂操作（伸缩操作）。

（2）工作斗摆动操作

将工作斗摆动操作杆按标牌箭头方向扳，使工作斗右摆动或左摆动。

(3）紧急停止操作

接通紧急停止操作杆时，上部的动作均会停止，但发动机不会停止。在下述情况时需紧急停止操作：

① 工作斗上的作业人员为避免危险情况需停止工作臂的动作；

② 操作控制出现失控的情况。

(4）小吊臂操作

如图4-7和图4-8所示，有的绝缘斗臂车设有工作斗小吊臂，其操作可参考相关使用手册。

图4-7　斗内小吊臂操作

图4-8　斗内小吊臂装置

(5）辅助装置的操作

将进油、回油及泄油接头与液压工具的相应接头可靠连接，把辅助操作杆转换到"工具"位置。

4. 下部操作（转台处操作盘的操作）

(1）工作臂的操作

在工作臂收回到托架上的状态下，不可进行工作臂的回转操作，转台处工作臂操作盘如图4-9所示。

（2）紧急停止操作

使用紧急停止操作杆进行紧急停止操作。接通紧急停止操作杆时，上部及下部操作的全部动作均停止。

（3）应急泵的操作

转台处应急泵的操作按钮，如图4-10所示。

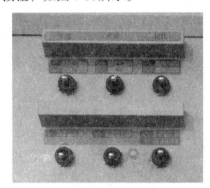

图4-9　转台处工作臂操作盘

图4-10　转台处应急泵的操作按钮

4.1.5　绝缘斗臂车的保养、维护及检查

1. 绝缘斗臂车的保养

运输期间的保养。工作斗必须回复到行驶位置。带吊臂的绝缘斗臂车，吊臂应卸掉或完全缩回，上、下臂均应回复到各自独立的支撑架上，且必须固定牢靠，以防止在运输过程中由于晃动受到撞击而损坏。绝缘斗臂车在行进过程中，两臂的液压操作系统必须切断，以防止工作斗的液压平衡装置来回摆动。

绝缘斗臂车暴露在露天环境时，由于雨水、路面灰尘、腐蚀和其他大气污染将会影响绝缘斗、臂的绝缘特性，降低其绝缘耐受水平。长时间的紫外线照射也会影响其绝缘性能。因此，在运输过程中应采用防潮保护罩进行防护。

2. 绝缘斗臂车的维护

（1）一般要求。必须建立一套绝缘斗臂车定期检查程序，包括详尽的外观检查和绝缘强度试验。

（2）清洁。如绝缘部件表面沾染了较小的污垢，可以用不起毛的布擦拭干净。如果绝缘部件异常脏污，可采用高压热水冲洗（水温不超过50℃，压力不超过690kPa）。

（3）涂硅。应先清洗和干燥绝缘表面后再行涂硅。

3. 检查

绝缘斗臂车的检查分为外观检查和功能检查。

绝缘部件的外观检查须在其外表面清洗后进行，上下臂、工作斗、吊臂等必须使用不起毛的布擦拭，如果需要，可以采用洁净布蘸少许异丙醇或其他合适的溶剂轻轻擦拭。外观检查主要针对结构损坏。结构损坏包含由于撞击而产生的绝缘破裂，以及玻璃纤维裸露、破洞以及沟痕，还有与树枝、电线杆等尖锐物体相撞击而产生的开裂痕迹，和由于超载而引起的上下臂连接处或靠近钢制连接件部分出现的裂缝、隆起等。

（1）开始工作前的检查

检查的目的是剔除绝缘斗臂车前期工作和库存期间可能遗留的缺陷和故障。主要由作业人员进行外观检查，同时辅以功能检查，重点倾听各部件是否有异常的噪声。

检查并确认绝缘斗臂车的绝缘部件的绝缘试验是否在有效期内。

①整车检查

外观检查和功能检查必须在正式开始工作前完成，并将检查结果以图表的形式记录下来。

外观检查：检查绝缘部件表面的损伤情况，如裂缝、绝缘剥落、深度划痕等。

功能检查：启动绝缘斗臂车后，在工作斗无人的情况下，采用下部控制系统操作绝缘臂伸缩、旋转及工作斗升降循环，检查是否有液体渗出、液压缸有无渗漏、异常噪声、工作失灵、漏油、不稳定运动或其他故障。

为了保证安全，应检查并操作各电源和紧急制动系统的灵活及可靠性，还应检验可视和音响报警装置。

对用于超高压及以上电压等级的绝缘斗臂车，其用于等电位作业部分时必须检查均压环，确认等电位点，并进行泄漏电流试验。

②工作斗检查

检查工作斗的底板是否有脏污或其他可能会损坏工作斗的物体，或在等电位作业的情况下会妨碍底板与导电鞋良好接触的物体。

检查工作斗的机械损伤，是否存在孔洞、裂缝或剥层等，并用蘸有适当溶剂的不起毛的软布将工作斗擦拭干净，并将斗内材料碎片和剥落物清除干净。

（2）每周检查

绝缘斗臂车应每周检查一次。

①整车检查。包括外观检查和功能检查。

②工作斗检查。内斗必须从外斗中取出，将污秽物清除干净。若存在损伤或剥落，必须查明是由机械损伤或化学因素引起的。任何机械损伤都会减少内斗的壁厚。当小于制造厂商推荐的壁厚最低值，在重新使用前，内外斗必须做电气试验。

（3）定期检查。定期检查周期，可根据生产厂商的建议和其他影响因素，如运行状况、保养程度、环境状况来确定。一般正常定期检查的最大周期为1年。

4.1.6　绝缘斗臂车预防性试验

绝缘斗臂车预防性试验项目包括绝缘工作斗工频耐压试验、绝缘工作斗泄漏电流试验、绝缘臂工频耐压试验、绝缘臂泄漏电流试验、整车的工频耐压试验、整车的泄漏电流试验、绝缘液压油击穿强度试验等，试验周期为半年。

4.2 旁路作业车及负荷开关车

采用旁路作业设备实施配网不停电作业的方法在国内外配电线路中得到了广泛应用并在提高供电可靠性方面取得了良好的效果。由于一套旁路作业设备拥有数量众多的旁路电缆、开关、滑轮、连接金具等,以往实施旁路作业的所有部件不能实现定制管理,运输途中容易磕碰导致部分设备损坏,另外因受道路交通限制,极大限制了旁路作业设备在应急抢修和配电线路不停电作业工作中的应用。旁路作业车的研制成功和推广应用,很好地解决了作业现场很多操作困难的问题,加上旁路作业负荷开关车的配合,大大缩短了旁路作业设备应用过程中的准备工作时间,为大力开展配电线路不停电作业提供有利条件。

如图4-11所示是旁路作业车,如图4-12所示为负荷开关车。

图4-11　旁路作业车

图4-12　负荷开关车

4.2.1 旁路作业车及负荷开关车的基本结构

1. 旁路作业车的基本结构

旁路作业车主要由车辆平台、电缆收放装置、部件收纳箱等组成，如图4-13所示。车辆平台包括车辆底盘、厢体（车厢）结构等，是旁路作业车的运输载体。电缆收放装置主要由环形轨道、三联电缆卷盘、卷盘驱动机构、起吊装置等组成。部件收纳箱用于定置存放（除旁路柔性电缆之外）旁路负荷开关、转接电缆、电缆连接器等旁路作业设备部件。旁路作业车采用分舱设计，有独立的驾驶室、部件收纳箱、电缆收放装置、工具箱等。

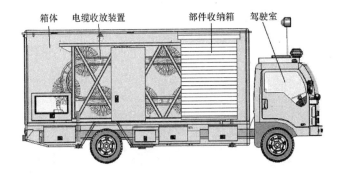

图4-13 带电作业旁路作业车整体结构图

2. 负荷开关车的基本结构

采用工程车底盘，车内配备一组电液驱动电缆卷盘装置和一台旁路负荷开关，其结构如图4-14所示。

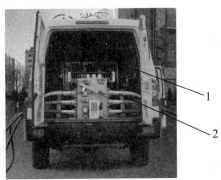

1—电缆卷盘装置；2—旁路负荷开关
图4-14 负荷开关车结构

4.2.3 旁路作业车的主要技术要求

1. 工作条件

旁路作业车在下列环境条件下应能正常工作：
（1）海拔不超过1000m；
（2）环境温度：-40~40℃；
（3）相对湿度不大于95%（25℃时）。
特殊使用条件时，应按照使用要求进行设计。

2. 功能要求

（1）定置装载旁路柔性电缆。整车为厢式工程车，在车厢内配置电缆收放装置，电缆收放装置主要由环形轨道、三联电缆卷盘、卷盘驱动机构、起吊装置等组成。电缆收放装置应定置装载不少于18盘旁路柔性电缆，如图4-15。

图4-15 电缆收放装置

（2）部件收纳箱。用于定置存放旁路负荷开关、转接电缆、电缆连接器等全部旁路作业设备部件。分类置放各种旁路作业部件并设计专用工装卡具可靠固定，存放小型部件的压型模应有数量标识，实现各部件的定置管理，防止工具在运输中互相磕碰和颠簸。

（3）手动收、放旁路柔性电缆功能。三联电缆卷盘横向并列安装在环形轨道内，通过电动或液压机构驱动每组卷盘，可按顺序逐个移动到车厢尾部指定收放旁路柔性电缆位置，每组卷盘在行驶状态应自动锁紧防止其窜动。一组卷盘装置在收放旁路柔性电缆位置应根据工作需要具有分别进行连续、点动、三相同时及单相收放功能。电缆卷盘应有定位锁紧功能，防止车辆在行驶过程中电缆卷盘移动和自转。电缆收放操作应通过配置的有线遥控操作装置实现，控制线缆长度不小于3m。机构设计时应有足够的检修空间，便于维护。

（4）现场快速拆解电缆卷盘的功能。配置的随车起吊装置可将电缆卷盘吊放到车厢外，一组电缆卷盘可快速拆分为3个单体卷盘以便于运送和卷盘的检修，起重臂额定起重量不得小于500kg。

（5）夜间作业现场照明功能。驾驶室外顶部安装车载升降式照明装置，配备全方位转向云台，用于夜间作业现场提供照明。照明电源宜采用车辆底盘蓄电池直流24V电源，照明灯具宜采用LED灯等节能灯具，照度满足现场工作要求。

（6）车辆存放支撑功能。车厢底部应配置4处液压垂直伸缩支腿，支腿伸出后应使轮胎不承载，并能承受整车和货载总质量，液压伸缩支腿的控制系统应安装在便于操作的位置。

（7）扩展功能。随着技术的进步和成熟，可增加新的功能。

（8）改装。旁路作业车采用已定型汽车整车进行改装。

3. 辅助系统

（1）电气系统。电路系统应设电源总开关，并将其布置在操作人员便于操作使用的位置。

（2）照明系统。旁路作业车的照明包括车辆本体照明、工作照明和应急照明。

4.2.3 旁路作业负荷开关车主要技术要求

由于旁路电缆及连接器的额定电流为200A，因此旁路负荷开关的额定电流不小于200A即可，旁路负荷开关如图4-16所示。按照200A的额定电流，结合10kV三相负荷开关的技术要求，其性能参数见表4-1。

图4-16 旁路负荷开关

表4-1 旁路负荷开关性能参数

序号	参数		标准	序号	参数		标准
1	额定电压（kV）		12	11	冲击耐受电压（kV）	对地	75
2	额定电流（A）		200			相间	75
3	额定频率（Hz）		50			断口间	85
4	额定开断负荷电流（A）		200	12	3s热稳定电流（kA）		16
5	额定开断负荷电流次数		20	13	动稳定电流（kA）		40
6	额定断开充电电流（A）		20	14	开关导通的接触电阻（μΩ）		<200
7	额定断开充电电流次数		20	15	三相分段不同期性（ms）		<5
8	开断时间（ms）		20	16	机械寿命		3000
9	关合短路电流能力（kA）		40	17	年漏气率		<0.5%
10	工频耐受电压（kV）	对地	42	18	防护等级		IP68
		相间	42				
		断口间	48				

4.2.4 旁路作业车及负荷开关车的使用注意事项

1. 旁路作业车

（1）旁路作业车是转载旁路设备的专用运输车辆，成本较高，运输和使用时应严格防止其受到挤压和磕碰。

（2）旁路作业车应指定专人操作，未经培训、许可的人员不得操作旁路作业车。

（3）车辆停车保管和使用时，应及时支起液压辅助支腿，以免车轮承重变形损坏。

2. 负荷开关车

（1）为使旁路负荷开关与旁路柔性电缆方便连接，其外接接口采用插拔式接口，连接方便、可靠。开关的分合指示标志清楚、明显，无论采用何种方式合闸，合闸以后必须闭锁，严禁出现在不经操作和不加任何措施情况下出现分闸的可能。它还具有压力显示和失压闭锁装置。

（2）旁路负荷开关具备核相功能，在分闸位置时可对开关两侧电压进行核相，避免旁路回路组装时相序错误合闸造成相间短路。否则，应具备验电核相功能的端子，端子电压不大于800V。核相装置或核相仪必须具有明显的同相或异相的指示信号、音响警报信号等。

（3）当拉开电源侧旁路负荷开关时，旁路负荷开关断开的是空载旁路电缆的电容电流即充电电流。空载电缆的充电电流与电缆的长度、线径、电压等有关。按照中国电科院高压研究所的试验分析，当电缆线路单芯（相）截面积不大于300mm²、长度不大于3km时，断接的最大单相稳态电容电流为5A。一般情况下架空线路旁路作业线路长度都比较短，旁路柔性电缆的截面积也较小，一般为35~50mm²，在这种情况下，拉开电源侧旁路负荷开关时断开的充电电流远小于20A。因此旁路开关的寿命不能简单地用断开负荷电流或充电电流次数衡量，当然也不等同于机械寿命，其寿命还与使用、保养和维护有关。

4.2.5 旁路作业车及负荷开关车的试验和存放

1. 车辆的试验

（1）带电作业旁路作业车试验包括型式试验、出厂试验和验收试验。型式试验和出厂试验由厂家进行。

（2）验收试验由用户与厂家共同完成。包括外观检查、电源切换试验、电缆收放、开关切换、照明系统和车辆存放支撑功能等多个方面。

（3）此外负荷开关车的旁路负荷开关应定期进行断开状态下断口间的工频耐压试验，以及闭合状态下的相地、相间工频耐压试验。使用前应采用绝缘电阻检测仪检查开关相地、相间、断口间的绝缘电阻，确保其值均大于700MΩ。

2. 车辆的存放

（1）旁路作业车和负荷开关车应存放在专用的库房内，库房存放的条件应满足防盗、防潮和通风良好的条件，并配备相应的消防设施。

（2）存放在库房内的车辆应将起门窗、抽屉等活动部件处于关闭状态，所有设备处于牢固的固定或绑扎状态。

（3）车辆应按机动车说明书进行定期的维护和保养。

4.3 移动箱变车

移动箱变车也称负荷转移车，就是装有一台箱式变电站的移动电源，箱变的高低压侧分别安装一组高压负荷开关和低压空气开关。通过负荷转移实现对杆上配电变压器的不停电检修，也可以从高压线路临时取电给低压用户供电。

4.3.1 移动箱变车的分类、结构和功能

1. 移动箱变车的分类

（1）按汽车产品分类。移动箱变车按照GB/T3730.1—2001《汽车和挂车类型的术语和定义》进行分类，属于特种专用作业车。按照GB/T 17350—2009《专用汽车和专用挂车术语、代号和编制方法》进行分类，属于箱式汽车。如图4-17所示为移动箱变车。

（2）按车载配置设备分类。移动箱变车按车载配置设备分为基本型和扩展型。

①基本型。开展较简单的配电线路及电缆临时供电作业项目。

②扩展型。开展较复杂的配电线路及电缆临时供电作业项目。

图4-17 移动箱变车示例图

2. 移动箱变车的基本结构

移动箱变车主要由车辆平台、车载设备、辅助系统等组成。

（1）车辆平台。车辆平台包括车辆底盘、箱体（车厢）结构等，是移动箱变车的运输载体。

（2）车载设备。车载设备主要包括变压器、旁路负荷开关、旁路柔性电缆、低压配电屏等。

（3）辅助系统。移动箱变车的辅助系统主要包括电气、照明、接地、液压、安全保护等系统。

移动箱变车内部结构如图4-18所示。移动箱变车旁路作业工作示意图如图4-19所示。

图4-18 移动箱变车内部结构

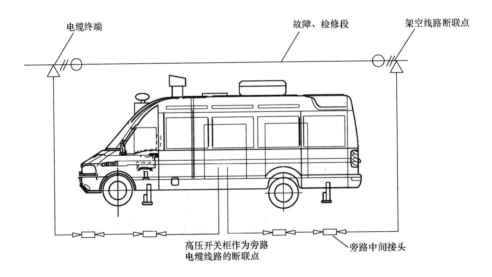

图4-19 移动箱变车旁路作业工作

3. 移动箱变车的基本功能

移动箱变车应具备输送、转换电能的不间断供电能力，其主要功能见表4-2。随着技术的进步和成熟，可增加新的功能。

第4章 带电作业特种车辆

表4-2 移动箱变车的主要功能

设备名称	序号	功能/项目	基本型	扩展型
旁路柔性电缆卷盘	1	手动卷缆	●	●
	2	机械或液压卷缆	○	○
低压电缆卷盘	1	手动卷缆	○	●
	2	机械或液压卷缆	○	●
相位检测	1	高压侧相位检测	●	●
	2	低压侧相位检测	●	●
	3	自动相位检测	○	○
低压翻相	1	手动翻相	●	●
	2	自动翻相	○	○
高低压侧出线	1	高压侧出线快速接口	○	●
	2	低压侧出线快速接口	○	○
旁路负荷开关及环网柜	1	旁路负荷开关应具备可靠的安全锁定机构	●	●
	2	配备至少一进二出的环网柜	○	○
高低压保护	1	高压保护	○	●
	2	低压保护开关额定值＞变压器容量的三分之二	○	●
辅助设备	1	液压垂直伸缩液压支撑	●	●
	2	应急照明	○	○

注：●表示应具备的功能；○表示看具备的功能。

图4-20所示为从架空线路临时取电给移动箱变车供电，如图4-21所示为从环网柜临时取电给移动箱变车供电。

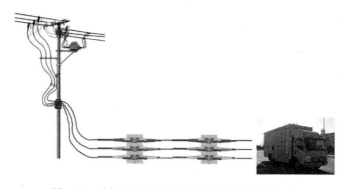

图4-20 从架空线路临时取电给移动箱变车供电

图4-21 从环网柜临时取电给移动箱变车供电

4.3.2 移动箱变车的技术要求

1. 工作条件

移动箱变车在下列环境条件下应能正常工作：

①海拔高度不超过1000m；

②环境温度在-40~40℃；

③相对湿度不大于95%（25℃时）。

特殊条件使用时，应按照使用要求进行设计。

2. 功能要求

（1）整体要求

①运输。移动箱变车应具有良好的机动性、抗震动、抗冲击、防尘等性能，满足可靠运输车载设备要求。

②改装。移动箱变车应采用已定型汽车整车进行改装。移动箱变车的改装应符合GB/T 1332—1991《载货汽车试验规程》、GB/T 13043—2006《客车定型试验规程》、GB/T 13044—1991《轻型客车定型试验规程》、QC/T 252—1998《专用汽车定型试验规程》等汽车改装技术标准的要求。

③生产。移动箱变车的生产除应符合本文件的规定外，还需遵守国家颁布的有关法律。

（2）车载设备

①一般要求

a. 维护检验。车载设备应按照相关管理规定或其说明书进行定期校准、维护或检验。

b. 性能和参数。车载设备的性能和参数除满足要求外，还应符合相关技术标准或规程的规定。

c. 接线方式。高压侧接线为一组进线与两组出线，出线一组用于连接变压器，另一组可用于转供负荷。低压侧出线为两组负荷（一主一备）输出。

d. 抗震性。车载设备元件或部件应安装牢固，有良好的抗震性。车载设备的抗震性能应符合 GB 4798.5—2007《电工电子产品应用环境条件 第5部分：地面车辆使用》的有关规定。

②配电变压器。配电变压器应符合 GB 50150—2006《电气装置安装工程 电气设备交持试验标准》的规定，容量可采用 250~630kVA 等规格的三相油浸直冷线圈无励磁调压配电变压器或干式变压器。

③旁路负荷开关。旁路负荷开关应符合 Q/GDW 249—2009《10kV 旁路作业设备条件》的规定，全绝缘全密封并能与环网柜、分支箱互连，具备良好的操作性能（机械寿命≥3000 次循环）和灭弧性，具备可靠的安全锁定机构。

④旁路柔性电缆。旁路柔性电缆应符合 Q/GDW 249—2009《10kV 旁路作业设备条件》的规定，可弯曲，能重复使用。

⑤旁路连接器。旁路连接器包括进线接头装置、终端接头、中间接头、T 型接头等其应符合 Q/GDW 249—2009《10kV 旁路作业设备条件》的规定。连接接头要求结构紧凑、对接方便，并有牢固、可靠的防止自动脱落锁口，在对接状态能方便改变分离状态。

⑥旁路电缆连接附件。旁路电缆连接附件包括可触摸式终端肘型电缆插头、可分离式电缆接头、辅助电缆、引下电缆等，应符合 Q/GDW 249—2009《10kV 旁路作业设备条件》的规定。型号与柔性电缆、带电作业消弧开关、箱式变压器、环网柜、分支箱和高低压进线柜匹配。

⑦低压配电屏。低压配电屏应符合 GB 7251.1—2005《低压成套开关设备和控制设备 第1部分：型式试验和部分型式试验成套设备》的规定，将低压电路所需的开关设备、测量仪表、保护装置和辅助设备等，按一定的接线方式布置安装在金属柜内。

⑧低压柔性电缆。低压柔性电缆应符合 GB 75941—1987《电线电缆橡皮绝缘和橡边套 第一部分：一般规定》的规定，可弯曲，重复使用。

⑨环网柜。环网柜应符合 GB 11022—2011《高压开关设备和控制设备标准的共用技术要求》的规定，应分为负荷开关室（断路器）、母线室、电缆室和控制仪表室等金属封闭的独立隔离室，其中负荷开关室（断路器）、母线室、电缆室均有独立的泄压通道。

⑩配置要求。包括基本型移动箱变车的典型设备配置和扩展型移动箱变车典型设备配置，按国家有关规范标准执行。

（3）辅助系统

①电路及控制。电路系统应设电源总开关并布置在操作人员便于操作使用的位置

②照明系统。移动箱变车的照明包括车辆本体照明、工作照明和应急照明。

③接地系统。移动箱变车应有专用的集中接地点，并具有明显的接地标志。接地电阻均应大于4Ω，保护接地和工作接地要相距5m及以上。

接地线应有足够的截面和长度，主接地回路接地线的截面应满足热容量和导线电压降的要求。

④液压系统。移动箱变车在车库停放或是在机组工作时，液压系统可为保护轮胎及车桥提供支撑，四支液压支腿带有锁定装置，每腿均能独立操作。

⑤安全保护、警示、防护。

a. 安全保护：移动箱变车的液压、机械、电动等运动部件，对承重、传动等安全有明显影响时，应有限位闭锁保护装置，闭锁装置应动作灵活、可靠。

可人工移动的可动部件，对运输、固定等有明显安全影响时，应有限位锁紧装置，锁紧装置应方便工人操作，动作应灵活、限位应可靠。

b. 警示：移动箱变车应有声光报警装置，并可由车上操作人员进行控制。设备区可根据带电检测需要安装烟雾、有毒气体等报警器。

c. 防护：移动箱变车宜配备常用的安全工器具、防护用具。驾驶室、设备区等不同功能区域应配备消防器材。消防器材应安装牢固，存取方便。

4.3.3 移动箱变车的试验和存放

1. 移动箱变车的试验

（1）带电作业专用车试验包括型式试验、出厂试验和验收试验。型式试验和出厂试验由厂家进行。

（2）验收试验由用户与厂家共同完成。包括外观检查、行驶试验、整车绝缘性能试验、电缆盘收放试验、车载设备试验等项目。

2. 移动箱变车的存放

移动箱变车属于带电作业特种车辆，在存放时应满足下列要求。

（1）带电作业专用车宜存放在车库内，减少太阳直接暴晒或雨淋，远离高温热源。

（2）长期停驶车辆时，应关闭电源总开关，切断全车的电路系统。

（3）当需要开启车辆前，应打开电源总开关，恢复全车的电路系统。

4.4 带电作业工具车

随着电力技术的不断发展及社会对供电安全性和稳定性的要求不断提高，带电作业在各供电企业中所起的作用越来越大。在配电线路带电作业和电力应急抢险的任务中，带电作业专用工具车有效解决了带电作业工器具的运输和保管问题，使带电作业工具能够在使用前保持正常状态，保证了带电作业的可靠性和安全性，提高了作业响应速度、工作效率和标准化水平。

4.4.1 带电作业工具专用车的基本类型、结构和功能

1. 带电作业工具专用车的基本类型

带电作业工具专用车按照使用特点和范围可以分为以下三种类型。

（1）Ⅰ型——输电线路（包括变电站）带电作业专用车，适合存放输电线路（包括变电站）作业使用的绝缘工具、金属工具和个人防护用具。可以根据工具所存放的带电作业工具的种类、尺寸和数量，调节空间及工具存放舱的温度范围。

（2）Ⅱ型——配电线路带电作业专用车，适合存放配电线路作业使用的绝缘工具、绝缘遮蔽用具和个人防护用具。可以根据工具所存放的带电作业工具的种类、尺寸和数量，调节空间及工具存放舱的温度范围。

（3）Ⅲ型——混合型带电作业专用车，适合存放输电线路（包括变电站）作业使用的绝缘工具、金属工具和个人防护用具，以及适合存放配电线路作业使用的绝缘工具、绝缘遮蔽用具和个人防护用具。可以根据存放工具的种类，分区隔离以适应不同的温度范围要求。

目前，配电线路带电作业（Ⅱ型）专用车已经广泛用于配电线路（包括10kV电缆线路不停电作业）带电作业操作，输电线路带电作业专用车（Ⅰ型）也有应用。

如图4-22所示为配电线路带电作业专用车，如图4-23所示为配电线路应急旁路工具车。

图4-22 配电线路带电作业专用车

图4-23 配电线路应急旁路工具车

近年来配电线路带电作业专用车更新很快,已经向多功能、智能化方面发展,而且充分利用了现代机电一体化及测控技术,集绝缘工具存放环境监控、现场抢修过程及现场环境监测、现场照明、专用电源、3G无线远程数据传输、报警功能、车载定位及导航等功能于一体,能够最大限度地满足配电线路带电作业和现场抢修工作的需要。如图4-24所示为陕西华安电力科技有限公司研制生产的智能型带电作业工具专用车。

图4-24 智能型带电作业工具专用车

2. 带电作业工具专用车的基本结构

带电作业工具专用车主要是由车辆平台、车载设备、辅助系统和工具存放舱构成。如图4-25所示为带电作业专用车结构。

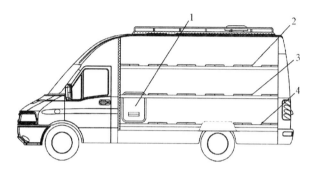

1—发电机；2—硬质绝缘工具存放区；3—绝缘防护用具、软质绝缘工具存放区；
4—仪表、仪器、电动工具、金属工具存放区

图4-25 带电作业专用车结构

① 车辆平台。以工程车或客车为基础改装而成，它包括车辆底盘、箱体等部分，它是带电作业工具专用车的运输载体。

② 车载设备。主要包括了车载发电机及电源转换装置，能够保证车载电源与市电相互切换，其次是除湿机、烘烤加热设备及排风设备等。

如图4-26所示为带电作业工具专用车所采用的便携式发电机。

图4-26 车载便携式发电机

③ 辅助系统。包括控制系统、车载GPS导航系统、防火烟雾报警探测及短信报警装置、车顶安装有高杆照明及野外视频监控系统、车顶配备微型气象系统、车载电台、GPS定位系统、车载电脑、热餐装置等。如图4-27所示为带电作业工具专用车视频监控终端设备。如图4-28为升降式车载照明系统。

图4-27 车载视频监控终端设备

图4-28 升降式车载照明系统

④工具存放舱。存放工具的舱体,如图4-29所示,具有调温调湿功能,可以满足绝缘工具、绝缘防护用具的防潮及存放温度要求。

图4-29 专用车工具存放舱

3. 带电作业工具专用车的基本功能

能够满足独立完成输配电线路一般性带电作业检修维护和应急处理等的任务要求;能够规范配电带电作业管理,在带电作业工作中,使用配电带电作业工具专用车；能够确保带电作业工具的可靠性，从而确保带电作业的安全性，通过带电作业工具专用车提供带电作业现场的辅助作业。

4.4.2 带电作业工具专用车的主要技术要求

1. 一般要求

①工具存放舱基本要求

a. 空间。工具存放舱的最小尺寸见表4-3所示。

表4-3　工具存放舱的最小尺寸　　　　　　　　　　　　mm

类型	长	宽	高
Ⅰ型	2400	1850	1750
Ⅱ型	1600	1850	1750
Ⅲ型	3200	1850	1750

b. 工具存放舱为封闭式，应有保温隔热层，舱门附有密封条。工具存放舱应有接地装置。

c. 工器具存放舱的消防、照明和装修材料应该应该满足 GB 25725—2010《带电作业工具专用车》（GB25725-2010）相关要求。

②工器具存放架

应该满足 GB 2725—2010《带电作业工具专用车》相关要求。如图4-30所示为带电作业工具专用车内部结构分布。

图4-30　带电作业工具专用车内部结构分布

2. 温度、湿度要求

①温度

a. 带电作业工具应根据工具类型分区存放，各存放区可有不同的温度要求。金属

工具的存放不做温度要求。

b. 硬质绝缘工具、软质绝缘工具、检测工具、屏蔽用具的存放区，温度应控制在 5~40℃内。

c. 绝缘遮蔽用具、绝缘防护用具的存放区的温度，应控制在 10~28℃之间。

d. Ⅰ型车的温度范围应控制在 5~40℃。

e. Ⅱ、Ⅲ型车的温度范围应控制在 10~28℃。

②湿度

工具存放舱湿度不大于60%。

3. 设备要求

除湿设备、烘干设备、通风设备、温湿度控制设备、报警设备、视频监控设备和其他设备应该满足 GB 25725—2010《带电作业工具专用车》相关要求。

4.4.3 带电作业工具专用车的试验和存放

1. 带电作业工具专用车的试验

（1）带电作业工具专用车试验包括型式试验、出厂试验和验收试验。型式试验和出厂试验由厂家进行。

（2）验收试验由用户与厂家共同完成。包括外观检查、电源切换试验、整车绝缘性能试验、工具舱密封性试验、工具舱存放温度和湿度试验、工具舱温度湿度调控试验、烘干设备电源自动切断及复位试验和手动操控试验等。

2. 带电作业专用车的存放

带电作业专用车宜存放在车库内，减少太阳直接暴晒或雨淋，远离高温热源。

第5章

安全防护用具

输配电线路带电作业用的安全防护用具主要有绝缘防护用具、绝缘遮蔽用具和屏蔽用具。前两类主要用于配电线路带电作业，其中绝缘防护用具主要是用于操作人员安全防护，绝缘遮蔽用具主要用于进行带电作业时，对带电体进行有限遮蔽，防止操作人员触电；屏蔽用具主要用于输电线路带电作业中，在输电线路带电作业时用于屏蔽高压电磁场对人体的影响，分流通过人体的工频电流，防护人体免受高压电场及电磁波的危害。

5.1 绝缘防护用具

配电线路带电作业常用的绝缘防护用具主要有绝缘手套、绝缘袖套、绝缘服、绝缘鞋、防刺穿手套、绝缘安全帽等，它们是配电线路带电作业最常用的绝缘防护用具。

5.1.1 带电作业用绝缘手套

1. 带电作业用绝缘手套的基本结构和功能

（1）基本结构

①手套由合成橡胶制成，手套可以加衬，防止化学腐蚀，降低臭氧对手套产生的老化影响。

②按结构可以分为大拇指基线、手腕、平袖口、卷边袖口、中指弯曲中点高度几部分。

（2）基本功能

带电作业用绝缘手套，是指在高压电气设备或装置上进行带电作业时起电气辅助绝缘作用的手套，由合成橡胶或天然橡胶制成，主要对带电作业人员手部进行绝缘防护。目前绝缘等级最大为2级，即10kV以下使用。

2. 带电作业用绝缘手套的分类和技术规格

（1）分类

①按使用方法可分为常规绝缘手套和复合型绝缘手套，本章重点介绍常规型绝缘手套，常规型绝缘手套如图5-1所示。

1—大拇指基线；2—手腕；3—平袖口；4—卷边袖口；5—中指弯曲中点高度

图5-1 常规型绝缘手套

②按形状可分为分指绝缘手套、连指绝缘手套、长袖复合绝缘手套和圆弧型袖口绝缘手套。

③按照电气性能的不同，可分为0、1、2、3、4五级。适用于不同标称电压手套见表5-1。

表5-1 适用于不同标称电压的手套级别

编号	级别	AC^a/V
1	0	380
2	1	3000
3	2	10 000
4	3	20 000
5	4	35 000

注：AC^a在三相系统中指线电压。

④具有特殊性能的手套分为5种A、H、Z、R、C 5种类型，见表5-2。

表5-2 特殊性能的手套类型

编号	型号	特殊性能
1	A	耐酸
2	H	耐油
3	Z	耐臭氧
4	R	耐酸、油、臭氧
5	C	耐超低温

注：R类兼有A、H、Z的性能。

(2) 常规型绝缘手套技术参数

①常规型绝缘手套的长度。常规型不同级别的绝缘手套的长度见表5-3。

表5-3 常规型不同级别的绝缘手套长度

编号	型号	长度/mm			
1	0	280	360	410	460
2	1	—	360	410	460
3	2	—	360	410	460
4	3	—	360	410	460
5	4	—	—	410	460

②常规型常规型绝缘手套的厚度。为保持适当的柔软性，手套平面（表面不加肋时）的最大厚度分为0、1、2、3、4五级，厚度有1.00mm、1.50mm、2.30mm、2.90mm和3.60mm几种。

3. 常规型绝缘手套的检查内容

常规型绝缘手套在使用前必须进行检查，其内容如下。

(1) 手套内外侧表面应通过检查确定无有害的和有形的表面缺陷。

(2) 如某双手套中的一只可能不安全，则这双手套不能使用，应将其返回进行试验。

(3) 使用前，应做漏气试验检查，先对绝缘手套进行充气并挤压，置于面部，检查是否漏气，漏气试验检查如图5-2所示。

图5-2 绝缘手套漏气试验检查

4. 普通型绝缘手套的使用方法

(1) 使用时，在绝缘手套的最外层应使用机械防护手套（如羊皮手套），绝缘手套的使用如图5-3所示。

图5-3 绝缘手套使用示例图

（2）避免用绝缘手套直接用力按压尖锐物体。

5. 常规型绝缘手套的保管注意事项

（1）绝缘手套应分件包装，手套应贮存在专用箱内，避免阳光直射或存放在人造热源附近，防止雨雪浸淋，防止挤压和尖锐物体碰撞，否则易造成刺破或划伤。

（2）禁止手套与油、酸、碱或其他有害物质接触，并距离热源1m以上。贮存环境温度宜为10~21℃。

（3）当手套被弄脏时应用肥皂和水清洗，彻底干燥后涂上滑石粉。如果有焦油和油漆这样的混合物黏附在手套上，应采用合适的溶剂擦去。

6. 常规型绝缘手套试验

常规型绝缘手套的常规试验内容如下。

（1）外观检查包括目视检查、尺寸检查、厚度检查、工艺及成型检查、标志检查、包装检查。

（2）机械试验包括拉伸强度及扯断伸长率试验、拉伸永久变形试验、抗机械刺穿试验、耐磨试验、耐切割试验、抗撕裂试验。

（3）电气试验包括交流验证试验、交流耐受试验、直流验证试验、直流电压试验。

（4）电气试验周期为：预防性试验每半年一次，检查性试验每年一次，两次试验间隔为半年。绝缘手套试验如图5-4所示。

图5-4 绝缘手套试验

5.1.2 绝缘袖套

1. 绝缘袖套的基本结构和功能

(1) 基本结构

绝缘袖套分为浸制和模压两种。浸制橡胶绝缘袖套与绝缘手套工艺相同,模压是采用模具压合成型。

绝缘袖套扣、袖套如图5-5、5-6所示。

图5-5 绝缘袖套扣

图5-6 绝缘袖套

(2) 基本功能

带电作业用绝缘袖套是配合同等级的绝缘手套使用的,用于保护肩臂部位;袖套扣配合袖套使用,要求具有良好的电气性能和较高的机械性能,并具有良好的绝缘性能。

2. 绝缘袖套的分类和技术规格

(1) 分类

①绝缘袖套按外形分为直筒式和曲肘式两种。

②特殊性能的绝缘袖套可分为A、H、Z、S和C类五种,分别具有耐酸、耐油、耐臭氧、耐低温性能。

③按耐压级别:分为0、1、2、3四级。

(2) 技术规格

①厚度:绝缘袖套应具有足够的弹性且平坦,合成橡胶最大厚度必须满足表5-4的规定。

表5-4 合成橡胶最大厚度

编号	级别	厚度（mm）
1	0	1.00
2	1	1.50
3	2	2.50
4	3	2.90

②绝缘袖套采用无缝制作，袖套与袖套扣连接所留的小孔必须采用非金属加固边缘，直径一般为8mm。

③袖套内外表面均匀，光滑，有规则，无小孔、裂纹、局部隆起，切口处无夹杂导电异物，无折缝、凹凸波纹和铸造标志等。

3. 绝缘袖套的检查内容

绝缘袖套在使用前必须进行检查，其内容有外观检查（袖套内外侧表面应进行目测检查，表面均匀）、尺寸检查、厚度检查、工艺及成型检查、标志检查、包装检查等。

4. 绝缘袖套的使用方法

（1）使用前，做外观检查。

（2）将绝缘袖套和绝缘扣可靠连接。

（3）避免被尖状物或锋利物划伤。

5. 绝缘袖套的保管注意事项

（1）绝缘袖套不能折叠，折痕会引起橡胶被氧化，降低绝缘性能，绝缘袖套的存放如图5-7所示。

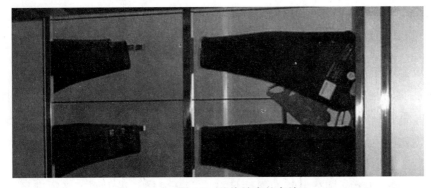

图5-7 绝缘袖套的存放

(2) 袖套要存放在正确尺寸的存放袋里，平坦放置，一个存放袋放入一副袖套。

(3) 绝缘袖套应分件包装，要注意防止阳光直射或存放在人造热源附近，尤其要避免直接碰触尖锐物体，造成刺破或划伤。

(4) 禁止袖套与油、酸、碱或其他有害物质接触，并距离热源1m以上。贮存环境温度宜为10~21℃之间。

(5) 当袖套被弄脏时应用肥皂和水清洗，彻底干燥后涂上滑石粉。如果有焦油和油漆这样的混合物黏附在手套上，应采用合适的溶剂擦去。

(6) 使用中袖套变湿或者洗了之后要进行彻底干燥，但是干燥温度不能超过65℃。

6. 绝缘袖套的试验内容

绝缘手套的常规试验内容如下。

(1) 机械试验。包括拉伸强度及伸长率试验、抗机械刺穿试验、拉伸永久变形试验。

(2) 电气耐压试验。包括型式试验、抽样试验和出厂例行试验。其中型式试验和抽样试验，电压持续时间为3min；出厂试验，电压持续时间为1min。

(3) 绝缘袖套的预防性试验周期为每半年进行一次。

5.1.3 绝缘服

绝缘服是由绝缘材料制成的服装，是保护带电作业人员接触带电体时免遭电击的一种人身安全防护用具。

1. 绝缘服的基本结构和功能

(1) 基本结构

①绝缘服由ERV树脂材料或合成橡胶制成，具有防止机械磨损、化学腐蚀和臭氧的作用，下面主要介绍ERV树脂材料制成的绝缘服，如图5-8所示。

②绝缘衣的结构分为衣袖、袖口收紧带，衣身、纽扣。

③绝缘裤的结构分为裤带、腰部松紧带。

(2) 带电作业用绝缘服基本功能

对人体除头部、手、足外实现绝缘遮蔽，以保护带电作业人员接触带电体时免遭电击。

图5-8 绝缘服例图

2. 绝缘服的分类和技术规格

(1) 分类

整套绝缘服包括绝缘衣和绝缘裤，绝缘衣分为：普通绝缘上衣、网眼绝缘上衣和绝缘披肩。

（2）技术规格

①绝缘衣、绝缘裤的型号分为小号、中号、大号、加大号。

②绝缘服的表面应平整、均匀、光滑，无小孔、局部隆起、夹杂异物、折缝、空隙等，结合部分应采取无缝制作的方式。

③表面拉伸强度平均值应不小于9Mpa，最低值应不低于平均值的90%。

④表面抗机械刺穿力平均值应不小于15N，最低值应不低于平均值的90%。

⑤表面抗撕裂拉断力平均值应不小于150N，最低值应不低于平均值的90%。

⑥电气性能应满足表5-5的要求。

表5-5 绝缘服的电气性能

交流电压试验	整衣层向验证电压（kV）	20
	整衣层向耐受电压（kV）	30
	沿面工频耐受电压（kV）	100
电阻率测量	内层材料体积电阻系数（Ω·cm）	$\geqslant 1\times 10^{15}$

3. 绝缘服的检查内容

绝缘服在使用前必须进行检查，其内容有外观检查，其重点是工艺及成型检查，对内外侧表面应进行目视检查，表面应平整、均匀、光滑，无小孔、局部隆起、夹杂异物、折缝、空隙等，结合部分应采取无缝制作的方式；标志检查；包装检查等。

4. 绝缘服的使用方法

（1）每次使用前都要对绝缘服的内外表面进行外观检查。

（2）发现绝缘服存在可能影响安全性能的缺陷时应禁止使用，并应对绝缘服进行试验。

（3）避免被尖状物或锋利物划伤。

绝缘服穿着如图5-9所示。

5. 绝缘服的保管注意事项

（1）绝缘服不能折叠，折痕会引起橡胶被氧化，降低绝缘性能。

（2）绝缘服应逐一悬挂在干燥、通风良好的带电作业工具库房专用不锈钢金属架上，如图5-10所示。

图5-9 绝缘服穿着

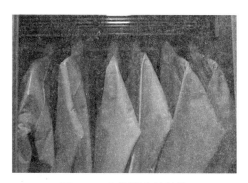

图 5-10　绝缘服存放保管

（3）绝缘服禁止储存在蒸汽管、散热器或其他人造热源附近；禁止储存在阳光直射的地方；尤其要避免直接碰触尖锐物体，造成刺破或划伤。

（4）禁止绝缘服与油、酸、碱或其他有害物质接触，并距离热源1m以上，贮存环境温度宜为10～21℃之间。

（5）当绝缘服被弄脏时应用肥皂和水清洗，彻底干燥后涂上滑石粉。如果有焦油和油漆这样的混合物黏附在其表面，应采用合适的溶剂擦去。

（6）使用中绝缘服变湿或者洗了之后要进行彻底干燥，但是干燥温度不能超过65℃。

6. 绝缘服的试验内容

绝缘服的常规试验内容如下。

（1）外观检查包括工艺及成型检查、标志检查、包装检查。

（2）机械试验包括拉伸强度及伸长率试验、抗机械刺穿试验、表面抗撕裂试验。

（3）绝缘服的预防性试验项目包括标志检查、交流耐压试验或直流耐压试验，试验周期为每半年进行一次。绝缘服预防性试验如图5-11所示。

图 5-11　绝缘服预防性试验

5.1.4 带电作业用绝缘鞋（靴）

1. 带电作业用绝缘鞋（靴）的基本结构和功能

（1）基本结构

基本绝缘鞋（靴）由鞋底、鞋面、鞋跟、靴筒等组成。绝缘鞋（靴）如图5-12所示。

图5-12　绝缘鞋（靴）

（2）基本功能

①绝缘鞋（靴）是配电线路带电作业时使用的辅助安全用具。

②带电作业用绝缘鞋（靴）在高压电气设备或装置上进行带电作业时起电气辅助绝缘作用。它用合成橡胶或天然橡胶制成，主要对带电作业操作人员脚部进行绝缘防护，要求具有良好的电气性能，并具有良好的绝缘性能。目前绝缘等级最大为2级，即10kV及以下使用。

2. 带电作业用绝缘鞋（靴）的分类和技术规格

（1）分类

①按系统电压可分为3～10kV（工频）绝缘鞋（靴）和0.4kV以下绝缘鞋（靴）。

②按材质分为布面绝缘鞋、皮面绝缘鞋（靴）、胶面绝缘鞋（靴）。

（2）技术规格

①绝缘鞋（靴）宜用平跟，外底应有防滑花纹。

②绝缘鞋（靴）只能在规定的范围内作辅助安全用具使用。

3. 带电作业用绝缘鞋（靴）的检查

绝缘鞋（靴）在使用前必须进行检查，其内容如下。

（1）外观检查，绝缘鞋（靴）内外侧表面应平整，无裂纹和孔洞等表面缺陷。

（2）如某双绝缘鞋（靴）中的一只可能不安全，则这双鞋不能使用，应将其返回进行试验。

（3）标志检查，其试验合格证应完整，且在有效期内。

4. 带电作业用绝缘鞋（靴）的使用方法

（1）穿戴正确。绝缘鞋（靴）应在进入绝缘台或绝缘斗前穿好后进入，穿好后不准在地面或其他尖锐物上行走和踩踏。

（2）绝缘鞋（靴）的型号与作业人员的脚码相适应，不可过大或过小。

（3）绝缘鞋（靴）凡有破损、鞋底防滑齿磨平、外底磨透露出绝缘层或预防性检验不合格时，均不得使用。

（4）使用中鞋（靴）变湿或者清洗之后要进行彻底干燥，但是干燥温度不能超过65℃。

5. 带电作业用绝缘鞋（靴）的保管注意事项

（1）存放地应干燥通风，堆放时离开地面和墙壁20cm以上，离开一切发热体1m以上，严禁与油、酸、碱或其他腐蚀性物品存放在一起。

（2）当鞋被弄脏时应用肥皂和水清洗，彻底干燥后涂上滑石粉。如果有焦油和油漆这样的混合物黏附在鞋上，应采用合适的溶剂擦去。

6. 带电作业用绝缘鞋（靴）的试验

（1）机械性能试验。包括拉伸强度及扯断伸长率试验、耐磨性能试验、邵氏A硬度试验、围条与鞋帮黏附强度试验、鞋帮与鞋底剥离度试验、耐折性能试验。

（2）电气试验。包括交流验证电压试验和泄漏电流试验。

（3）各种绝缘鞋（靴）预防性检验周期不应超过6个月。

5.1.5 带电作业用防机械刺穿绝缘手套

1. 带电作业用防机械刺穿绝缘手套的基本结构和功能

（1）基本结构

①手套由合成橡胶制成，手套可以加衬，以防止机械磨损、化学腐蚀和臭氧氧化。

②主要部分由大拇指基线、手腕、平袖口、卷边袖口、中指弯曲中点高度组成，如图5-13所示。

（2）基本功能

带电作业用防刺穿绝缘手套是在高压电气设备或

图5-13 防刺穿手套

置上进行带电作业时起电气辅助绝缘作用的手套,主要对带电作业操作人员手部进行绝缘防护,要求具有良好的电气性能和较高的机械性能。

2. 带电作业用防机械刺穿绝缘手套的分类和技术规格

(1) 分类

①按使用方法可分为复合型绝缘手套、长袖复合型绝缘手套。

②按形状可分为分指绝缘手套和连指绝缘手套。

③按照电气特性的不同,规定了三种等级的手套:00、0和1级。适用于不同标称电压的防机械刺穿手套见表5-6。

表5-6 适用于不同标称电压的防机械刺穿手套级别

编号	级别	交流有效值(V)	直流(V)
1	00	500	750
2	0	1000	1500
3	1	3000	11250

注:在三相系统中指线电压。

③具有特殊性能的手套分为A、H、Z、P、C5种类型见表5-7。

表5-7 特殊性能的手套类型

编号	型号	特殊性能
1	A	耐酸
2	H	耐油
3	Z	耐臭氧
4	R	耐酸、油、臭氧
5	C	耐超低温

注:P类兼有A、H、Z的性能。

(2) 技术规格

①带电作业用防刺穿绝缘手套自身具备机械保护性能,可以不用配合机械防护手套使用,并具有良好的绝缘性能。

②手套袖口可以制成带卷边的或不带卷边的。

③绝缘手套的长度。不同级别的手套的长度标准见表5-8。

表5-8　常规型不同级别的绝缘手套的长度

编号	级别	长度（mm）			
1	00	270	360	—	800
2	0	270	360	410	460	800
3	1	—	—	410	460	800

注：复合绝缘手套长度偏差允许±20mm。

④带电作业用防刺穿绝缘手套的厚度。为保持适当的柔软性，手套平面（表面不加肋时）的最大厚度见表5-9。

表5-9　合成橡胶最大厚度

编号	级别	厚度（mm）
1	00	1.80
2	0	2.30
3	1	

注：级别"1"对应的厚度数值未确定。

3. 带电作业用防机械刺穿绝缘手套的检查内容

（1）手套内外侧表面应通过检查确定无有害的和有形的表面缺陷。

（2）为改善紧握性能而设计的手掌和手指表面，不应视为表面缺陷。

（3）如某双手套中的一只可能不安全，则这双手套不能使用，应将其返回进行试验。

4. 带电作业用防机械刺穿绝缘手套的使用方法

（1）使用前，做漏气试验检查，先对绝缘手套进行充气并挤压，置于面部，检查是否漏气。

（2）避免用绝缘手套直接用力按压尖状物体。

5. 带电作业用防机械刺穿绝缘手套的保管注意事项

（1）绝缘手套应分件包装，手套应贮存在专用箱内，避免阳光直射或存放在人造热源附近，防止雨雪浸淋、挤压和尖锐物体碰撞，否则易造成刺破或划伤。

（2）禁止手套与油、酸、碱或其他有害物质接触，并距离热源1m以上。贮存环境温度宜为10～21℃之间。

（3）当手套被弄脏时应用肥皂和水清洗，彻底干燥后涂上滑石粉。如果有焦油和油漆这样的混合物黏附在手套上,应采用合适的溶剂擦去。

(4)使用中手套变湿或者洗了之后要进行彻底干燥,但是干燥温度不能超过65℃。

(5)不要将手套不必要地暴露于热、光之中,也不要与油、油脂、松脂或弱酸接触。

6. 防机械刺穿绝缘手套的试验内容

绝缘手套的常规试验内容如下。

(1)机械试验。包括耐磨试验、抗机械刺穿试验、抗撕裂试验、拉伸强度及扯断伸长率试验、抗切割试验。

(2)电气性能试验。包括交流验证试验、交流耐受试验、受潮后的泄漏电流试验。

(3)绝缘手套预防性试验周期为每半年进行一次。

5.1.6 绝缘安全帽

绝缘安全帽是用来保护电气作业人员头部,防止其在带电作业时被电击、撞伤或被坠物打击伤害的防护用具。

1. 带电作业用绝缘安全帽的基本结构和功能

(1)基本结构

绝缘安全帽由帽壳、帽衬、下颏带、后箍等组成,绝缘安全帽无透气孔。如图5-14和如图5-15所示,为绝缘安全帽内外部结构。

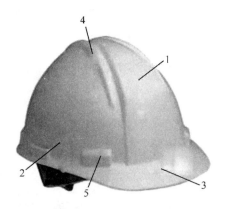

1—帽衬;2—连接孔;3—帽衬接头;4—托带、后箍;6—后箍调节器;7—下颏带;
8—吸汗带;9—衬垫;10—帽箍;11—锁紧卡;12—护带

图5-14 绝缘安全帽内部结构

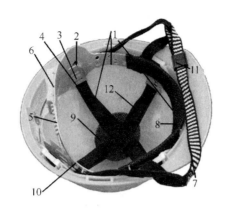

1—帽壳；2—帽沿；3—帽舌；4—顶筋；5—插座

图5-15 绝缘安全帽外部结构示例图

（2）基本功能

在带电作业时保护作业人员头部，能够屏蔽电弧，缓冲减震和分散应力，避免受到电击或机械伤害。

2. 带电作业用绝缘安全帽的分类和技术参数

（1）分类

带电作业绝缘安全帽主要分为美制、欧制、日制和国产，颜色主要有白色和黄色。

（2）技术参数

带电作业用绝缘安全帽采用高密度复合聚酯材料，除具有符合安全帽检测标准的机械强度，还应符合相关配电带电作业电气检测标准，其电介质的强度必须满足20kV/3 min的试验要求。

3. 绝缘安全帽的检查内容

（1）新的绝缘安全帽，首先检查是否有劳动部门允许生产的证明及产品合格证；再看是否破损、厚薄是否均匀，缓冲层及调整带和弹性带是否齐全有效；最后检查安全帽上商标、型号、制造厂名称、生产日期和生产许可证编号是否完好。

（2）绝缘安全帽在使用前必须进行外观检查，检查安全帽的帽壳、帽箍、顶衬、下颏带、后扣（或帽箍扣）等组件应完好无损，帽壳与顶衬缓冲空间在25～50mm之间。试验合格证完好，且在试验有效期内。

（3）定期检查，检查有无龟裂、下凹、裂痕和磨损等情况，发现异常现象要立即更换，不准再继续使用；任何受过重击、电击、有裂痕的绝缘安全帽不论有无损坏现象，均应报废；试验检测其绝缘性能。

绝缘安全帽壳不能有透气孔，应避免与普通安全帽混淆。

4. 绝缘安全帽的使用方法

（1）戴绝缘安全帽前应将帽后调整带按自己头型调整到适合的位置（头部稍有约束感，但不难受的程度，以不系下颏带低头时安全帽不会脱落为宜）。佩戴安全帽必须系好下颏带，下颏带应紧贴下颏，松紧以下颏有约束感，但不难受为宜。然后将帽内弹性带系牢。缓冲衬垫的松紧由带子调节，人的头顶和帽体内顶部的空间垂直距离一般在25～50mm之间，至少不要小于32mm。

（2）安全帽戴好后，应将后扣拧到合适位置（或将帽箍扣调整到合适的位置），锁好下颏带，防止工作中前倾后仰或其他原因造成滑落。不要把绝缘安全帽歪戴，也不要把帽沿戴在脑后方。否则，会降低安全帽对于冲击的防护作用。

（3）绝缘安全帽的下颏带必须扣在颏下并系牢，松紧要适度。这样不至于被大风吹掉或者是被其他障碍物碰掉或者由于头的前后摆动使安全帽脱落。

（4）严禁使用只有下颏带与帽壳连接的绝缘安全帽，也就是帽内无缓冲层的绝缘安全帽。

（5）严禁不规范使用安全帽，在现场作业中，作业人员不得将安全帽脱下搁置一旁或当坐垫使用，不得不系扣带或者不收紧，不得将扣带放在帽衬内。

（6）平时使用绝缘安全帽时应保持整洁，不能接触火源，不要任意涂刷油漆。

5. 绝缘安全帽的保管注意事项

（1）绝缘安全帽不应贮存在酸、碱、高温、日晒、潮湿等处所，更不可和硬物放在一起。

（2）绝缘安全帽应放在恒温、恒湿的专用带电作业工具库房内，并分类放置，避免与普通安全帽混合放置。

（3）安全帽的适用温度为-10～+50℃，应摆放在透风干燥处，并避免接触火（热）源和侵蚀性物质，以免影响其正常使用寿命。应尽量缩短安全帽的库存期，最好随买随用，购置出厂时间不长的产品，以免因库存时间过长而影响实际有效使用时间。

6. 绝缘安全帽的试验内容

（1）绝缘安全帽出厂试验包括冲击吸收性能试验（经低温、高温、淋水预处理后做冲击试验，传递到头模上的力不超过4900N），耐穿刺性能试验，电绝缘性能试验，阻燃性能试验，侧向刚性试验，抗静电性能试验。

（2）绝缘安全帽要进行定期试验，机械试验和电气试验应每年进行一次，合格后方可继续使用。

5.2 绝缘遮蔽用具

绝缘遮蔽用具由绝缘材料制成，主要用于配电线路带电作业中遮蔽带电导体或非带电导体的绝缘保护用具。常用的绝缘遮蔽用具有绝缘遮蔽罩、绝缘挡板、绝缘套管、绝缘毯、绝缘垫等。

5.2.1 绝缘遮蔽罩

绝缘遮蔽罩设置于作业人员与被遮蔽物之间，防止作业人员与带电体发生直接接触，起遮蔽或隔离的保护作用。

1. 绝缘遮蔽罩的基本结构和功能

下面主要介绍10kV配电线路带电作业用绝缘遮蔽罩（以下简称遮蔽罩）。

（1）基本结构

①遮蔽罩采用环氧树脂材料、橡胶材料、塑料材料、聚合材料等绝缘材料制成。

②遮蔽罩可以是硬壳的，也可以是软质的，应适于被遮蔽物，有阻碍人体直接接触带电体或接地体的功能，其长度一般不应超过1.5m。除满足必要的电气特性要求外，其尺寸要减到最小。

③遮蔽罩的保护区应有清晰、明显且牢固的标记。

④所有遮蔽罩应能用绝缘杆来装设，应设有提环、孔眼、挂钩等部件。

绝缘遮蔽罩如图5-16所示。

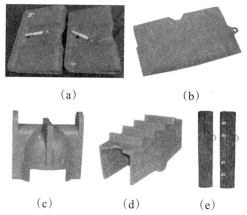

（a）硬质跌落式熔断器绝缘遮蔽罩；（b）软质跌落式熔断器遮蔽罩；
（c）针式绝缘子绝缘遮蔽罩；（d）横担绝缘遮蔽罩；（e）电杆用绝缘遮蔽罩

图5-16 绝缘遮蔽罩示例图

(2) 基本功能

在10kV配电线路带电作业中，遮蔽罩不起主绝缘作用，它只适用于在带电作业人员发生意外短暂碰撞时即擦过接触时，起绝缘遮蔽或隔离的保护作用。

2. 绝缘遮蔽罩的分类和技术规格

(1) 分类

①按照遮蔽对象的不同分为硬壳型、软型或变形型，也可分为定型或平展型。

②根据遮蔽罩的用途可分为导线遮蔽罩、针式绝缘子遮蔽罩、耐张装置遮蔽罩、悬垂装置遮蔽罩、线夹遮蔽罩、棒形绝缘子遮蔽罩、横担遮蔽罩、电杆遮蔽罩、套管遮蔽罩、跌落式熔断器遮蔽罩、避雷器遮蔽罩等，也可根据被遮物体专门设计。

③特殊性能的遮蔽罩分为A、H、C、W、P五种。

(2) 技术规格

①遮蔽罩应由吸湿性小的绝缘材料制成，绝缘材料应能满足在一定高温和低温情况下所要求的电气和机械性能。

②按照遮蔽罩电气性能可分为0、1、2、3、4五级。

③绝缘遮蔽罩的主体表面应光滑，其内表面与外表面均不允许有小孔、接缝裂纹、浮泡、破口、不明杂物、磨损擦伤、明显机械加工痕迹等表面缺陷。

④遮蔽罩在结构上应有提环、筒眼、挂钩等部件。

⑤遮蔽罩上应设有一个或多个闭锁部件，防止在使用中，或在外力作用下突然滑落。

3. 绝缘遮蔽罩的检查内容

为了确保遮蔽罩电气和机械特性的完整，在每次使用工具之前，应进行仔细的外观检查和试装配，内容如下。

(1) 遮蔽罩在经贮存和运输之后应无损伤，工具的绝缘表面应无孔洞、撞伤、擦伤和裂缝等。

(2) 遮蔽罩表面洁净干燥。

(3) 遮蔽罩的可拆卸部件或各组件经装配后完整无缺。

(4) 遮蔽罩应能正确操作，工具应转动灵活无卡阻，锁位功能正确等。

4. 绝缘遮蔽罩的使用方法

(1) 仅限于10kV及以下电力设备的带电作业，如图5-17所示为横担遮蔽罩和绝缘子遮蔽罩。

图 5-17 绝缘遮蔽罩的使用

（2）遮蔽罩不起主绝缘作用，但允许偶尔短时擦过接触，主要还是限制人体活动范围。

（3）遮蔽罩应与人体安全保护用具并用。

（4）在遮蔽作业过程中绝缘遮蔽罩之间或与其他遮蔽物体的边沿重叠部位不得少于150mm。

（5）每个遮蔽罩遮蔽的范围不能超出遮蔽罩保护区的保护范围。

5. 绝缘遮蔽罩的保管注意事项

（1）绝缘遮蔽罩不能折叠、挤压，折痕会引起橡胶被氧化，降低绝缘性能。

（2）遮蔽罩应存放在正确尺寸的存放袋内，分件包装，平坦放置。

（3）遮蔽罩禁止储藏在蒸汽管、散热管或其他人造热源附近，禁止贮藏在阳光、灯光或其他光源直射的环境下，尤其要避免贮藏或挪动时直接碰触尖锐物体，造成刺破或划伤。

（4）禁止遮蔽罩与油、酸、碱或其他有害物质接触，并距离热源1m以上。贮存环境温度宜为10~21℃。

（5）对潮湿的遮蔽罩应进行彻底干燥，但干燥处理的温度不能超过65℃。

6. 绝缘遮蔽罩的试验内容

绝缘遮蔽罩的试验内容如下。

（1）机械试验。遮蔽罩的机械试验为低温条件下的试验。遮蔽罩进行机械性能方面的试验包括模拟装配试验、软形绝缘遮蔽罩低温折叠试验、硬质绝缘遮蔽罩低温耐冲击试验。

（2）电气试验。包括工频耐压和泄漏电流试验、认证试验、耐臭氧试验、特殊性能试验。

(3)预防性试验。绝缘遮蔽罩的预防性试验应逐只进行,试验周期为每半年进行一次。

5.2.2 绝缘挡板

1. 绝缘挡板的基本结构和功能

(1)基本结构

①绝缘隔板一般为环氧树脂玻璃钢材料制造,如图5-18所示。

②绝缘隔板主要由浸渍纸、棉布、无碱玻璃纤维布、浸渍酚醛、环氧树脂等材料组成,经浸透加压、烘干、打磨、固化而成的硬质板状绝缘材料。

(2)基本功能

①绝缘挡板应具有很高的绝缘性能、防腐性及较高的耐热性,起临时绝缘和隔离带电部件,限制带电作业人员活动范围,提高对邻相绝缘水平的作用。

②在带电作业中,绝缘挡板也可置于拉开隔离开关的动、静触头之间,以防止隔离开关自行落下后误送电。

③环氧树脂具有良好的机械性能和稳定的耐压性,具有阻挡电弧、阻断电压及部分电流的作用。

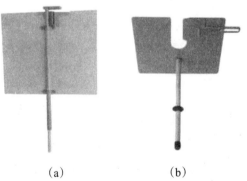

(a) 竖向用绝缘挡板;(b) 横向用绝缘挡板

图5-18 绝缘挡板示例图

2. 绝缘挡板的分类和技术规格

(1)分类

绝缘挡板主要分为带手柄绝缘挡板和系绳式绝缘挡板。

(2)技术规格

①环氧树脂绝缘板外观、表面应平整光滑,均采用整板制作,颜色均匀,不允许

有杂质和其他明显的缺陷，允许有轻微的擦伤，但边缘应切割整齐，断面没有分层和裂纹。

②环氧树脂绝缘隔板的耐高温 180～200℃，具有较高的机械性能和介电性能，较好的耐热性和耐潮性，并具有良好的机械加工性能。

③环氧树脂绝缘隔板颜色一般为米黄色，具有良好的耐油性和耐腐蚀性。

3. 绝缘挡板的检查内容

为了确保绝缘挡板电气和机械特性的完整，在每次使用之前，应进行仔细的检查。其外观、表面应平整光滑，颜色均匀，无杂质和其他明显的缺陷，边缘整齐，断面没有分层和裂纹等；绝缘挡板表面应洁净干燥；绝缘挡板各组件完整无缺；绝缘挡板应能正确操作，锁位功能正确等。

4. 绝缘挡板的使用方法

（1）仅限于10kV及以下电力设备的带电作业，绝缘挡板使用如图5-19所示。

（2）绝缘挡板不起主绝缘作用，应与人体安全保护用具并用。

（3）每个绝缘挡板隔离的范围不能超出挡板的保护范围。

（4）安装应牢固可靠，安装后便于拆卸。

图5-19　绝缘挡板使用

5. 绝缘挡板的保管注意事项

（1）绝缘挡板不能挤压，应平坦放置。

（2）禁止绝缘挡板与油、酸、碱或其他有害物质接触，并距离热源1m以上。贮存环境温度宜为10～21℃之间。

（3）对潮湿的绝缘挡板应进行彻底干燥，但干燥处理的温度不能超过65℃。

6. 绝缘挡板的试验内容

绝缘挡板的试验内容如下。

（1）机械试验。遮蔽罩进行机械性能方面的试验包括模拟装配试验和低温耐冲击试验。

（2）电气试验。包括：工频耐压和泄漏电流试验、认证试验等。

（3）预防性试验绝缘挡板的预防性试验应逐只进行，试验周期为每半年进行一次。

5.2.3　绝缘毯

1. 绝缘毯的基本结构和功能

（1）基本结构

①绝缘毯的形状可以是平展式的，也可以是开槽式的，也可以专门设计以满足特殊用途的需要。平展式绝缘毯，如图5-20、5-21所示，开槽式绝缘毯如图5-22所示。

图5-20　平展式树脂绝缘毯

图5-21　平展式橡胶绝缘毯

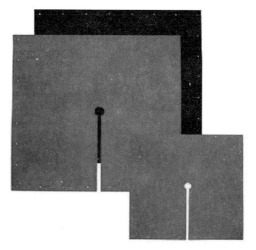

图 5-22 开槽式橡胶绝缘毯

②绝缘毯一般采用环氧树脂、橡胶等绝缘材料制成。

（2）基本功能

①绝缘毯应具有很高的绝缘性能，具有良好的防腐性及较高的耐热性，起临时绝缘和保护电气作业人员在带电作业时避免误触带电体，隔离带电部件，并提高对邻相绝缘水平的作用。

②具有稳定的耐压性，能阻挡电弧、阻断电压及部分电流的作用。

2. 绝缘毯的分类和技术规格

（1）分类

①绝缘毯分为树脂绝缘毯和橡胶绝缘毯。

②具有特殊性能和多重特殊性能的绝缘毯分为A、H、Z、M、S、C6种类型。

（2）技术规格

①绝缘毯按电气性能分为0、1、2、3四级。

②绝缘毯应采用绝缘的橡胶类和塑胶类材料，采用无缝制作工艺制成。绝缘毯上的孔眼必须用非金属材料加固边缘，其直径通常为8mm。

③绝缘毯上、下表面应不存在有害的不规则性，绝缘毯的保护区应有清晰、明显且牢固的标记。

3. 绝缘毯的检查内容

为了确保绝缘毯电气和机械特性的完整，在每次使用之前，应对绝缘毯的两面进行仔细检查。其外观、表面应平整光滑，颜色均匀，无杂质，无针孔、无裂纹，无割伤和其他明显的缺陷；绝缘毯表面应洁净干燥；绝缘毯应具有试验合格证，且在试验有限期内。

4. 绝缘毯的使用方法

（1）仅限于10kV及以下电力设备的带电作业，如图5-23所示。

图5-23 绝缘毯的使用

（2）不起主绝缘作用，但允许偶尔短时擦过接触，应与人体安全保护用具并用。

（3）每个绝缘毯的隔离的范围不能超出保护区的保护范围。

（4）绝缘毯安装好后应用绝缘毯夹可靠固定。

（5）在遮蔽作业过程中绝缘毯之间或与其他遮蔽物体的边沿重叠部位不得少于150mm。

5. 绝缘毯的保管注意事项

（1）绝缘毯应逐一贮藏于有足够强度的包装袋内。小心地放置绝缘毯以确保其不被挤压和折叠，尤其要避免直接碰触尖锐物体，造成刺破或划伤。禁止贮藏在蒸汽管、散热管或其他人造热源附近，并距离热源1m以上。禁止贮藏在阳光、灯光或其他光源直射的条件下；禁止与油、酸、碱或其他有害物质接触，贮存环境温度宜为10~21℃之间。

（2）对潮湿的绝缘毯应进行干燥处理，但干燥处理的温度不能超过65℃。

6. 绝缘毯的试验内容

绝缘毯的试验内容如下。

（1）机械试验。绝缘毯机械性能方面的试验包括拉伸强度及伸长率试验、抗机械刺穿试验、拉伸永久变形试验、抗撕裂试验。

（2）电气试验。交流电压试验包括交流电压认证试验和交流耐压试验。

（3）预防性试验。绝缘毯的预防性试验周期为每半年进行一次。

5.2.4 绝缘垫

1. 绝缘垫的基本结构和功能

（1）基本结构

绝缘垫采用橡胶类绝缘材料制作，上表面采用皱纹状或菱形花纹状等防滑设计以增强表面防滑性能，背面可采用布料或其他防滑材料，如图5-24所示。

图5-2-8　绝缘垫

（2）基本功能

①绝缘垫由橡胶类绝缘材料制成，敷设在地面或接地物体上以保护作业人员免遭电击。

②具有稳定的耐压性，能阻挡电弧、阻断电压及部分电流的作用。

2. 绝缘垫的分类和技术规格

（1）分类

①按电压等级分可以分为5kV绝缘胶垫、10kV绝缘胶垫、15kV绝缘毯、20kV绝缘胶板、25kV绝缘胶板、30kV绝缘胶板和35kV绝缘胶板。

②按颜色可以分为赤色、玄色、绿色绝缘胶垫。

（2）技术

①绝缘垫按电气性能分为0、1、2、3四级。

②绝缘垫应采用无缝制作工艺制成。

③绝缘垫上、下表面应不存在有害的不规则性。

④绝缘垫凹陷的直径不大于1.6m，边缘光滑，当凹陷处的反面包敷在拇指上扩展时，正面不应有可见痕迹。凹陷在5个以下，且任意两个凹陷之间的距离不大于15mm。当拉伸时，凹槽或模型趋向于平滑的表面。表面上由杂质形成的凸块不影响材料的延展。

3. 绝缘垫的检查内容

为了确保绝缘垫电气和机械特性的完整,在每次使用之前,应对绝缘垫的两面进行仔细检查。其外观、表面应平整光滑,颜色均匀,无杂质,针孔、裂纹、割伤和其他明显的缺陷;绝缘垫表面应洁净干燥;绝缘垫具有试验合格证,且在试验有限期内。

4. 绝缘垫的使用方法

(1) 绝缘垫仅限于10kV及以下电力设备的带电作业,如图5-25所示。

(2) 绝缘垫不起主绝缘作用,但允许偶尔短时擦过接触,应与人体安全保护用具并用。

图 5-25 绝缘垫的使用

5. 绝缘垫的保管注意事项

(1) 绝缘垫应储存在专用箱内,避免阳光直射,雨雪浸淋,防止挤压和尖锐物体碰撞。

(2) 禁止贮藏在蒸汽管、散热管或其他人造热源附近,并距离热源1m以上。禁止贮藏在阳光、灯光或其他光源直射的条件下;禁止与油、酸、碱或其他有害物质接触,贮存环境温度宜为10~21℃之间。

(3) 对潮湿的绝缘垫应进行干燥处理,但干燥处理的温度不能超过65℃。

6. 绝缘垫的试验内容

绝缘垫的试验内容如下。

(1) 机械试验。包括抗机械刺穿试验、防滑试验。

(2) 电气试验:包括交流电压验证试验和交流耐压试验,交流耐压试验周期为每半年进行一次。

(3) 预防性试验绝缘垫的预防性试验项目包括标志检查。

5.3 屏蔽用具

输电线路带电作业常用的屏蔽防护用具主要有屏蔽服和防静电服。

5.3.1 屏蔽服

1. 屏蔽服的基本结构和功能

（1）基本结构特点

成套的屏蔽服装应包括上衣、裤子、帽子、袜子、手套、鞋及其相应的连接线和连接头。

（2）基本功能

带电作业屏蔽服又叫等电位均压服，是采用均匀的导体材料和纤维材料制成的服装。其作用是在穿用后，使处于高压电场中的人体外表面各部位形成一个等电位屏蔽面，从而防护人体免受高压电场及电磁波的危害。

2. 屏蔽服的分类和技术规格

（1）分类

屏蔽服分两种类型。用于交流110（66）kV～500kV、直流±500kV及以下电压等级的屏蔽服装为Ⅰ型，用于交流750kV电压等级的屏蔽服装为Ⅱ型。Ⅱ型屏蔽服装必须配置面罩，整套服装为连体衣裤帽。Ⅰ型屏蔽服如图5-26所示，Ⅱ型屏蔽服如图5-27所示。

图5-26　Ⅰ型屏蔽服

图5-27　Ⅱ型屏蔽服

（2）技术参数

①屏蔽服装应有较好的屏蔽性能、较低的电阻、适当的通流容量、一定的阻燃性及较好的服用性能。屏蔽服装各部件应经过两个可拆卸的连接头进行可靠的电气连接。

②用于制作屏蔽服的衣料，其电阻不得大于800mΩ；衣料应具有一定的耐火花能力，在充电电容产生的高频火花放电时而不烧损，仅碳化而无明火蔓延。经过耐火花试验2min以后，衣料碳化破坏面积不得大于300mm^2。

③用于制作屏蔽服的衣料，其熔断电流不得小于5A。

④衣料与明火接触时，必须能够阻止明火的蔓延。试样的碳长度不得大于300mm；且烧坏面积不得大于100mm^2，烧坏面积不得扩散到试样边缘。

⑤多次洗涤后，衣料的电气和耐燃性能无明显降低。

⑥衣料必须耐磨损，使衣服具有一定的耐用价值。经过500次摩擦试验后，衣料电阻不得大于1Ω。

⑦对导电纤维类衣料，衣料的径向断裂强度不得小于343N，纬向断裂强度不得小于294N，经、纬向断裂伸长率均不得小于10%；对于导电涂层类衣料，衣料的径向短裂强度不得小于245N，纬向断裂强度不得小于245N，经纬向断裂伸长率均不得小于10%。

⑧衣料应具有较大的透气量，以达到穿戴者舒适的目的。透过衣料的空气流量不得小于35L/（m^2·s）。

⑨全套屏蔽服连接良好后，任意两个最远端点之间的电阻均不得大于20Ω，手套、袜子、衣、裤单件任意两个最远端之间的电阻不得大于15Ω，导电鞋的电阻不得大于500Ω。

⑩必须确保帽子和上衣之间的电气连接良好。帽子必须通过屏蔽效应试验。帽子的屏蔽效应试验在整套衣服的屏蔽性能试验一起进行。对Ⅰ型屏蔽服装，帽子的保护盖舌和外伸边沿必须确保人体外露部位（如面部）不产生不舒适感，并应确保在最高使用电压的情况下，人体外露部位的表面场强不得大于240kV/m；用于750kV电压等级的Ⅱ型屏蔽服装必须配置屏蔽面罩，面罩采用导电材料和阻燃纤维编织，视觉良好，其屏蔽效率不得小于20db；对屏蔽服通以规定的工频电流，并经一定时间的热稳定以后，检查屏蔽服任何部位的温升不得超过50℃。

3. 屏蔽服的检查内容

屏蔽服在使用前必须进行仔细检查，有无空洞、划破、折断等缺陷；各部位连接良好，牢固可靠。用专用仪表对屏蔽服进行检测时，全套屏蔽服连接良好后任意两个最远端点之间的电阻均不得大于20Ω，手套、袜子、衣、裤单件任意两个最远端之间的电阻不得大于15Ω，导电鞋的电阻不得大于500Ω，屏蔽衣料的屏蔽效率不得小于40db。

4. 屏蔽服的使用方法

（1）穿着时必须将衣服、帽、手套、袜、鞋等各部分的多股金属连接线按照规定次序连接良好，并且不能和皮肤直接接触，屏蔽服内应穿阻燃内衣，如图5-28所示。

（2）严禁通过屏蔽服短接接地电流、空载电路和耦合电容器的电容电流。

（3）屏蔽服使用后必须妥善保管，不得与水气和污染物质接触，以免损坏，影响电气性能。

图5-28　屏蔽服的使用

5. 屏蔽服的保管注意事项

（1）屏蔽服应该尽量在通风干燥的地方存放。屏蔽服的保管如图5-29所示，应该放在干燥、通风良好的工器具库房内，并放置在专用工器具盒内上架存放。

（2）屏蔽服在不穿的时候尽量挂着，或裹成圆筒状放在专用的箱子内，以防止功能性纤维断裂。

（3）屏蔽服不应与其他化学物品放置在一起。

图5-29　屏蔽服的存放保管

6. 屏蔽服的试验内容

屏蔽服的试验内容如下。

（1）衣料效率试验。包括屏蔽效率试验、衣料电阻试验、衣料熔断电流试验、耐电火花试验、耐燃试验、耐磨试验、断裂强度和断裂伸长试验、成品试验。

（2）屏蔽服成品试验。包括上衣和裤子电阻试验、手套和短袜电阻试验、鞋子电阻试验、帽子屏蔽效应试验、屏蔽面罩屏蔽效率试验、整套屏蔽服电阻试验、整套衣服内部电场强度试验、整套衣服内流经人体电流试验、整套衣服通流经人体电流容量试验等。

（3）屏蔽服预防性试验。包括成衣电阻试验、整套服装的屏蔽效率试验，试验周期为半年。

5.3.2　防静电服

1. 防静电服的基本结构和功能

（1）基本结构

①带电作业用防静电服是由专用的防静电洁净面料制作。面料采用专用涤纶长丝，经向或纬向嵌织导电纤维。

②根据级别要求提供不同款式，并采用导电纤维缝制，使服装各部分保持电气连续性。袖管裤管为特有的双层结构，内层使用导电或防静电螺纹，从而满足高级别无尘环境的要求。

③整套高压防静电服包括上衣、裤、帽、手套和鞋，如图5-30所示。

图5-30　防静电服

（2）基本功能

①防静电服是防止衣物的静电集聚，用防静电织物面料缝制，具有高效、永久的

防静电、防尘性能，还具有薄滑，织纹清晰的特点。

②能有效地防护人体免受高压电场及电磁波。

③用于输电线路和变电站登塔（构架）作业人员和需防静电感应的地面作业人员。

2. 防静电服的分类和技术规格

（1）分类

防静电服有分体式、连体式、套头式等多种款式，袜套、帽子、口罩可以自由组合。按照导电成分在纤维中的分布情况又可将导电纤维分为导电成分均一型、导电成分覆盖型和导电成分复合型3类。

（2）技术规格

①屏蔽效率不得小于28db，衣料电阻不得大于300Ω，鞋的电阻不得大于500Ω。

②衣料应具有较大的透气量，以达到穿戴者舒适的目的。透过衣料的空气流量不得小于35L/（$m^2·s$）。

③防静电服全部使用防静电织物，不使用衬里，必须使用衬里（衣袋、加固布等）时，衬里的露出面积占全部防静电服内面露出面积的20%以下；超过20%（如防寒服或特殊服装）时，应做成面罩与衬里为可拆式。

3. 防静电服的检查内容

（1）重点检查防静电服上有无附加或佩戴任何金属物件。

（2）穿用一段时间后，应对防静电服进行检验，若静电性能不符合要求，则不能再使用。

（3）外观无破损、斑点、污物以及其他影响绝缘性能的缺陷。

（4）服装上一般不得使用金属附件，必须使用（纽扣、拉链等）时，金属附件不得直接外露。

4. 防静电服的使用方法

（1）凡是在正常情形下，爆炸性气体混杂物持续地、短时光频繁地涌现或长时间存在的场合及爆炸性气体混杂物有可能呈现的场合，可燃物的最小点燃能量在0.25 mJ以下时，应穿用防静电服。

（2）不准在易燃易爆场合穿脱防静电服。

（3）不准在防静电服上附加或佩戴任何金属物件。

（4）穿用防静电服时，还应与防静电鞋配套应用，同时地面也应是防静电地板并有接地系统。

(5) 高压静电防护服不得作为等电位屏蔽服使用。

5. 防静电服的保管注意事项

(1) 新缝制的洁净防静电服可直接进行洗涤，而回收穿过的洁净防静电服发现油污，应仔细去除油污再进行洗涤程序。洗涤后，在专用的洁净空气循环系统中进行干燥，洗涤干燥后，在洗涤专用的洁净室内叠好，装入洁净的聚酯袋或口袋内，根据要求可双层包装，也可进行真空塑封。

(2) 防静电服应该尽量存放在通风干燥的工器具库房内。并放置在专用工器具盒内上架存放。

6. 防静电服的试验内容

防静电服的试验内容主要是整套防护服装的屏蔽效率试验，试验周期为半年。

第6章

带电作业常用仪器仪表

6.1 火花间隙检测装置

多年来，国内外对输配电线路瓷质绝缘子的检测进行了大量研究，提出了许多检测方法，如电压分布测量法、红外成像法、电场测量法、超声波测量法等，研制了大量检测仪器，如光电式检测杆、自爬式检测仪、超声波检测仪、直流绝缘子检测器等，但真正被广泛应用于生产实践的还是火花间隙检测装置。火花间隙检测装置有结构简单、加工容易、使用方便、重量轻、便于携带等优点，因此被广泛应用于输配电线路带电作业中。

1. 火花间隙检测装置的基本结构和功能

（1）基本结构

①如图6-1所示，火花间隙检测装置由支撑板、接触电极固定架、火花电极、接触电极和连接头等组成。

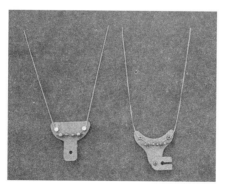

1—支撑板；2—接触电极固定架；3—火花电极；4—接触电极；5—连接头
图6-1 火花间隙检测器示例图

②火花电极、接触电极和固定架可用中碳钢制成，接触电极采用直径为3mm的圆钢制作，前段倒角。

③支撑板可用环氧层压玻璃布板制成。

(2) 基本功能

火花间隙检测装置主要用于带电检测零值瓷质绝缘子，以判断瓷质绝缘子的好坏。

2. 火花间隙检测装置的分类和技术规格

(1) 分类

①我国目前使用的火花间隙检测装置可分为固定式和可变式两种类型。固定式就是在检测过程中，间隙是固定不变的。可变式是在检测过程中可变动间隙距离，可粗略估计绝缘子的分布电压。

②按照火花电极的型式可分为球—球电极（如图6-1右侧装置）和尖—尖电极（如图6-1左侧装置）。

(2) 技术规格

①支撑板要求符合GB/T 1303.1—2009《电气用热固性树脂工业硬质层压板 第1部分：定义、命名和一般要求》的要求，厚度为3~5mm，要便于安装固定架和连接头。

②两电极组装后，其中心线应在一条轴线上，其最大偏移不得超过0.5mm。

③使用时装在相应电压等级的伸缩式绝缘杆上。

④检测各电压等级绝缘子串的低、零值绝缘子时，相应的电极间隙距离见表6-1。

表6-1 检测各电压等级绝缘子串相应的电极间隙距离

系统标称电压（kV）	63	110	220	330	500
火花电极间隙距离（mm）	0.4	0.5	0.6	0.6	0.6

3. 火花间隙检测装置的检查内容

火花间隙检测装置在使用前必须进行检查，内容如下。

(1) 使用前应对火花间隙检测装置进行检测，保证操作灵活，测量准确。

(2) 根据检测设备的电压等级调整火花电极间隙距离。

如图6-2所示为火花间隙检测装置检查。

图 6-2　火花间隙检测装置检查

4. 火花间隙检测装置的使用方法

（1）针式及少于 3 片的悬式绝缘子不得使用火花间隙检测器进行检测。火花间隙检测法实际是用试短接一片绝缘子的方法来判断绝缘子的绝缘性能。当绝缘子串中绝缘子片数少于 3 片时，如果有一片为零值，则进行检测时将极易引起接地短路故障，并烧坏工器具，造成设备事故。

（2）当检测 35kV 及以上电压等级的绝缘子串时，如发现同一串中零值绝缘子片数达到表 6-2 规定的片数，应立即停止。

表 6-2　一串中允许零值绝缘子片数

电压等级（kV）	63(66)	110	220	330	500
绝缘子串片数	5	7	13	19	28
零值片数	2	3	5	4	6

（3）带电检测绝缘子时，应在干燥天气进行。

（4）检测顺序应从导线端开始逐片向横担端进行。

（5）应注意区别检测装置靠近导线时的放电声与火花间隙的放电声，以免误判。

（6）测量过程中，应确保作业人员与带电体的最小安全距离和配合火花间隙检测装置使用的绝缘杆的有效绝缘长度。

（7）检测过程中，如遇有零值绝缘子应复测 2～3 次。使用火花间隙检测装置时，两金属探针应轻轻接触瓷质绝缘子钢帽和钢脚。作业人员应注意避免测量间隙变形。

如图 6-3、6-4 所示为使用火花间隙检测装置检测绝缘子。

图6-3　使用火花间隙检测装置检测绝缘子示例图1

图6-4　使用火花间隙检测装置检测绝缘子示例图2

5. 火花间隙检测装置保管注意事项

火花间隙检测装置应存放在带电作业工器具专用的库房内，并上架定置存放。如图6-5所示为存放火花间隙检测装置的工具柜。

图6-5　存放火花间隙检测装置的工具柜

6. 火花间隙检测装置预防性试验

火花间隙检测装置预防性试验的周期为一年，试验类型如下。

（1）间隙调整与放电试验。

（2）工频耐压试验。

（3）操作冲击耐压试验。

6.2 绝缘电阻测试仪

绝缘诊断是检测电气设备绝缘缺陷或故障的重要手段。绝缘电阻测试仪（兆欧表）是测量绝缘电阻的专用仪表。目前，绝缘电阻测试仪可分为数字式的和指针式的两种。数字式绝缘电阻测试仪目前已发展到数字显示智能型水平。指针式绝缘电阻测试仪又分为电子式和手摇式，目前生产现场多使用指针式绝缘电阻测试仪。

6.2.1 手摇式绝缘电阻测试仪

绝缘电阻测试仪即兆欧表，俗称摇表，是用来检测电气设备的绝缘电阻的一种便携式仪表。它的计量单位是兆欧（MΩ），故称兆欧表。

1. 手摇式绝缘电阻测试仪的基本结构和功能

（1）如图6-6所示，一般的兆欧表主要是由手摇发电机、比率型磁电系测量机构以及测量电路等组成。

（2）手摇式绝缘电阻测试仪有三个接线柱，上端两个较大的接线柱上分别标有"接地"（E）和"线路"（L），在下方较小的一个接线柱上标有"保护环"（或"屏蔽"）(G)。

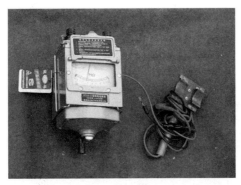

1—线路接线柱；2—接地接线柱；3—屏蔽端钮；4—表盖；5—刻度盘；
6—提手；7—发电机摇柄；8—测试电极
图6-6 手摇式绝缘电阻测试仪

2. 技术参数

（1）测量额定电压在500V以下的设备或线路的绝缘电阻时，选用500V或1000V手摇式绝缘电阻测试仪。

(2）测量额定电压在500V以上的设备或线路的绝缘电阻时，选用1000～2500V手摇式绝缘电阻测试仪。

（3）测量绝缘子时，应选用2500～5000V兆欧表。一般情况下，测量低压电气设备绝缘电阻时可选用0～200MΩ量程的手摇式绝缘电阻测试仪。

（4）手摇式绝缘电阻测试仪按准确度等级分为5级：1.0、2.0、3.0、5.0、10.0、20.0。

3. 手摇式绝缘电阻测试仪的检查内容

兆欧表在使用前必须进行检查，内容如下。

（1）检查手摇式绝缘电阻测试仪的外观是否良好，表面是否脏污。

（2）检查手摇式绝缘电阻测试仪连接线是否完好，连接是否牢固可靠。

（3）检查手摇式绝缘电阻测试仪的出厂合格证和校验合格证，检查确认绝缘电阻测试仪在检验合格周期内。

（4）对手摇式绝缘电阻测试仪进行开路试验。将两连接线开路，摇动手柄指针应指在无穷大处。

（5）对手摇式绝缘电阻测试仪进行短路试验。慢摇兆欧表摇柄，把两连接线短接一下，指针应指在零处。要注意，将两连接线断开后，方可停止摇动兆欧表摇柄。

如图6-7为手摇式绝缘电阻测试仪的检查，如图6-8所示为手摇式绝缘电阻测试仪的开路试验。

图6-7　手摇式绝缘电阻测试仪的检查

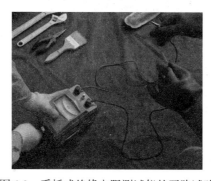

图6-8　手摇式绝缘电阻测试仪的开路试验

4. 手摇式绝缘电阻测试仪的使用方法

（1）将手摇式绝缘电阻测试仪的"接地"接线柱（即E接线柱）通过测试线可靠地接地（一般接到某一接地体上），将"线路"接线柱（即L接线柱）通过测试线接到被测物上。

（2）连接好后，顺时针摇动手摇式绝缘电阻测试仪，转速逐渐加快，保持在约120转/分后匀速摇动，当转速稳定，表的指针也稳定后，指针所指示的数值即为被测物的绝缘电阻值。

（3）实际使用中，E、L两个接线柱也可以任意连接，即E可以与接被测物相连接，L可以与接地体连接（即接地），但G接线柱决不能接错。

（4）使用注意事项。

①使用前应作开路和短路试验。使L、E两接线柱处在断开状态，摇动手摇式绝缘电阻测试仪，指针应指向"∞"；将L和E两个接线柱短接，慢慢地转动，指针应指向在"0"处。这两项都满足要求，说明手摇式绝缘电阻测试仪是好的。

②测量电气设备的绝缘电阻时，必须先切断电源，然后将设备进行放电，以保证人身安全和测量准确。

③手摇式绝缘电阻测试仪测量时应放在水平位置，并用力按住手摇式绝缘电阻测试仪，防止在摇动中晃动，摇动的转速为120转/分钟。

④测试线应采用多股软线，且要有良好的绝缘性能，两根测试线切忌绞在一起，以免造成测量数据的不准确。

⑤测量完后应立即对被测物放电，在摇表的摇把未停止转动和被测物未放电前，不可用手去触及被测物的测量部分或拆除导线，以防触电。

如图6-9所示为手摇式绝缘电阻测试仪的使用。

图6-9　手摇式绝缘电阻测试仪的使用

5. 手摇式绝缘电阻测试仪的保管注意事项

（1）绝缘电阻测试仪应存放保管在通风良好、干燥、清洁的环境内，并上架定置存放。

（2）做好仪器仪表的维护保养工作。

（3）仪器仪表的技术资料及说明书要妥善保管。

如图6-10所示为存放手摇式绝缘电阻测试仪的工具柜。

图6-10　存放手摇式绝缘电阻测试仪的工具柜

6. 手摇式绝缘电阻测试仪的预防性试验

手摇式绝缘电阻测试仪预防性试验主要是定期对手摇式绝缘电阻测试仪进行校验。手摇式绝缘电阻测试仪的检定周期不得超过两年。

6.2.2　电子式绝缘电阻测试仪

1. 电子式绝缘电阻测试仪的基本结构和功能

（1）基本结构

电子式绝缘电阻测试仪主要由接地端、屏蔽端、接线端、仪表刻度盘、发光显示管、调零装置、测试键、波段开关和测试电极组成，各部分功能如下。

①地端。接于被试设备的外壳或地上。

②屏蔽端。接于被试设备的高压护环，以消除表面泄漏电流的影响。

③线路端。高压输出端口，接于被试设备的高压导体上。

④双排刻度线。上档为绿色——5000V/2GΩ～200GΩ，10 000V/4GΩ～400 GΩ。下档为红色——5000V/0～4000 MΩ，10 000V/0～8000 MΩ。

⑤绿色发光二极管。发光时读绿档（上档）刻度。

⑥红色发光二极管。发光时读红档（下档）刻度。

⑦机械调零。调整机械指针位置，使其对准∞刻度线。

⑧测试键。按下开始测试，按下后如顺时针旋转可锁定此键。

⑨波段开关。可实现输出电压选择、电池检测、电源开关等功能。

⑩测试电极。一端接到仪表，另一端与被测物体接触。

（2）基本功能

绝缘电阻测试仪可用于测量变压器、互感器、发电机、高压电动机、电力电容器、电力电缆、避雷器等绝缘电阻。

电子式绝缘电阻测试仪的外形如图6-11所示。

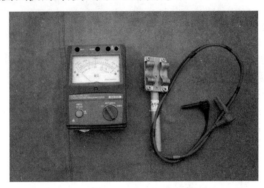

1—地端；2—线路端；3—屏蔽端；4—双排刻度线；5—绿色发光二极管；6—红色发光二极管；
7—机械调零；8—测试键；9—波段开关；10—测试电极

图6-11 电子式绝缘电阻检测仪

2. 电子式绝缘电阻测试仪的技术参数

（1）输出电压：500V、1000V、2500V、5000V。

（2）绝缘电阻：200GΩ。

（3）精度：0.05。

3. 电子式绝缘电阻测试仪的检查内容

电子式绝缘电阻测试仪在使用前必须进行检查（如图6-12所示），内容如下。

（1）检查电子式绝缘电阻测试仪的外观良好。

（2）检查电子式绝缘电阻测试仪的出厂合格证和校验合格证，检查确认电子式绝缘电阻测试仪在检验合格周期内。

（3）校验仪表指针是否在无穷大上，否则需调整机械调零螺丝。

（4）检查仪表电池是否良好。

（5）检查各接头状况是否良好。

图 6-12 绝缘电阻测试仪的检查

4. 电子式绝缘电阻测试仪的使用方法

(1) 将高压测试线一端(红色)插入 LINE 端,另一端接于或使用挂钩挂在被试设备的高压导体上,将绿色测试线一端插入 GUARD 端,另一端接于被试设备的高压护环上,以消除表面泄漏电流的影响。将另外一根黑色测试线插入地端(EARTH)端,另一端接于被试设备的外壳或地上。在接线时,特别注意 LINE(红色)与 GUARD(绿色)的接法,不要让其短路。

(2) 转动波段开关 BATT.CHECK 挡,按下测试键,仪表开始检测电池容量。

(3) 转动波段开关,选择需要的测试电压。

(4) 按下或锁定测试键开始测试。这时测试键上方高压输出指示灯发亮并且仪表内置蜂鸣器每隔 1 秒钟响一声,代表 LINE 端有高压输出。注意,测试过程中,严禁触摸探棒前端裸露部分以免发生触电危险。

(5) 当绿色发光二极管亮时,则读外圈绝缘电阻值(高范围);红色发光二极管亮时,则读内圈绝缘电阻值。测试完后,松开测试键,仪表停止测试,等待几秒钟,不要立即把测试电极移开。这时仪表将自动释放测试电路中的残存电荷。注意,试验完毕或重复进行试验时,必须将被试物短接后对地充分放电(仪表也有内置自动放电功能,不过时间较长)。

(6) 需连续进行第二次测量时,可按(3)~(5)步骤执行。注意:如长期不进行测试,需将电池仓中的电池拿出来,以免电池液渗漏损坏仪表。

(7) 对于两节及以上的被试品,例如避雷器、耦合电容可采用图 6-13 所示的接线进行测量。图中将屏蔽端接到被测避雷器上一节法兰上,这样,由上方高压线路等所引起的干扰电流由屏蔽端子屏蔽掉,而不经过测试主回路,从而避免了干扰电流的影响。对最上节的避雷器,可将其上法兰接仪表地端(EARTH)后再接地,使干扰电流直接入地。但后者不能将干扰完全消除掉。

如图6-14所示为绝缘电阻测试仪的使用。

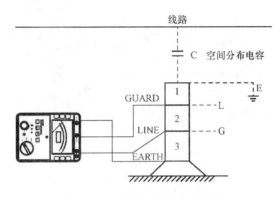

图6-13 两节以上被试品接线

图6-14 电子式绝缘电阻检测仪的使用

5. 电子式绝缘电阻测试仪的保管注意事项

（1）电子式绝缘电阻测试仪应存放保管在通风良好、干燥、清洁的环境内，并上架定置存放。

（2）做好仪器仪表的维护保养工作。

（3）仪器仪表的技术资料及说明书要妥善保管。

如图6-15所示为绝缘电阻测试仪存放工具柜。

图6-15 绝缘电阻检测仪的存放工具柜保管

6. 电子式绝缘电阻测试仪预防性试验

电子式绝缘电阻测试仪预防性试验主要是定期对电子式绝缘电阻测试仪进行校验。电子式绝缘电阻测试仪的检定周期不得超过2年。

6.2.3 其他常用的绝缘电阻测试仪

在生产现场常用的绝缘电阻测试仪有很多型号，如图6-16和6-17所示的两种型式所示，其结构、功能、检查和使用方法见相关技术手册。

图6-16 快速绝缘电阻测试仪

图6-17 数字式绝缘电阻测试仪

6.3 钳形电流表

1. 钳形电流表的基本结构和功能

（1）基本结构

钳形电流表，简称钳形表，其工作部分主要是由一只电磁式电流表和穿心式电流互感器组成。穿心式电流互感器铁芯制成活动开口，且成钳形，故名钳形电流表。如图6-18所示为钳形电流表结构图。

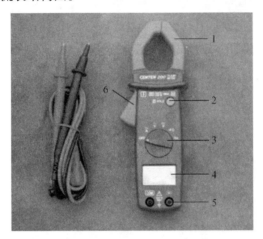

1—卡口；2—资料锁住按钮；3—挡位选择；4—数字显示屏；5—接线端子；6—扳手

图6-18 钳形电流表结构图

（2）基本功能

钳形电流表是一种不需断开电路就可直接测量电路交流电流的便携式仪表，在电气检修中使用非常方便。

2. 钳形电流表的分类和技术规格

（1）分类

①按显示方式分有指针式和数字式。

②按功能分主要有交流钳形电流表、多用钳形表、谐波数字钳形电流表、泄漏电流钳形表和交直流钳形电流表等几种。

③从测量电压分有高压钳形表和低压钳形表。

④根据其结构及用途分为互感器式和电磁系两种。常用的是互感器式钳形电流

表，由电流互感器和整流系仪表组成。它只能测量交流电流。电磁系仪表可动部分的偏转与电流的极性无关，因此，它可以交直流两用。

（2）技术规格

钳形电流表准确度主要有2.5、3.0、5.0级等几种。

3. 钳形电流表的检查内容

钳形电流表在使用前必须进行检查，检查内容如下。

（1）检查钳形电流表是否有出厂合格证和校验合格证，是否在试验合格有效期内。

（2）使用前，检查钳形电流表有无损坏，指针是否指向零位。如发现没有指向零位，可用小螺丝刀轻轻旋动机械调零旋钮，使指针回到零位上。

（3）检查钳口上的绝缘材料有无脱落、破裂等损伤现象，闭合后有无明显的缝隙，若有则必须待修复之后方可使用。

（4）检查钳形电流表包括表头玻璃在内的整个外壳，不得有开裂和破损现象，因为钳口绝缘和仪表外壳的完好与否，直接关系着测量安全问题，还涉及仪表的性能问题。

（5）对于多用型钳形电流表，还应检查测试线和表棒有无损坏，要求导电性能良好、绝缘完好无损。

（6）对于数字式钳形电流表，还需检查表内电池的电量是否充足，不足时必须更新。

如图6-19所示为钳形电流表的检查。

图6-19　钳形电流表的检查

4. 钳形电流表的使用方法

（1）测量时，应先估计被测电流大小，选择适当量程。若无法估计，可先选较大量程，然后逐挡减少，转换到合适的挡位。转换量程挡位时，必须在不带电情况下或者在钳口张开情况下进行，以免损坏仪表。

（2）测量时，被测导线应尽量放在钳口中部，钳口的结合面如有杂声，应重新开合一次，仍有杂声，应处理结合面，以使读数准确。另外，不可同时钳住两根导线。

（3）用高压钳形表测量时，应由两人操作，测量时应戴绝缘手套，站在绝缘垫上，不得触及其他设备，以防止短路或接地。

（4）测量5A以下电流时，为得到较为准确的读数，在条件许可时，可将导线多绕几圈，放进钳口测量，其实际电流值应为仪表读数除以放进钳口内的导线根数。

（5）测量时应注意身体各部分与带电体保持安全距离，低压系统安全距离为0.1～0.3m。测量高压电缆各相电流时，电缆头线间距离应在300mm以上，且绝缘良好，具备测量操作条件。观测表计时，要特别注意保持头部与带电部分的安全距离，人体任何部分与带电体的距离不得小于钳形表的整个长度。

（6）测量低压可熔保险器或水平排列低压母线电流时，应在测量前将各相可熔保险或母线用绝缘材料加以保护隔离，以免引起相间短路。当电缆有一相接地时，严禁测量，防止出现因电缆头的绝缘水平低发生对地击穿爆炸而危及人身安全事故的发生。

（7）每次测量前后，要把调节电流量程的切换开关放在最高挡位，以免使用时，因未经选择量程就进行测量而损坏仪表。

如图6-20所示为钳形电流表的使用。

图6-20　钳形电流表的使用

5. 钳形电流表的保管注意事项

（1）钳形电流表应存放保管在通风良好、干燥、清洁的环境内，并上架定置存放。

（2）做好仪器仪表的维护保养工作。

（3）仪器仪表的技术资料及说明书要妥善保管。

(4)对钳形电流表应做到定期保养。保养时,首先检查表计外观各部件、各部分有无损伤,采用柔软的毛刷或棉布将浮尘擦掉,然后蘸少许无腐蚀性的溶剂或清水仔细擦净并擦干,保养时可将电池盒打开检查或擦洗,但不得将表壳打开。

(5)钳形电流表若内部装有电池的必须将其取出,防止电池漏液而损坏表计。

(6)保管钳形电流表的室内光线不可太强,严禁阳光直射到表计上,因为其塑料外壳在阳光的光化作用下将会加速老化,而液晶显示屏则须防止阳光直射。如图6-21所示为存放钳形电流表的工具柜。

图6-21 存放钳形电流表的工具柜

6. 钳形电流表的预防性试验

钳形电流表预防性试验类型主要是定期对钳形电流表进行校验,校验周期通常为3个月,或者连续使用表计的次数较多后均应进行校验。

6.4 高压核相仪

在电力系统环网和双电源电力网建设或检修中，对于闭环点断路器两侧电源进行核相检查是非常重要的试验项目，否则可能发生相间短路，后果不堪设想。

1. 高压核相仪的基本结构和功能

（1）基本结构

①高压核相仪由三个部分组成：两个电压检测和发射的联合装置（包括绝缘杆），一个无线接收装置。如图6-22所示为高压核相仪的结构。

②两个发射装置分别将采样到的电压数据信号通过高频发射装置发送至接收端，接收端根据发射器发射的信号进行相应的处理，并在显示屏上显示相位关系和两线路的频率。

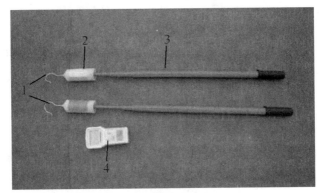

1—接触电极；2—发射器；3—绝缘杆；4—接收器
图6-22 高压核相仪的结构

（2）基本功能

对电力线路进行核相操作。

2. 高压核相仪的分类和技术规格

（1）分类

①根据核相仪的工作原理，高压核相仪可分为电容型核相仪和电阻型核相仪。电容型核相仪用于探测和指示基于电流通过杂散电容接地相位关系的设备，有双杆无连接引线和单杆记忆系统两种。电阻型核相仪用于探测和指示基于电流通过电阻元件的相位关系的设备，如双杆核相仪。

②根据核相仪的类型,可分为单体核相仪和分体核相仪。

(2)技术规格

①使用范围:6~35kV。

②发射器之间最大距离:20m。

③绝缘杆:2根共4节,使用时将两节接合使用,绝缘杆由环氧树脂构成。

3. 高压核相仪的检查内容

高压核相仪在使用前必须进行检查,检查内容如下。

(1)检查高压核相仪的外观是否良好。

(2)检查高压核相仪的出厂合格证和校验合格证,检查确认高压核相仪在检验合格周期内。

(3)检查绝缘杆耐压试验报告,如果没有绝缘杆耐压试验合格报告则不能使用,禁止雨天用高压核相仪作业。

(4)检查仪表电池是否良好,使用前最好对其进行充电。

(5)核相前,在同一电网上检测核相仪是否正常,一人操作一人监护。

图6-23所示为高压核相仪检查,图6-24所示为高压核相仪绝缘杆检查。

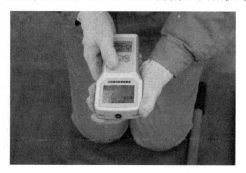

图6-23 高压核相仪检查

图6-24 高压核相仪绝缘杆检查

4. 高压核相仪的使用方法

（1）使用前将发射器和绝缘杆连接起来。

（2）核对相位。将核相器主杆（带表杆）金属钩接于带电甲侧线路任一相（此时仪表有指示），将另一杆（副杆）接于乙侧线路任一相，如果此时仪表指示为零或近于零时表明被测线路两线相位为同相。如此时表指针指示为两个输出之间的电位差（线电压），则两相位异相。

（3）验电。按核相器组装好，主杆接高压线，副杆接地或接铁横担上，也可接另一导线，如果指针指示较大时则标明线路有电，反之则无电。

（4）查找接地故障时。在中性点不直接接地的电力系统中，可用来查找接地点，将主杆接于接地事故的线路上，副杆接于接地线及金属构架或横担上，此时仪表指示为零，当断开带事故某一设备时，仪表指示较大，此时接地消除。

5. 高压核相仪的保管注意事项

（1）高压核相仪应存放保管在通风良好、干燥、清洁的环境内，并上架定置存放。

（2）做好仪器仪表的维护保养工作。

（3）仪器仪表的技术资料及说明书要妥善保管。

如图6-25所示为存放高压核相仪的工具柜。

图6-25 存放高压核相仪的工具柜

6. 高压核相仪预防性试验

高压核相仪预防性试验类型如下。

（1）定期将高压核相仪送检。

（2）定期对高压核相仪进行校验。

（3）对高压核相仪的绝缘杆要定期做耐压试验，高压核相仪的耐压试验的周期为半年。

6.5 风速仪、温湿度仪

由于输配电线路带电作业对风速、温度、湿度等有着严格的要求,每次进行带电作业之前,都需要对现场的风速和温湿度进行较为准确的检测。

6.5.1 风速仪

1. 风速仪的基本结构和功能

(1) 基本结构

风速仪由风速测量传感器和显示表两部分组成。

(2) 基本功能

风速计将流速信号转变为电信号的一种测速仪器。如图6-26所示为风速仪的结构。

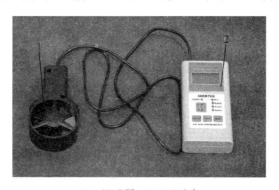

1—传感器;2—显示表
图6-26 风速仪的结构

2. 风速仪的分类

按照风速仪的原理,可将风速仪分为风杯式风速仪、螺旋桨式风速仪、热线风速仪、数字风速仪和声学风速仪。现在带电作业最常用的是数字风速仪。

3. 风速仪的检查内容

风速仪在使用前必须进行检查,检查内容如下。

(1) 检查风速仪的外观是否良好。

(2) 检查风速仪的出厂合格证和校验合格证,检查确认风速仪在检验合格周期内。

(3) 检查风速仪电池是否良好。

(4) 检查各接头状况是否良好。

4. 风速仪的使用方法

(1) 将传感器插在显示表上，一手举起传感器，一手操作显示表读数。

(2) 人保持姿势不变，转动一定角度，再次测量风速。

(3) 重复以上步骤，多测几个点后取最大风速。

(4) 在使用中，如遇风速仪散发出异常气味，或发出声音，或冒烟，或有液体流入风速内部，请立即关机取出电池。否则有被电击、发生火灾和损坏风速仪的危险。

(5) 不要将探头和风速仪本体暴露在雨中。否则，可能有被电击、发生火灾和伤及人身的危险。

(6) 不要触摸探头内部传感器部位。

(7) 禁止在可燃性气体环境中使用风速仪。

(8) 不要摔落或重压风速仪。否则会导致风速仪故障或损坏。

(9) 不要在风速仪带电的情况下触摸探头的传感器部位。否则会影响测量结果或导致风速仪内部电路的损坏。

如图6-27所示为风速仪的使用。

图6-27 风速仪的使用

5. 风速仪的保管注意事项

(1) 风速仪应存放保管在通风良好、干燥、清洁的环境内，并上架定置存放。

(2) 仪器仪表的技术资料及说明书要妥善保管。

(3) 风速仪长期不使用时，请取出内部的电池。否则电池可能漏液，导致风速仪损坏。

（4）不要将风速仪放置在高温、高湿、多尘和阳光直射的地方。否则气导致内部器件的损坏或风速仪性能的变坏。

（5）不要用挥发性液体来擦拭风速仪。否则可能气导致风速仪壳体变形变色。风速仪表面有污渍时，可用柔软的织物和中性洗涤剂来擦拭。

如图6-28所示为存放风速仪的工具柜。

图6-28　存放风速仪的工具柜

6. 风速仪的预防性试验

风速仪预防性试验类型主要是定期对风速仪进行校验，校验周期为半年。

6.5.2　温湿度仪

1. 温湿度仪的基本结构和功能

（1）基本结构

温湿度仪由温湿度探棒和显示表两部分组成。

（2）基本功能

温湿度仪，是一种用于测量瞬时温度、湿度和平均温度、湿度的仪器，具有温湿度测量、显示、记录、实时时钟、数据通信和超限报警等功能。如图6-29所示为温湿度仪的结构。

2. 温湿度仪的技术参数

（1）量程范围：湿度为0%～100%R.H.，温度为-20～+50℃。

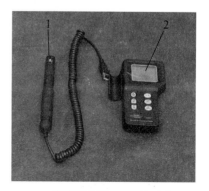

1—温湿度探棒；2—显示表
图6-29　温湿度仪的结构

(2) 分辨率：0.1% R.H.，0.1℃，0.1℉。

(3) 准确度：湿度为±3%，温度为±1℃。

(4) 显示：双显示温度及湿度。

3. 温湿度仪的检查内容

温湿度仪在使用前必须进行检查，检查内容如下。

(1) 检查温湿度仪的外观是否良好。

(2) 检查温湿度仪的出厂合格证和校验合格证，检查确认温湿度仪在检验合格周期内。

(3) 检查温湿度仪电池是否良好。

(4) 检查各接头状况是否良好。

4. 温湿度仪的使用方法

(1) 将温湿度探棒连接到显示表上，一手举起温湿度探棒，一手操作显示表读数。

(2) 在使用中，如遇温湿度仪散发出异常气味，或发出声音，或冒烟，或有液体流入温湿度内部，请立即关机取出电池。否则有被电击、发生火灾和损坏温湿度仪的危险。

(3) 不要将探头和温湿度仪本体暴露在雨中。否则，可能被电击、发生火灾和伤及人身的危险。

(4) 不要触摸探头内部传感器的部位。

(5) 不要摔落或重压温湿度仪。否则将会导致温湿度仪故障或损坏。

如图 6-30 所示为温湿度仪的使用。

图 6-30 温湿度仪的使用

5. 温湿度仪的保管注意事项

(1) 温湿度仪应存放保管在通风良好、干燥、清洁的环境内，并上架定置存放。

（2）做好仪器仪表的维护保养工作。

（3）仪器仪表的技术资料及说明书要妥善保管。

（4）温湿度仪长期不使用时，请取出内部的电池。否则电池可能漏液，导致温湿度仪损坏。

（5）不要将温湿度仪放置在高温、高湿、多尘和阳光直射的地方。否则会导致内部器件的损坏或温湿度仪性能的变坏。

如图6-31所示为温湿度仪存放的工具柜。

图6-31　存放温湿度仪的工具柜

6. 温湿度仪的预防性试验

温湿度仪预防性试验类型主要是每年对温湿度仪进行校验，校验周期一般不得超过一年。

第7章

安全及绝缘手工工具

7.1 安全工具

输配电线路带电作业常用的安全工具有携带式短路接地线、安全带、脚扣、升降板、绝缘扳手、拔销钳、剥皮器、断线钳等。安全工具的检查、使用和保管是输配电线路带电作业的重要内容。

7.1.1 携带式短路接地线

1. 携带式短路接地线的基本结构和功能

（1）基本结构

①如图7-1所示，携带式短路接地线由导线端线夹、短路线、汇流夹、接地线、接线鼻、接地端线夹以及接地操作棒等组成。

②导线端和接地端的线夹采用优质铝合金压铸，与其配套的紧固件金属均经镀镍处理。

③接地操作棒采用绝缘性能与机械强度俱佳的高级环氧树脂、玻璃纤维等制作而成，同时在操作手柄处加装硅橡胶护套，使绝缘更为安全可靠。

④短路线与接地线采用多股优质软铜线绞合而成，并外覆柔软、耐高温的透明绝缘护层，可以防止使用中对接地铜线的磨损，确保作业人员在操作中的安全。

⑤接线鼻和汇流夹与接地铜线及外护套采用压接新工艺，软连接可有效地防止使用时连接处铜线断股，提高了接地线的可靠性和使用寿命。

（2）基本功能

携带式短路接地线可防止邻近高压带电设备产生感应电压对人体的危害，还可用于放尽处于强电场中导体上的电荷,以及释放断电设备的剩余电荷。

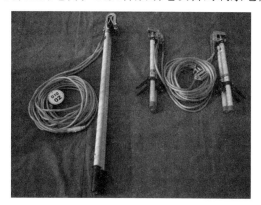

1—接地操作棒;2—导线端线夹;3—汇流夹;
4—短路线;5—接线鼻;6—接地线;7—接地端线夹

图 7-1　携带式短路接地线

2. 携带式短路接地线的分类和性能参数

(1) 分类

①按照使用环境可以分为室内母排型接地线(JDX-NL)和室外线路型接地线(JDX-WS)。

②按照电压等级可分为 0.4kV、10kV、35kV、110kV、220kV、500kV 接地线等。

③携带式短路接地线接地短路线标称截面积有:16mm^2、25mm^2、35mm^2、50mm^2、70mm^2、95mm^2,可根据电压等级及额定短路电流大小进行选择。

(2) 性能参数

10~110kV 携带式短路接地线性能参数见表 7-1。

表 7-1　10~110kV 携带式短路接地线性能参数表

电压 kV	参数 绝缘杆部分 长度(mm)	握手部分 长度(mm)	金属接头 长度(mm)	节数	杆径 (mm)	标称截面 (mm^2)	总长 (mm)
10kV	700	300	50	1	30	25	1050
35kV	900	600	50	1	30	25	1550
110kV	1300	700	140	2	30	35.67	2140

3. 携带式短路接地线的检查内容

携带式短路接地线在使用前必须进行检查,检查内容如下。

(1) 首先检查接地线规格是否同工作线路电压等级一致,严禁使用不符合规格的接地线。

（2）检查携带式短路接地线试验合格证是否在有效期内。

（3）检查携带式短路接地线的连接器（线卡或线夹）装上后接触应良好，并应有足够的夹持力，以防短路电流幅值较大时，由于接触不良而熔断或因电动力的作用而脱落。

（4）检查携带式短路接地铜线和三根短接铜线的连接是否牢固，一般应由螺丝拴紧后，再加焊锡焊牢，以防因接触不良而熔断。

（5）检查携带式短路接地线软铜线塑料外套是否破损，软铜线是否有磨损和断股现象。

如图7-2所示为便携式短路接地线的检查。

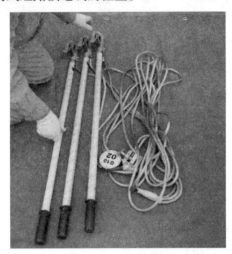

图7-2 携带式短路接地线的检查

4. 接地线的使用方法

（1）使用时，携带式短路接地线的连接器（线卡或线夹）装上后接触应良好，并有足够的夹持力，以防短路电流幅值较大时，由于接触不良而熔断或因电动力的作用而脱落。

（2）携带式短路接地线使用时必须注意装拆顺序，装设接地线必须先连接接地夹，后连接导线夹，且必须接触良好，拆除携带式短路接地线的顺序与此相反。

（3）装设携带式短路接地线必须由两人进行，装、拆携带式短路接地线时应穿戴绝缘手套。

（4）携带式短路接地线在每次装设以前应经过详细检查，损坏的携带式短路接地线应及时修理或更换，禁止使用不符合规定的导线作接地线或短路线。

（5）用于释放电荷时，接地线接地端必须使用专用线夹固定接地体上，严禁用缠绕的方法进行接地或短路。

（6）携带式短路接地线和工作设备之间不允许连接刀闸或熔断器，以防它们断开时，设备失去接地，使检修人员发生触电事故。

（7）装设的携带式短路接地线的最大摆动范围与带电部分保持足够的安全距离。

5. 携带式短路接地线的保管注意事项

（1）携带式短路接地线应存放保管在通风良好、干燥的环境内，并上架定置存放。

（2）每组携带式短路接地线均应编号，存放位置亦应编号。携带式短路接地线号码与存放位置号码必须一致，以免在复杂的系统中进行部分停电检修时，发生误拆或忘拆地线而造成的事故。

如图7-3所示为存放接地线的工具柜。

图7-3　存放接地线的工具柜

6. 携带式短路接地线的预防性试验

携带式短路接地线的预防性试验每年进行1次，试验内容如下。

（1）成组携带式短路接地线的电压降试验。

（2）多股软导线疲劳试验。

（3）线鼻与多股软导线的紧握力试验。

（4）接地线端线夹超紧固力试验。

（5）接地操作棒的机械强度试验。

（6）短路电流试验。

7.1.2 安全带

1. 安全带的基本结构和功能

（1）基本结构

如图7-4所示，全身式安全带由带子、绳子和金属配件组成。根据作业性质的不同，其结构形式也有所不同，主要有围杆作业安全带、悬挂作业安全带等形式。

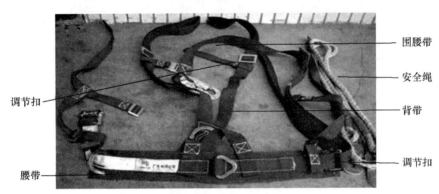

图7-4　全身式安全带各部分

安全带采用锦纶、维尼纶、蚕丝等材料制成。但因蚕丝原料少、成本高，故目前多以锦纶为主要材料。金属配件用普通碳素钢或铝合金钢。

（2）基本功能

安全带是高处作业人员预防坠落伤亡的防护用品。它广泛用于在架空输配电线路杆塔上进行施工安装、检修作业中，为防止作业人员从高空摔跌，必须使用安全带。围杆作业安全带主要用于线路检修和巡视作业；悬挂安全带主要用于建筑、安装等作业中。

如图7-5所示为围杆作业安全带和悬挂作业安全带两种安全带的结构，图7-6所示为三点式、五点式安全带实物图。

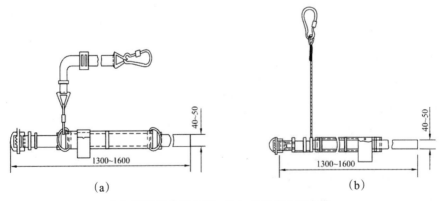

（a）围杆作业安全带；（b）悬挂作业安全带
图7-5　安全带的结构

(a) (b)

(a) 五点式安全带;(b) 三点式安全带

图 7-6 安全带

2. 安全带的分类和技术规格

(1) 分类

①安全带根据使用条件可以分为围杆作业安全带、区域限制安全带、坠落悬挂安全带三大类,见表 7-2。

表 7-2 安全带根据使用条件分类

分类	部件组成	挂点装置
围杆作业安全带	系带、连接器、调节器(调节扣)、安全绳、围杆带(围杆绳)	杆柱
区域限制安全带	系带、连接器(可选)、安全绳、调节器、连接器、围杆带(围杆绳)	挂点
	系带、连接器(可选)、安全绳、调节器、连接器、围杆带(围杆绳)、滑车	导轨
坠落悬挂安全带	系带、连接器(可选)、缓冲器、安全绳、连接器、围杆带(围杆绳)、速差自控器	挂点
	系带、连接器(可选)、缓冲器(可选)、安全绳、连接器、围杆带(围杆绳)、自锁器	导轨

②安全带按结构形式不同又可以分为三大类,见表 7-3。

表 7-3 安全带按结构形式分类

分类	适用范围
腰带式安全带	变电站检修试验、110kV 及以下变电站构架作业、35kV 及以下配电线路高空作业
半身式安全带	220kV 变电站构架作业和 110kV 线路高空作业以及其他必须使用区域限制安全带的工作
全身式安全带	220kV 及以上输电线路或工作高度在 30 m 以上的高空作业、500kV 及以上的变电站构架作业以及其他必须使用坠落悬挂安全带的工作

（2）技术规格

①主带必须是一整根，其宽度不小于40mm，长度为1300～1600mm。

②护腰带宽度不小于80mm，长度为600～700mm，辅带宽度应不小于20mm。

③安全绳（包括未展开的缓冲器）有效长度不应大于2m，特殊需要超过2m的应增装缓冲器或防坠器。当在高空作业，活动范围超出安全绳保护范围时，必须配合速差式自控器使用。有两根安全绳（包括未展开的缓冲器）的安全带，其单根有效长度应不大于1.2m。

3. 安全带的检查内容

（1）安全带使用前，必须做外观检查，如发现破损、变质及金属配件有断裂者，应禁止使用，平时不用时，也应每月做一次外观检查。

（2）安全带在使用前必须检查其试验合格证是否在有效期内，超过有效试验期的安全带不能使用。

（3）安全带在使用前必须进行受力的冲击试验，以检验安全带承力情况，如图7-7所示。

图7-7 安全带冲击试验

4. 安全带的使用方法

（1）安全带应高挂低用或水平拴挂。高挂低用就是将安全带的绳挂在高处，作业人员在悬挂点下方工作；水平拴挂就是使用单腰带时，将安全带系在腰部，绳的挂钩挂在和安全带同一水平的位置，人和挂钩保持差不多等于绳长的距离。切忌低挂高用，并应将活梁卡子系紧。

（2）在低温环境中使用安全带时，要注意防止安全带变硬割裂。

（3）使用频繁的安全绳应经常做外观检查，发生异常时应及时更换新绳，并注意加绳套的问题。

（4）不能将安全带打结使用，以免发生冲击时安全绳从打结处断开，应将安全挂钩挂在连接环上，不能直接挂在安全绳上，以免发生坠落时安全绳被割断。

（5）安全带上的各种部件不得任意拆掉，更换新绳时要注意加绳套，带子使用期为3～5年，发现异常应提前报废。

（6）安全带使用和存放时，应避免接触高温、明火和酸类物质，以及有锐角的坚硬物体和化学药物。

如图7-8所示为安全带的穿戴方法，如图7-9所示是在杆上作业时安全带的使用方法。

图7-8　安全带的穿戴方法

图7-9　杆上作业安全带的使用

5. 安全带的保管

（1）安全带应贮藏在干燥、通风的环境内，并编号上架定置存放，不准接触高温、明火、强酸、强碱和尖利的硬物，也不要暴晒。搬动时不能用带钩刺的工具，运输过程中要防止日晒雨淋。

（2）安全带应该经常保洁，可放入温水中用肥皂水轻轻擦，然后用清水漂净，然后晾干。

如图7-10所示为安全带的上架存放。

图7-10　安全带的上架存放

6. 安全带的预防性试验

安全带的预防性试验主要是静负荷试验，试验周期为1年（其中牛皮带的试验周期为半年）。

7.1.3　脚扣

1. 脚扣的基本结构和功能

脚扣，又叫铁脚，是攀登电杆的工具。脚扣一般采用高强无缝管制作，经过热处理，具有重量轻、强度高、韧性好、可调性好、轻便灵活、安全可靠、携带方便等优点，是电工攀登不同规格的混凝土电杆或木质电杆的理想工具。

2. 基本类型和性能参数

（1）基本类型

脚扣分两种：一种是扣环上制有铁齿，供登木杆电杆使用，如图7-11(a)所示；另一种在扣环上裹有橡胶，供登混凝土电杆用，如图7-11(b)所示，图7-11(c)为脚扣实物图。脚扣攀登电杆速度较快。

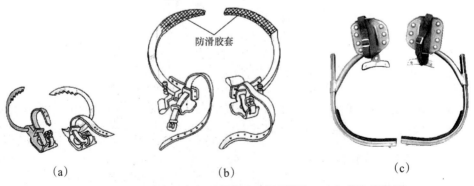

(a)　　　　　　　　　(b)　　　　　　　　　(c)

(a) 登木质电杆用脚扣；(b) 登混凝土电杆用脚扣；(c) 脚扣实物图
图7-11　脚扣

（2）性能参数

登杆脚扣性能参数见表7-4。

表7-4 登杆脚扣性能参数表

产品名称	规格型号	高度m	材质	可选配件
登杆脚扣	JK-T-250	8	高碳钢	脚扣带
登杆脚扣	JK-T-300	10	高碳钢	脚扣带
登杆脚扣	JK-T-350	12	高碳钢	脚扣带
登杆脚扣	JK-T-400	15	高碳钢	脚扣带
登杆脚扣	JK-T-450	18	高碳钢	脚扣带
登杆脚扣	JK-T-500	21	高碳钢	脚扣带

3. 脚扣使用前的检查

（1）脚扣在使用前必须检查其金属部件是否有磨损、锈蚀和裂纹，活动扣在扣体内滑动是否灵活，脚背绑带是否完好等。

（2）脚扣使用前必须认真检查其试验合格证是否在有效期内，超过试验合格有效期的脚扣不能使用。

（3）脚扣使用前必须对脚扣进行人体载荷冲击试验，检验脚扣各部位是否牢固可靠。

4. 脚扣的使用方法

脚扣主要用于攀登电杆，其中木杆电杆脚扣主要适用于电力、邮电线路混凝土电杆或钢管塔的登高。混凝土电杆脚扣适用于电力、邮电线路混凝土电杆的登杆。

如图7-12所示为运用脚扣登杆，如图7-13所示为运用脚扣下杆。

图7-12 运用脚扣登杆（上杆）

图7-13 运用脚扣下杆（下杆）

5. 脚扣的保管注意事项

脚扣应保管在干燥、通风的库房内，必须编号上架定置存放。

如图7-14所示为登杆工器具（脚扣、升降板）的保管。

图7-14 登杆工器具的保管

6. 脚扣的预防性试验

脚扣的预防性试验为静负荷试验，试验周期为一年。

7.1.4 升降板

1. 升降板的基本结构和功能

升降板，又称三角板、蹬板、踏板或踩板。它由铁钩、麻绳、木板组成，如图7-15所示。升降板是电工攀登电杆及杆上作业的一种工具。绳钩至木板的垂直长度与作业人的高度相适应，一般应以作业人员手臂长为宜。

图 7-15 升降板实物图

2. 升降板的性能参数

脚踏板是选用质地坚韧的木材，如水曲柳、柞木等，制成 30~50mm 厚的长方形踏板，再将白棕绳的两端系在踏板两端。白棕绳采用 16mm 直径的三股白棕绳。升降板的木板和白棕绳均应能承受 300kg 的重量。

3. 升降板使用前的检查

（1）升降板使用前必须认真检查其踏板有无开裂或腐朽，绳索有无腐蚀或断股现象，如发现要及时处理。否则登杆时容易出现滑落伤人事故。

（2）使用前必须认真检查其试验合格证是否在有效期内，超过试验合格有效期的升降板不能使用。

（3）使用前必须对升降板进行人体载荷冲击试验，检验升降板各部位是否牢固可靠。如图 7-16 所示为对升降板进行载荷冲击试验。

图 7-16 升降板载荷冲击试验

4. 升降板的使用方法

升降板的使用方法要掌握得当，否则发生脱钩或下滑，就会造成人身伤亡事故。升降板在杆上作业，比较灵活舒适，适用于长时间的杆上作业，能降低疲劳程度。

如图7-17所示为使用升降板登杆。

图7-17　使用升降板登杆

5. 升降板的保管注意事项

升降板应保管在干燥、通风的库房内，必须上架定置存放。

6. 升降板的预防性试验

升降板的预防性试验为静负荷试验，试验周期为半年。

7.2 绝缘手工用具

输配电线路绝缘手工用具主要有绝缘螺栓刀、绝缘扳手、绝缘手钳、绝缘导线剥皮器和手动断线钳等,它们是带电操作人员最常用的个人工器具。

7.2.1 绝缘螺钉旋具

1. 绝缘螺钉旋具的结构和功能

绝缘螺钉旋具又称为改锥、起子,是在普通螺丝刀的旋杆金属部分套有绝缘套管,其头部有一字型和十字型两种;绝缘螺丝刀是用来旋紧或松开头部带沟槽的螺丝钉的工具。绝缘螺钉旋具实物图如图7-18所示。

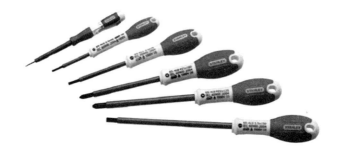

图7-18 绝缘螺钉旋具实物图

2. 绝缘螺钉旋具的类型和性能参数

(1)基本类型

绝缘螺钉旋具有绝缘一字螺钉旋具、绝缘十字螺钉旋具、绝缘快速螺钉旋具、绝缘弯头螺钉旋具等几种类型。

(2)性能参数

绝缘螺钉旋具一般以旋杆直径×旋杆长度作为规格参数,常用旋杆直径有3mm、5mm、7mm、9mm等,常用的旋杆长度有50mm、100mm、150mm和300mm等。

3. 绝缘螺钉旋具使用前的检查

(1)检查绝缘螺钉旋具的手柄部分是否破裂和损坏,检查其绝缘套管是否松动和破裂。

(2) 检查绝缘螺钉旋具的刀口是否变钝和损坏，对刀口损坏严重、变形、手柄裂开或损坏的应报废。

4. 螺钉旋具的使用方法

(1) 应根据旋紧或松开的螺钉头部的槽宽和槽形选用适当的绝缘螺钉旋具。

(2) 不能用较小的螺钉旋具去旋拧较大的螺丝钉。

(3) 绝缘十字螺钉旋具用于旋紧或松开头部带十字槽的螺丝钉。

(4) 绝缘弯头螺钉旋具用于空间受到限制的螺丝钉头。

(5) 不要用绝缘螺钉旋具旋紧或松开握在手中工件上的螺丝钉，应将工件夹固在夹具内，以防伤人。

(6) 不可用锤击绝缘螺钉旋具手把柄端部的方法撬开缝隙或剔除金属毛刺及其他的物体。

(7) 绝缘螺钉旋具不可当撬棒使用，不可用手锤击打螺钉旋具手把柄，也不可在螺钉旋具手把柄与刀口处用扳手或钳子来增加扭力，以防螺钉旋具弯曲损坏。

(8) 不能用绝缘螺钉旋具工具玩耍、打闹，以免伤人。

(9) 使用绝缘螺钉旋具时，姿势要正确，用力要适当。

绝缘螺钉旋具的正确使用方法如图7-19所示。

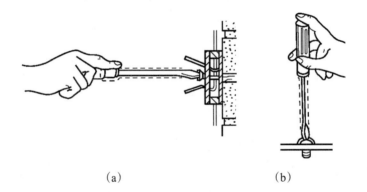

(a) 较大旋具用法；(b) 较小旋具用法
图7-19　螺钉旋具的使用方法

5. 绝缘螺钉旋具的保管注意事项

绝缘螺钉旋具应存放在干燥、通风良好的工器具库房里面，可以放在专用工器具盒内或编号上架定点放置。

6. 绝缘螺钉旋具的预防性试验

普通绝缘螺钉旋具绝缘部分的耐压为500V，高压绝缘螺钉旋具绝缘部分的耐压可以达到1000V或1500V。绝缘螺钉旋具预防性试验包括外观检查和绝缘部做高压工频耐压试验，试验周期为1年。

7.2.2 绝缘扳手

1. 绝缘扳手的结构和功能

绝缘扳手是用于旋紧六角形、正方形螺钉和各种螺母的工具。绝缘扳手采用工具钢、合金钢或可锻铸铁制成。绝缘活动扳手主要由呆扳唇、活络扳唇、蜗轮、轴销，手柄等构成，如图7-20所示。绝缘活动扳手是一种旋紧或起松有角螺丝或螺母的工具，转动活络扳手的蜗轮，就可以调节扳口的大小。绝缘呆扳手的扳口固定成形不可以调节，只适用于同种规格的螺钉，如图7-21所示。

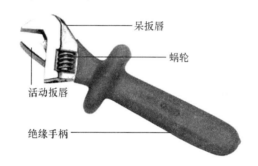

图7-20 绝缘活动扳手

图7-21 绝缘呆扳手

2. 绝缘扳手的基本类型和技术规格

输配电线路带电作业常用的绝缘扳手一般有活扳手、呆扳手和套筒扳手等几种类型。

绝缘活扳手有200mm、250mm和300mm（英制8in、10in、12in）三种。使用时要根据螺母的大小，选用适当规格的绝缘扳手，以免扳手过大，损伤螺母，或螺母过大，损伤扳手。

3. 绝缘扳手使用前的检查

（1）检查绝缘扳手的手柄部分是否破裂和损坏，绝缘部分是否老化。

（2）检查绝缘活动扳手扳口是否可以灵活调节。

4. 绝缘扳手的使用方法

（1）使用时应根据螺钉、螺母的形状、规格及工作条件选用规格相适应的绝缘活动扳手操作。

（2）绝缘活扳手开口尺寸可在一定的范围内调节，所以在开口尺寸范围内的螺钉、螺母一般都可以使用。

（3）绝缘活扳手不可用大尺寸的扳手去旋紧尺寸较小的螺钉，这样会因扭矩过大而使螺钉折断；应按螺钉六方头或螺母六方的对边尺寸调整开口，间隙不要过大，否则将会损坏螺钉头或螺母，并且容易滑脱，造成伤害事故；应让固定钳口受主要作用力，要将扳手柄向作业者方向拉紧，不要向前推，扳手手柄不可以任意接长；不应将扳手当锤击工具使用。

5. 绝缘扳手的保管注意事项

绝缘扳手应存放在干燥、通风良好的工器具库房里面，并编号上架定点放置。如图7-22所示绝缘扳手的存放。

图7-22 绝缘扳手的存放

6. 绝缘扳手的预防性试验

绝缘扳手绝缘部分应定期做高压工频耐压试验，试验周期为1年。

7.2.3 钢丝钳

1. 钢丝钳的结构和功能

钢丝钳,俗称卡钳、手钳,又称电工钳,是输配电线路带电作业使用的安全用具之一。它是用来钳夹和剪切的工具,其结构如图 7-23 所示,由钳头和钳柄组成。钳头有四口,即钳口、齿口、刀口和铡口;钳柄套有绝缘套管。如图 7-24 所示为钢丝钳实物图。

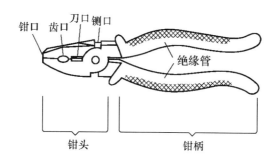

图 7-23 钢丝钳的结构

图 7-24 钢丝钳实物图

2. 钢丝钳的技术规格

常用的钢丝钳规格有 150mm、175mm、200mm 三种,普通钢丝钳绝缘手柄耐压为 500V,耐高压绝缘钢丝钳工作电压可以达到 1000V。

如图 7-25 所示为 1000V 高压绝缘钢丝钳实物图。

图 7-25 1000V 耐高压钢丝钳

3. 钢丝钳使用前的检查

（1）钢丝钳在使用前应检查绝缘手柄绝缘套管是否磨损、碰裂，以防在工作中钳头触碰到带电部位，使钳柄带电而造成意外事故。

（2）使用前应扳动手柄，检查其钢丝钳钳口开合是否灵活可靠。

4. 钢丝钳的使用方法

（1）使用钢丝钳，要使钳头的刀口朝内侧，即朝向自己，便于控制钳口部位；用小指伸在两钳柄中间，用以抵住钳柄，张开钳头。

（2）在使用中还需注意的切勿用刀口去钳断钢丝，以免刀口损伤。绝缘柄钢丝钳的刀口可用来切剪电线、铁丝。剪8号镀锌铁丝时，应用刀刃绕表面来回割几下，然后只需轻轻一扳，铁丝即断。

用电工钢丝钳剪切带电导线时，不得用钳口同时剪切相线和零线，或同时剪切两根相线，那样均会造成线路短路。

（3）钳头不可代替手锤作为敲打工具。

（4）绝缘柄钢丝钳使用中切忌乱扔、砸或使用手柄矫直钢线，以免损坏绝缘塑料管。

（5）用绝缘柄钢丝钳缠绕抱箍固定拉线时，用钳子齿口夹住铁丝，以顺时针方向缠绕。

5. 钢丝钳的保管注意事项

钢丝钳应存放在干燥、通风良好的工器具库房里面，可以放在专用工器具盒内或上架定点放置。

6. 钢丝钳的预防性试验

高压绝缘钢丝钳预防性试验包括外观检查和绝缘部做高压工频耐压试验，试验周期为1年。

7.2.4 拔销器

1. 拔销器的结构和功能

拔销器又称为拔销钳，它的钳口是空心的，其他部分同钢丝钳类似，其功能主要是用于线路工作中摘取绝缘子的弹簧销。

2. 拔销器的基本类型

拔销器可分停电和带电两种类型。停电的使用的多半是用手握式的，有普通型和"Z"型两种。带电作业使用的有手握和绝缘操作杆两大类，手握式与停电时使用的报销钳一样，绝缘操作杆有直线拔销器和耐张拔销器两种。如图7-26所示为几种常见拔销器。

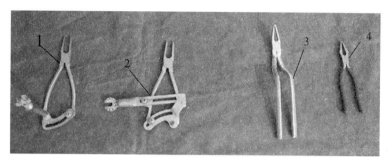

1—耐张绝缘子拔销钳；2—直线绝缘子拔销钳；
3—220kV手握式拔销钳；4—10kV手握式拔销钳
图7-26 拔销钳

3. 拔销器使用前的检查

（1）耐张、直线绝缘子拔销器使用前要检查钳口是否变形、磨损，闭合是否完整，钳柄调节螺丝是否松动。

（2）手握式拔销钳要检查钳口是否变形、磨损，绝缘护套是否破裂和老化。

4. 拔销器的使用方法

（1）耐张、直线绝缘子拔销钳使用时必须同绝缘操作杆配合，绝缘杆需要选用相同电压等级的绝缘操作杆配合使用。

（2）耐张、直线绝缘子拔销钳适用于输电线路带电作业地电位作业法操作使用。在地电位情况下使用耐张、直线绝缘子拔销钳时，绝缘杆操作必须平稳，动作迅速。如图7-27所示为带电使用耐张绝缘子拔销钳更换耐张单片绝缘子操作。

图7-27 带电使用耐张绝缘子拔销器更换
耐张单片绝缘子操作

5. 拔销器的保管注意事项

拔销器应存放在干燥、通风良好的工器具库房里，并上架定点放置。如图7-28所示为拔销器存。

图7-28　拔销器的上架存放

6. 拔销器的预防性试验

耐张、直线绝缘子拔销器预防性行试验主要是外观检查绝缘杆预防性试验，绝缘杆预防性试验主要是工频耐压试验；手握式拔销钳预防性试验主要是外观检查和绝缘手柄工频耐压试验，试验周期均为1年。

7.2.5　绝缘导线剥皮器

1. 绝缘导线剥皮器的结构和功能

绝缘导线剥皮器又称为绝缘导线剥皮钳或剥线钳，主要用于剥除绝缘导线绝缘层，目前主要用于10kV绝缘导线和高压电缆绝缘层剥除操作。美国、日本、意大利、中国等国家均有生产厂家，其外形结构方式多样。国产BXQ-Z-40A旋切式剥皮器结构由固定旋钮、进刀旋钮、固定钳头、圆柱体切刀、滑动钳头和手柄组成，如图7-29所示。刀片由合金钢制造，刀片锋利、耐用，刀头可以拆卸、修磨，剥皮厚度有进给刻度，操作直观便利。如图7-30所示为BXQ-Z-40A旋切式剥皮器。

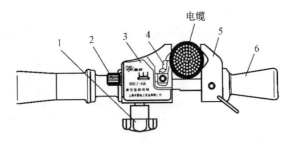

1—固定旋钮；2—进刀旋钮；3—圆柱体切刀；4—滑动钳头；5—手柄
图7-29　BXQ-Z-40A旋切式剥线钳的结构

第7章 安全及绝缘手工工具

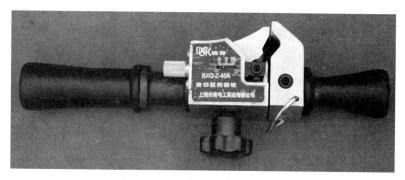

图 7-30 BXQ-Z-40A 旋切式剥皮器

2. 绝缘导线剥皮器的基本类型和性能参数

绝缘导线剥皮器按操作方式来分有手动和电动两类，按生产厂家制式来分有美制方式、日制方式、意制方式、合资方式和国产方式几种。

下面介绍几种常用绝缘导线剥皮器规格及性能参数，见表7-5。

表7-5 绝缘导线剥皮器的规格和性能参数

序号	型号	制式类别	性能参数	用途
1	NP400	日制	适用于直径从12~32mm，绝缘层厚度1.4~4mm的绝缘导线	剥除架空绝缘导线绝缘层
2	WS-50	美制	剥除线径为12.7~57.1mm，剥除厚度为8.5mm，定位剥除长度为150mm	适用于10kV主绝缘层、架空绝缘导线
3	TYX-300	合资	可剥除绝缘导线面积为25~300mm^2	适用于架空绝缘导线
4	AV6220	意制	适用电缆直径为25mm以上，切割厚度为5mm	适用于三芯电缆外皮的剥除
5	BXQ-Z-40A	国产	剥切直径为12~40mm，剥除厚度：终端剥切小于6mm，中段剥切小于4mm	主要用于绝缘导线、架空导线及电缆终端和中段快速剥切

如图7-31所示为NP400、WS-50、TYX-300和AV6220剥皮器。

· 179 ·

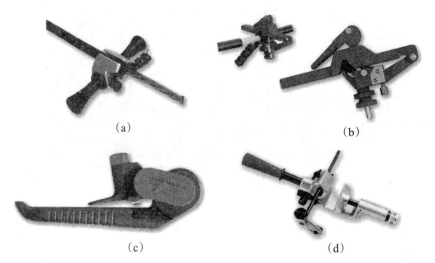

(a) NP400剥皮器；(b) AV6220剥皮器；(c) TYX-300剥皮器；(d) WS50剥皮器
图7-31 剥皮器

3. 绝缘导线剥皮器使用前的检查

（1）使用前根据导线直径选择相应剥皮器，切忌超越范围使用，以免刀刃受损。

（2）使用时间长，圆柱体切刀难免会钝，应及时用油石修磨，保证剥削质量。

（3）使用前应检查剥皮器各部分结构是否良好，旋钮是否灵活可靠，有无打滑失效情况。

4. 绝缘导线剥皮器的使用方法

国产BXQ-Z-40A剥皮器的使用方法如下。

（1）根据BXQ-Z-40A结构示意图，放松固定旋钮，拉开滑头钳头，卡住导线的剥皮处，然后旋紧固定旋钮。

（2）逆时针旋紧旋钮，同时观察刻度进给量（每格1mm），到适当深度，停止进给。

（3）旋紧剥皮器，进给方向从右到左，快速时向右略施加推力，使剥皮器向左倾斜。

（4）旋削至预定长度时，取消向左推力，使机身原地旋削，此时皮屑即自行断落。

（5）顺时针方向旋进旋钮，使圆柱体切刀退回（刻度退到0），然后，放松固定旋钮，拉开滑头钳头，取出导线，剥皮结束。

（6）在使用剥皮器时，进刀进给深度应适当，通常情况下，进给深度应留有0.5～1mm的绝缘层，以免进给过深导致导线及圆柱体切口的刃口受损害。

其他剥皮器请参考其使用说明书。

5. 绝缘导线剥皮器的保管注意事项

绝缘导线剥线钳应存放在干燥、通风良好的工器具库房内，并装在专用工器具盒内上架定点放置。如图7-32所示是BXQ-Z-40A剥皮器的存放。

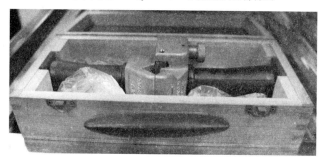

图7-32　剥皮器的保管

6. 绝缘导线剥皮器的预防性试验

绝缘导线剥皮器预防性试验包括外观检查和绝缘部分工频耐压试验，试验周期为1年。

7.2.6　齿轮断线钳

输配电线路带电作业所使用的电缆断线钳种类繁多，这里只介绍携带方便、适用于个人使用的齿轮式断线钳。

1. 齿轮断线钳的结构和功能

齿轮式断线钳主要用于切断电缆，包括钢芯铝绞线、钢绞线和绝缘导线。台湾生产的LK-300B型齿轮断线钳由绝缘手柄、动刀片、定刀片、盖板、止退旋钮等几部分组成，如图7-33所示。

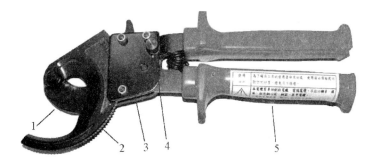

1—定刀片；2—动刀片；3—盖板；4—止退旋钮；5—绝缘手柄
图7-33　LK-300B齿轮断线钳的结构

2. 齿轮断线钳的型号和技术参数

齿轮断线钳有美制、日制和国内生产的，表7-6介绍了美制的两种型号齿轮断线钳的型号和技术参数。如图7-34所示为RCC-325型，RCC-500型齿轮断线钳。

表7-6　美制的两种齿轮断线钳的型号和技术参数

品牌	美国KUDOS	美国KUDOS
型号	RCC-325	RCC-500
适用范围	325mm²以下铜铝电缆	500mm²以下铜铝电缆
尺寸（长）	250mm	700mm
重量	0.65kg	1.67kg
特点	单手操作 浸泡式防滑手柄 不可用于切断硬铜线或是任何有钢芯的线缆 切断刀刃较宽不易变形，棘轮齿距较小切断省力	

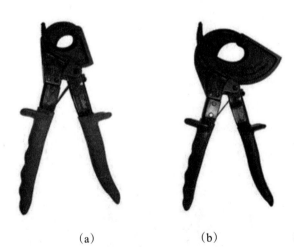

（a）RCC-325型；（b）RCC-500型

图7-34　齿轮断线钳

3. 齿轮断线钳使用前的检查

（1）使用前应检查断线钳齿轮刃口是否损坏，各部分螺丝是否松动。

（2）使用前应检查两个刀片是否错位或不能闭合，造成剪切困难，如果出现此类情况，只需要调整相应螺丝即可。

（3）检查断线钳齿轮转动部分是否灵活，如果出现转动困难，应及时在齿轮部分加油润滑。

（4）检查断线钳绝缘手柄是否破损，是否有裂口和老化，如果绝缘手柄出现此类情况，最好不要使用。

4. 齿轮断线钳的使用方法

（1）严禁超范围、超负荷使用。齿轮断线钳有其特定的额定强度，在使用时应该根据实际需要合理选择品种规格，不得以小代大，不得剪切硬度大于断线钳刀口的物品，以免造成崩刃或卷刃。

（2）不得把它当作普通钢制工器具替代其他工器具。

（3）剪切前将止动旋钮置于使用位置，根据剪切导线直径调整剪口，将导线放入剪口，并通过压、放绝缘手柄使齿轮转动，使得定刀片和动刀片逐渐闭合。

（4）使用完后要将已经闭合的动、定刀片置于打开状态，并及时清除弹簧、齿槽中的泥土杂物。

5. 齿轮断线钳的保管注意事项

齿轮断线钳应存放在干燥、通风良好的工器具库房内，并装在专用工器具盒内，上架定点放置。

6. 齿轮断线钳的预防性试验

齿轮断线钳的预防性试验包括外观检查和绝缘部分工频耐压试验，试验周期为1年。

第8章

带电作业工器具的保管和运输

带电作业在我国开展已有好几十年，带电作业行业积累了大量的实践经验。但在20世纪90年代，我国带电作业行业内还是出现了多起由于工具使用、管理不规范而造成的事故，使带电作业工作难以大范围的开展，究其原因主要是带电作业工具在存放、运输过程中受潮，使绝缘性能大幅下降甚至发生击穿闪络事故。尤其是我国南方地区，空气湿度较大，带电作业工具更易受潮，造成事故隐患。

本章重点介绍带电作业工器具的保管与运输基本内容。

8.1 带电作业工器具库房基本要求

带电作业工器具库房是长期保存带电作业工器具的专用房间，为保证工器具使用安全，工器具库房的环境条件、技术条件与设施、测控装置及库房信息管理系统必须满足相关规程规范的要求。

8.1.1 带电作业工具库房的一般要求

带电作业工具库房位置要求在防潮、空气通风条件良好，如果有条件的地方最好放置在楼层的一层，底层做高为0.8~1.2m的架空层。库房面积应根据工具数量及尺寸专门设计，也可以按表8-1参考面积进行设计。还应注意分区存放，并有区分标识说明，如电压等级、工具名称、规格等。工具存放空间与活动空间的比例为2∶1左右。库房的内空高度宜大于3.0m，若建筑高度难以满足时，一般应不低于2.7m。门窗要求能够防盗、防尘，封闭良好，同时还要求能够防火、隔湿。观察窗距地面1.0~1.2m为宜，窗玻璃应采用双层玻璃，每层玻璃厚度一般不小于8mm，以确保库房具有隔湿及防火功能。

表8-1　库房面积设计表

存放工具的电压等级kV	10~66	66~220	220~500
库房面积m²	20~60	50~150	60~200

在对库房的装修中，材料必须采用不起尘、阻燃、隔热、防潮、无毒的材料。位于一楼的库房地面应采用隔湿、防潮材料。工器具存放架一般应采用不锈钢等防锈蚀材料制作。

库房应该配备充足的消防器材，对于安装库房自动化管理系统的还应该要求消防的自动化报警系统；库房的用电线路最好能够独立，线径大小选择要考虑到库房各种系统的要求。同时满足防火要求。

绝缘斗臂车库的存放体积一般应为车体的1.5~2.0倍。顶部应有0.5~1.0m的空间，车库门可采用具有保温、防火的专用车库门，车库门可实行电动遥控，也可实行手动。

8.1.2　带电作业工具库房的技术条件与设施

带电作业工具库房的技术条件主要包括温度、湿度及通风条件。在库房内安装相应的加热、除湿及通风设备，保证满足库房内带电作业工器具的存储条件。

1. 湿度要求

库房内空气相对湿度应不大于60%。为了保证湿度测量的可靠性，要求在库房的每个房间内安装两个湿度传感器。工具库房的设计应满足GB/T 18037—2008《带电作业工具基本技术要求与设计导则》的规定。

2. 温度要求

带电作业工具及防护用具应根据工具类型分区存放，各存放区可有不同的温度要求。硬质绝缘工具、软质绝缘工具、检测工具、屏蔽用具的存放区，温度宜控制在5~40℃内；配电带电作业用绝缘遮蔽用具、绝缘防护用具的存放区的温度，宜控制在10~21℃之间；金属工具的存放不做温度要求。

另外，考虑到北方地区冬天室内外温差大，工具入库时易出现凝露问题，该地区的库房温度应根据环境温度的变化在一定范围内调控。若库房整体温度难以调整，工具在入库前也可先在可调温度的预备间暂存，在不会出现凝露时再入库存放。

为保证温度测量的可靠性，要求在库房的每个房间内安装两个温度传感器，为比较室内外温差，整套库房控制系统在室外安装一个温度传感器。

3. 除湿设施

库房内应装设除湿设备。除湿量按库房空间体积的大小来选择，一般按 0.05～0.2L/（d·m³）选配；对于北方地区，可按 0.05～0.15L/（d·m³）选配；对于南方地区，可按 0.13～0.2L/（d·m³）选配。在上述地区中，对湿度相对较高的区域，除湿机应按上限选配。

4. 烘干加热设施

库房内应装设烘干加热设备。建议采用热风循环加热设备，在能保证加热均匀的情况下也可采用红外线加热设备、不发光加热管、新型低温辐射管等。加热功率按库房空间体积的大小来选择，可根据当地的温度环境按 15～30W/m³ 选配。

加热设备在库房内应均匀分散安装，加热设备或热风口距工器具表面距离应不少于 30～50cm，热风式烘干加热设备安装高度以距地面 1.5m 左右为宜，低温无光加热器可安置于与地面平齐高度。车库的加热器安装在顶部或斗臂部位高度。加热设备内部风机应有延时停止装置。

5. 通风设施

库房内必须装设排风设备。每个房间必须安装两个排风扇，一个进气，一个排气，让空气循环，排风量可按每平方米 1～2m³/h 选配排风机。吸顶式排风机应安装在吊顶上，轴流式排风机宜安装在库房内净高度 2/3～4/5 高度的墙面上。出风口应设置百叶窗或铁丝窗，进风口应设置过滤网，预防鸟、蛇、鼠等小动物进入库房内。

6. 报警设施

应设有温度超限保护装置、烟雾报警器、室外报警器、防盗报警等报警设施。当库房温度超过 50℃时，温度超限保护装置应该能自动切断加热电源并启动室外报警器，要求温度超限保护装置在控制系统失灵时也应能正常启动，当库房内产生烟雾时，烟雾报警器和室外报警器应能自动报警，未经注册人员开启库房房门，防盗报警启动。

7. 库房设施的综合配置和选择

在除湿、烘干加热、通风设施的综合配置和选择上，主要应以能否满足温度、湿度要求以及调控要求来确定。

8. 绝缘斗臂车库

绝缘斗臂车库的通风、除湿、烘干装置要求与带电作业工具库房的要求相同。车库的加热器一般应安装在便于烘烤斗臂的部位或顶部，下部不需安装加热器。

如图8-1所示为绝缘斗臂车库。

图8-1 绝缘斗臂车库

8.1.3 监控功能与装置要求

自动化条件较高的带电作业库房，由传感装置、监测仪表、控制装置、计算机、开关柜等组成的监测系统应能根据设置要求对库房的各项参数实施实时监测。对工具库房的湿度、温度调控系统，应可根据监测的参数自动启动加热、除湿及通风装置，实现对库房湿度、温度的调节和控制。当调控失效并超过规定值时，应能报警及显示。

1. 功能要求

为了保证工具库房的温度、湿度环境能满足使用要求，应专设温湿度测控系统。温湿度测控系统应具备湿度测控、温度测控、库房温湿度设定、超限报警及库房温湿度自动记录，显示、查询、报表打印等功能，支持远程调控和查阅历史记录。

2. 监测要求

由传感器、测量装置、控制屏柜及其附件等组成的监测系统应对库房的温湿度实施实时监测并加以记录保存。

3. 调控要求

工具库房的湿度、温度调控系统，应可根据监测的参数自动启动加热、除湿及通风装置，实现对库房湿度、温度的调节和控制。当调控失效并超过规定值时，应能报警及显示；当库房温度超限时，温度超限保护装置应能自动切断加热电源。

为了有效保证测控系统的安全有效运行，控制系统需设置自动复位装置，以保证测控系统在受到外界干扰而失灵时能立即自动复位，进而恢复正常运行。

为了保证在测控系统完全失效或检修时除湿装置及加热装置等仍能投入工作，应在控制屏柜上设立手动/自动切换开关及相应的手动开关。

4. 元件及设备要求

库房内的设备、装置、元器件的技术性能和指标均应满足相关设备和元件标准的要求，以保证测控系统稳定、可靠、安全运行。

5. 显示和打印

测控系统应能存储库房一年时间的温湿度数据，具有全天任意时段的库房温湿度数据的报表显示、曲线显示、报表打印等功能，实时监测和记录库房的工作状态。

6. 远程监控要求

库房与车库应具有远程视频监控功能。测控系统具有Web发布功能，要求与企业局域网连接，实现远程信息共享及分级管理、库房温湿度参数设置等功能，实现对库房的环境温湿度变化情况的远程监控和工器具的远程管理等。

7. 主要测控元件的技术性能要求

（1）温度测控指标：范围–10℃～80℃，精度±2℃。

（2）湿度测控指标：范围30%～95%R.H.，精度±5%。

（3）温度传感器指标：量程–40℃～80℃，在–10℃～85℃范围内精度±0.5℃。

（4）湿度传感器指标：量程0%～100%R.H.，在10%～95%R.H.，范围内精度±3%。

如图8-2所示为带电作业工器具房监控系统。

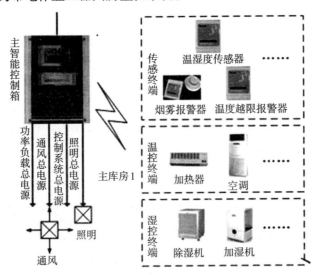

图8-2 带电作业工器具房监控系统

8.2 带电作业工器具库房管理系统

工具库房计算机管理系统应对工具贮存状况、出入库信息、领用手续、试验情况等信息进行实时记录。根据需要，工器具库房计算机管理系统还可具备在企业局域网上实施WEB发布及远程维护的功能。管理系统基本功能框图如图8-3所示。

8.2.1 带电作业工器具库房管理系统

带电作业工器具库房管理系统组成架构框图如图8-3所示。该系统框图可供各单位参考使用。各单位还可根据本单位管理的具体要求进行修改和补充。

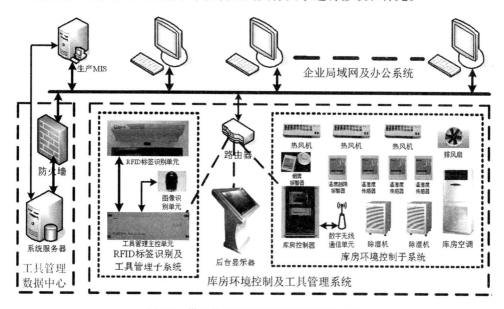

图8-3 带电作业工器具库房管理系统结构

8.2.2 带电作业工具管理系统

1. 系统组成构架框图

带电作业工器具管理系统组成架构图如图8-4所示。

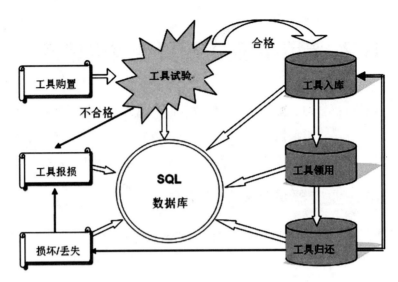

图 8-4 带电作业工器具库房管理系统结构

2. 工具出入库信息管理

（1）应建立带电工器具入库记录，包括工器具厂家、出厂日期、入库时间、试验情况、工具编号等。

（2）应建立带电工器具报废记录，主要包括工具名称、工具编号、入库时间、报废时间、报废原因等。

（3）对于带电工器具在工作中出现意外的脏污、损坏、丢失或其他情况必须填写相关报表给予说明，并经相关人员签字确认。

3. 工具借用管理

（1）带电工器具出、入库必须建立台账，主要记录借出、归还以及外观检查情况。

（2）工作人员从库中领用各种工具，选择工具方法多样，可以任意由分类、规格、工具名称、条形码或者简码来选择工具。

（3）在工作人员使用完工具后，需要进行此操作，完成对工具的归还，归还有三种状态，分别为归还、丢失、损坏。工器具归还后必须摆放在规定位置。

（4）当日不能够归还的绝缘工器具应保存在满足条件的场所。

4. 工具试验管理

定期对工器具进行试验，将大大加强工器具的安全性和可靠性。试验管理环节将是本平台最重要的一个环节，带电作业工器具试验分为：检查性试验、预防性试验和型式试验。试验后合格的工具可以继续使用,不合格的工具必须申请工具报损。根据不同类型的工具设置试验类型、试验周期、试验日期、试验编号,建立试验台账,提示各个工器具下次试验时间,超期工器具应报警等。

5. 综合查询

(1) 全智能一体化工具管理平台具有先进的查询系统,可以分别对库存状态、工具分布、工具入库、工具领用、工具报损、电气试验、机械试验进行查询,所有的数据都被保存在 SQL 数据库中,可以安全可靠高速地被调阅。

(2) 工器具盘点系统是有别于查询系统的系统,工器具盘点系统主要是为了让库房管理员清楚地知道某一个库房现存的工具情况。可以分别对每一个库房进行盘点。带电作业工器具库房综合查询系统如图 8-5 所示。

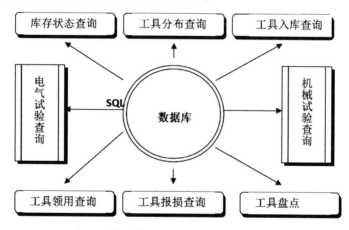

图 8-5 带电作业工器具库房综合查询系统

8.2.3 射频识别系统

RFID 射频识别是一种非接触式的自动识别技术,它通过射频信号自动识别目标对象并获取相关数据,识别工作无需人工干预,可在各种恶劣环境下工作。与此同时,还可识别高速运动物体并可同时识别多个标签,操作快捷方便。如图 8-6 所示为 RFID 射频系统。

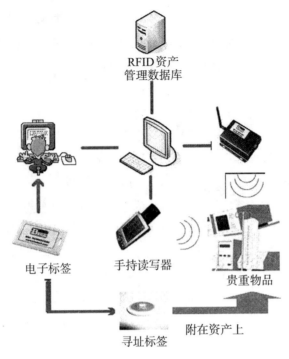

图 8-3-4 RFID 射频系统

RFID 射频识别系统是一种简单的无线系统，只有两个基本器件，主要用于控制、检测和跟踪物体。该系统由一个阅读器（或询问器）和很多标签（或应答器）组成。

由射频标签、识读器和计算机网络组成的自动识别系统。通常，识读器在一个区域发射能量形成电磁场，射频标签经过这个区域时检测到识读器的信号后发送存储的数据，识读器接收射频标签发送的信号，解码并校验数据的准确性以达到识别的目的。

RFID 在带电作业工器具库房管理中的应用：存放在带电作业工器具库房的工具，在工具上粘贴 RFID 标签后，能够在工作流程中的各个关键环节中被阅读器识别。

8.2.4 库房环境自动测控系统

1. 系统基本组成

系统充分采用智能分布控制、智能神经传感网络、PnP 即插即用等最新工业测控技术，硬件平台以嵌入式工业计算机为系统基层监控中心，立式带触摸的液晶屏作为人机对话平台，以 PnP 式湿度传感器、温度传感器、电量采集器、越限报警器及各种状态量采集器做输入设备，以自组网无线信号交换模块作为系统信息交换通道，以控制量输出模块及接触器等做输出装置，由拨号语言报警、蜂鸣器及指示灯等组成报警输出系统，在（前置控制计算机）Linux操作系统及环境测控软件、（后台机）Windows操作系统、Office套件、后台Web发布、视频监控、电话报警等专业软件的支持下，实现实时测量、数据采集、数据处理、设备控制、人机对话等基本功能。

它与传感器和控制设备间的通信采用国际通用免费ISM频段2.4G频率的Zigbee无线组网技术，可靠点到点通信距离无遮挡约30m。由于采用组网技术，每个终端均具备路由（级联）功能，因此只要任意两个终端距离不超过30m，所有网络内的设备均可通信。

后台客户端系统与智能控制箱之间采用网络通信，可远程查看库房内的环境参数、设备运行情况等，还可远程对库房参数进行配置及对库房设备进行远程控制，同一地点的多个智能控制箱通过路由器或交换机只连接一台后台客户端系统。

电话报警模块一般安装在后台客户端系统上，当某个库房发生异常时，后台客户端系统启动电话报警功能，电话号码最多可预设6个。

2. 系统结构

库房环境自动侦测系统结构如图8-7所示。

3. 系统独特的功能特点

前置机采用工业嵌入式主机（Linux平台）实现系统数据采集及系统控制，可执行系统最基本的功能，如保护、测量、控制等，按设定控制规则投切适当的加热机、除湿机、通风机，控制规则应灵活多样。

后台工作站采用Windows平台进行复杂数据处理及人机对话，能与内部局域网联网，支持通过局域网查询库房温湿度、工器具库存品种、数量、规格，领用和试验情况。系统具有优秀的防病毒特性，支持Watchdog timer及系统映射技术，程序跑飞或系统瘫痪后可迅速自行恢复，防止系统死机，保障系统的稳定性及可靠性。

手动/自动控制器有手动和自动切换开关，可实现手动或自动控制。在烟雾和高温报警状况下，可自动切断加热、除湿、空调等设备控制电源，并强制开启通风装置。

8.2.5 视频监控系统

带电作业工器具库房安装视频监控系统，使工作人员在不进入库房的情况下，可对库房内的情况进行实施监控，并可根据需要旋转角度。该视频监控系统可对库房进行实时录像。

本系统采用带云台红外摄像头，可以全方位360°旋转。在夜视状态下，本系统所采用的红外摄像机可以拍摄到黑暗环境下的影像。

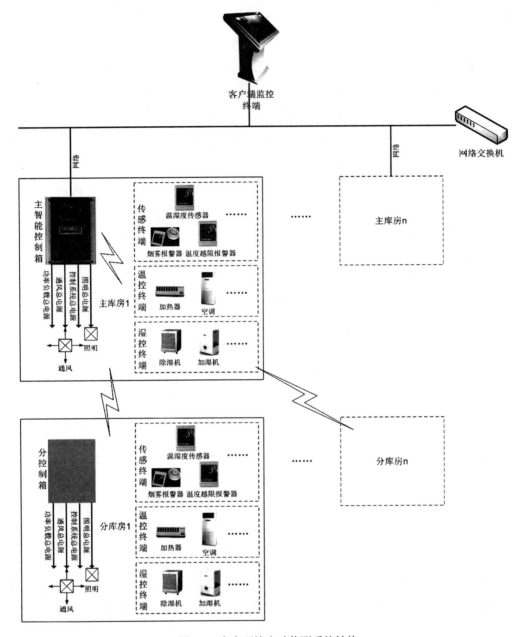

图 8-7 库房环境自动侦测系统结构

硬盘录像机（Digital Video Recorder，DVR），它是一套进行图像存储处理的计算机系统，具有对图像/语音进行长时间录像、录音、远程监视和控制的功能。DVR集合了录像机、画面分割器、云台镜头控制、报警控制、网络传输等五种功能于一身，用一台设备就能取代模拟监控系统一大堆设备的功能，而且在价格上占有优势。

视频资料存放分为有人员进出时段视频和库房无人时段视频，对视频资料可存储

3个月，对超期视频自动清除。

8.2.6 远程监控系统

带电作业工器具库房远程管理系统，使库房管理员或者单位领导在不进入库房的情况下，可对库房运行情况进行实时监控。以国际最流行的控制类软件Symantec pcAnywhere进行介绍。

1. Symantec pcAnywhere 用户界面

库房远程监控系统Symantec pcAnywhere用户界面如图8-8所示。

2. Symantec pcAnywhere 操作示意流程图

pcAnywhere操作示意流程图8-9所示。

3. Symantec pcAnywhere 功能一览

（1）以简易且安全的方式连接至远端装置。

（2）轻松进行跨平台作业。

图8-8　库房远程监控系统界面

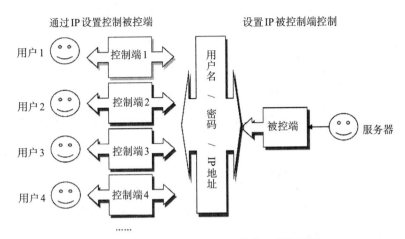

图8-9　Symantec pcAnywhere操作示意流程图

（3）管理电脑及迅速解决服务台问题。

（4）简易安全的远端连接能力，可以跨区域、跨平台地对库房的主控台轻松操作，温湿度监控、工具入库管理等都可以轻松实现。使用 Symantec pcAnywhere，远端连线既简易又安全。崭新的闸道功能可让主控端使用者寻找所需的被控端，即使被控端位于防火墙或路由器之后，或者没有固定或公开 IP 位址，也可轻松找到。基本的窗口界面专为新使用者设计，并且针对每项作业提供简易的图形选项。内建 FIPS 140-2 验证 AES 256 位元加密，可确保阶段作业的安全性。Symantec pcAnywhere 现在可以支持连接运行 Microsoft Windows、Linux、Mac OS X Universal 或 Microsoft Pocket PC 的计算机，并对其进行管理。

4. Symantec pcAnywhere 的主要特点

（1）简化的 GUI 和改进的连接选项。

（2）提供了强大的保护远程控制、文件传输、命令队列、远程管理、快速部署以及连接功能。

（3）提供了安全、高效、灵活的远程管理方法，可以排除故障并快速进行修复，纠正用户错误以及减少解决问题所需的时间。

（4）提供了管理本地和远程设备的功能，加大对整个 IT 基础架构的控制力度和范围。

（5）系统中的"连线精灵"软件会引导新使用者进行初次的用户端与被控端连线。

（6）强大的档案传送功能可让使用者跨平台上传及下载档案。

（7）强制的密码保护与登入加密能确保只有经过授权的使用者，才能存取 Symantec pcAnywhere 被控端。

（8）使用高达 AES 256 位元的 FIPS 140-2 验证加密。

（9）自动侦测频宽功能会自动将每一种连线类型的 pcAnywhere 效能最佳化。

8.3 带电作业工器具的存放方法

带电作业工器具必须在合格的带电作业工具专用库房内长期保存。工器具每次使用后应擦拭干净才能存入库房。带电作业工器具应按电压等级及工具类别分区存放，主要分类为：金属工器具、硬质绝缘工具、软质绝缘工具、屏蔽保护用具、绝缘遮蔽用具、绝缘防护用具、检测工具等。每件工器具都必须有唯一编号，标明相关机电性能参数。非合格工器具严谨存放在带电作业专用工具库房。

8.3.1 金属工器具

金属工器具的存放设施应考虑承重要求，应便于存取，可采用多层式存放架。同时要考虑到使用的频繁程度，对金属工具中的卡具要进行分类摆放，以电压等级作为分类的标准。螺钉等小件随大件存放。

8.3.2 硬质绝缘工具

硬质绝缘工具中的硬梯、平梯、挂梯、升降梯、托瓶架等可采用水平式存放架存放，每层间隔30cm以上，最低层对地面高度不小于50cm，同时应考虑承重要求，应便于存取。绝缘操作杆、吊拉支杆等的存放可采用垂直吊挂的排列架，每个杆件相距10~15cm，每排相距30~50cm。在杆件较长、不便于垂直吊挂时，可采用水平式存放架存放。在分层摆放时应该以电压等级进行分类摆的。大吨位绝缘吊拉杆可采用水平式存放架存放。

8.3.3 软质绝缘工具

绝缘绳索、软梯的存放可采用垂直吊挂的构架。绝缘绳索挂钩的间距为20~25cm，绳索下端距地面不小于30cm。软质工具盘绕要分型号、规格存放整齐。绝缘软梯盘绕后也可以放在水平式存放架存放。

8.3.4 滑车

对滑车和滑车组可采用垂直吊挂构架存放，也可以放在水平式存放架存放。根据滑车的大小、承受重量、类别分组定置定位存放。

8.3.5 检测仪器

验电器、相位检测仪、分布电压测试仪、绝缘子检测仪、干湿温度仪、风速仪、绝缘电阻表等检测用具应分件摆放，防止碰撞，可采用多层水平不锈钢构架存放。对仪器仪表内的干电池，不使用时需要取出存放；对仪器仪表内的锂电池，每月进行一次充电，以保证足够电量，电池老化及时更换新电池。

8.3.6 绝缘遮蔽用具

绝缘遮蔽用具，如绝缘毯、导线遮蔽罩、绝缘子遮蔽罩、横担遮蔽罩、电杆遮蔽罩等，应储存在有足够强度的袋内或箱内，再置放在多层式水平构架上；禁止储存在蒸汽管、散热管和其他人造热源附近，禁止储存在阳光直射的环境下。

8.3.7 绝缘防护用具

绝缘防护用具，如绝缘服、绝缘袖套、绝缘披肩、绝缘手套、绝缘靴等应分件包装，要注意防止阳光直射或存放在人造热源附近，存放严禁对折，尤其要避免直接碰触尖锐物体，造成刺破或划伤。

8.3.8 屏蔽用具

屏蔽用具如屏蔽服、导电手套、导电袜、导电鞋、屏蔽面罩等应分件包装，成套贮存在有足够强度的包装袋或箱内，再置放在多层式水平构架上。

屏蔽服用完后应装入专用箱内，严禁揉摺和重压，特别是对绝缘服上的圆康铜合金丝更应注意，以免折断造成使用时放电，对带电作业人员带来刺痛感。夏天使用完后，应将其浸泡在50~100倍（与屏蔽服重量比）的50~60℃热水中15min，以溶解衣服中的汗水。

8.3.9 绝缘斗臂车

绝缘斗臂车必须停放在专用车库内，车库具有自动除湿、烘干设备，烘干设备必须安装在绝缘臂高度，车辆不需加热。长时间停放须使用支腿支撑。绝缘斗臂车绝缘臂液压部分每两年必须检修、换油一次，行走部分保养按厂家要求执行。如图8-10所示为带电作业工器具的存放。

第8章 带电作业工器具的保管和运输

(a)

(b)

图 8-10 带电作业工器具的存放

8.4 带电作业工器具的保管和运输

输电线路带电作业比较偏远，工器具在运输途中的时间长、配电线路大部分在城区作业，但同类型的作业点多，在室外使用的时间也较长，为保证带电作业工器具能安全使用，对带电作业工器具的运输及保管提出了很严格的要求。

8.4.1 带电作业工器具的保管

1. 带电作业工具应存放于通风良好，清洁干燥的专用工具房内。库房内的相对湿度应不大于60%，硬质绝缘工具、软质绝缘工具、检测工具、屏蔽用具的存放区，温度宜控制在5～40℃。

2. 带电作业工具房应配备温度计、湿度计、除湿机、辐射均匀的加热器，足够的工具摆放架、吊架和灭火器。

3. 带电作业工器具应设专人保管，带电作业工具应统一编号、登记造册，并建立试验、检修、使用记录。各类工器具要有完整的出厂说明书、试验卡片或试验报告书。

4. 带电作业工器具有使用寿命，根据台账即时报废；根据使用情况，预计每年带电作业工器具采购计划。

5. 有缺陷的带电作业工具应及时修复，不合格的应予以报废，禁止继续使用。不合格的工器具严禁放在带电作业库房内。

6. 定期对充电设备进行复充电，以免电池亏电损坏。

7. 每月应对带电作业库房及带电作业工具车进行整理。对工器具进行烘干或进行外表检查及保养，发现问题应及时上报专责人员。

8. 对带电工器具进行定期预防性试验和检查性试验。

9. 高架绝缘斗臂车应存放在干燥通风的车库内，其绝缘部分应有防潮措施。

10. 潮湿地区或潮湿季节，带电作业工具需外出超过24小时，需配备专用带电作业工具库房车，专用带电作业工具库房车必须带有烘干除湿设备、温湿自动控制系统，并按带电作业库房及带电作业工具车标准执行。带电作业工具库房车必须专用，车载发电机、温湿控制系统必须处于良好状态，随时随地可以进行烘干除湿工作。如图8-11所示为带电作业工具库房车温湿控制系统界面。

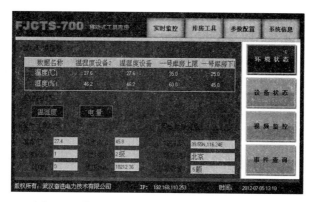

图 8-11　带电作业工具库房车温湿控制系统界面

8.4.2　带电作业工器具在运输过程中的注意事项

1. 带电作业工器具出库装车前必须用专用清洁帆布袋包装，带电作业工具在运输过程中，应该装在专用工具袋、工具箱或专用工具车内，以防受潮和损伤。有条件时可使用带电作业移动库房车进行工具的运输。

2. 带电作业工器具在运输过程中应该分类摆放，并采取必要的固定措施，以防途中车辆颠簸相互碰撞、磨损。铝合金工具、表面硬度较低的卡具、夹具及不宜磕碰的金属机具，运输时应有专用的木质和皮革工具箱，每箱容量以一套工具为限，零散的部件在箱内应予固定。金属工具应放下层摆放，绝缘工具放在上层固定，不得混放装运。如图 8-12 所示为绝缘工具的摆放。

图 8-12　绝缘工具的摆放

3. 绝缘硬梯、绝缘操作杆、绝缘托瓶架等物品应该在车厢内平放，工具袋应完好无损。车厢内不得存放腐蚀性、油漆及杂物等。

4. 外出连续工作时，还应配备烘干设备，每日返回驻地后，要对所带绝缘工器具进行一段时间的烘干，以备次日使用。

参 考 文 献

[1] 胡毅. 带电作业工具及安全工具试验方法[M]. 北京：中国电力出版社，2003.

[2] 华捷. 带电作业人员资质认证培训专用教材. 输电线路[M]. 北京：中国电力出版社，2011.

[3] 国家电网公司人力资源部. 生产技能人员职业能力培训通用教材·带电作业基础知识[M]. 北京：中国电力出版社，2010.

[4] 国家电网公司人力资源部. 生产技能人员职业能力培训专用教材·配电线路带电作业[M]. 北京：中国电力出版社，2010.

[5] 国家电网公司人力资源部. 国家电网公司生产技能人员职业能力培训专用教材·输电线路带电作业[M]. 北京：中国电力出版社，2010.

[6] 国家电网公司运维检修部 10kV电缆线路不停电作业培训教材[M]. 北京：中国电力出版社，2013.

Chapter 1

Overview of Tools and Instruments For Live Working

Live working refers to operation and maintenance work safely carried out by operating personnel on live transmission and distribution lines and electrical equipment by using professional tools and instruments with high insulation performance and mechanical strength. Tools and instruments for live working with reliable performance are necessary to ensure the safety of live working. With the popularization of live working methods, many practitioners have developed a wide range of tools and instruments for live working with different specifications. At present, with the completion and operation of several UHV AC 1000kV and UHV DC ± 800kV transmission lines, it is urgent to manufacture and use tools and instruments for live working in UHV operation environment, and new requirements are put forward for their materials, performance and preventive tests.

1.1 Development History of Tools and Instruments for Live Working

The development history of live working is also that of tools and instruments for live working. Tools and instruments for live working are developed, used and improved with the development of live working.

1.1.1 History of Live Working & Tools and Instruments in China

1. First Stage of Development of Tools and Instruments for Live Working

Live working technology in China started in the 1950s when Anshan Iron and steel factory, the largest iron and steel base in China at that time, had a very high dependence on power energy to the extent that even the power outage necessary for grid maintenance could not be realized. In order to solve the contradiction that power line maintenance required to cut off power while user power supply required not, the so-called "non-interruption maintenance technology" at that time came into being.

In 1953, workers of the former Anshan Electric Power Bureau began to study simple tools for live cleaning, replacement and disassembly of power distribution equipment and leads. On May 12, 1954, in the mass movement of technological innovation, Liu Changgeng, a technician of Anshan Electric Power Bureau, came up with a method of live removing and binding distribution porcelain bottle binding wires, showed tools he had developed, such as insulated tongs, insulated clip scissors, porcelain insulator cable clip, etc., and put forward technical analysis report and operation scheme of live replacing straight wire pole, porcelain insulator and cross arm of 3.3kV line.

Live replacement of items like cross arm, wooden pole and porcelain insulator of 3.3kV distribution line was successfully completed on the scene by using tools fabricated with wood similar to birch wood bar. His technical innovation was recognized by experts and passed the feasibility study.

The date, May 12, 1954, was recorded in the annals of Anshan Electric Power Bureau as the founding day of live working, and also recorded in the history of electric power development in China as the beginning of development of live working in China.

Live working was highly valued by leaders of Anshan Electric Power Bureau, and in order to consolidate and develop initial achievements of live working, Anshan Electric Power Bureau formally established a professional team of live working in June 1956, with Zhang Renjie as the team leader, and Liu Changgeng as the specific engineer. It was further developed to live replacement of wooden straight wire pole, cross arm and insulator of 44-66kV line.

The first generation of tools have the following characteristics: the main insulating tools are made of porcelain insulator and porcelain bushing, and the metal parts are made of ordinary steel and malleable cast iron, which makes the tools and instruments bulky and insufficient in universality. As shown in Fig. 1-1, the first generation of "rise and fall" tool is used for live replacement of a straight wooden pole of 66kV line, and in Fig. 1-2, the first generation of "insulated splint" tool is used for live replacement of a strain insulator string of 22-33kV line.

Fig. 1-1 Live Replacement of a Straight Wooden Pole of 66kV Line

Fig. 1-2 Live Replacement of a Strain Insulator String of 22-33kV Line

2. Second Stage of Development of Tools and Instruments for Live Working

In October 1957, the former Northeast Electric Power Bureau designed the first set of tools for live working for 220kV high-voltage transmission line, and successfully applied them to live working of 220kV high-voltage transmission lines. At the same time, the complete set of maintenance tools for 3.3-33kV wooden pole and tower line had also been improved and perfected, which laid a material and technical foundation for the implementation of non-interruption maintenance of lines at all voltage levels.

In 1958, the current Shenyang Central Test Institute started the experimental study on the maintenance with human body directly contacting with conductors. On the basis of learning from foreign experience, they solved the shielding problem of high-voltage electric field, and successfully carried out the first equipotential experiment of human body directly contacting with 220kV live conductors in the test field. The experimenter was Liu Dewei who completed the task of equipotential operation and conductor repair on 220kV line for the first time. The success of this equipotential live working experiment opened a new chapter of live working in China. Since then, equipotential operation technology has been widely used in live working in China.

Fig. 1-3 is the picture of 220kV equipotential experiment conducted by Anshan Electric Power Bureau in 1958, and Fig. 1-4 is the operation of removing lead wire of 220kV arrester conducted by Anshan Electric Power Bureau in 1958.

Fig. 1-3   220kV Equipotential Experiment

Fig. 1-4   220kV Operation of Removing Lead Wire of Arrester

Around 1959, Anshan Electric Power Bureau developed a set of tools and operation methods for non-interruption maintenance of transformer equipment on the basis of 3.3-220kV outdoor power transmission and distribution equipment. Till then, China's live working technology had developed into a 3.3-220kV comprehensive maintenance technology integrated with transmission, transformation and distribution.

In March 1958, the former Electric Power Bureau of Liaoning Province held a conference on non-interruption maintenance of lines in Anshan, after which the live working technology began to be popularized in power supply departments of Liaoning Province and northeast electric power system. On April 12, 1958, the People's Daily reported the success of live working technology experiment in Anshan Electric Power Bureau under the title of Major Technological Innovation in the Electric Power Industry - Non-interruption Maintenance of Electric Power Lines. From May 20 to July 1 of the same year, 46 technical representatives from power supply departments of Beijing, Shanghai, Hubei, Sichuan, Shaanxi and other provinces and cities received technical training on live working in Anshan. In that year, more than 5700 representatives from associate units visited Anshan to learn the live working technology.

The second generation of tools and instruments for live working used in 1957-1960 have the following characteristics: the main insulating tools are made of phenolic wood laminate, phenolic rubber paper tube, phenolic cardboard and silk, and the metal parts are made of ordinary steel and malleable cast iron. The tools and instruments in this stage are bulky, with certain universality.

3. Third Stage of Development of Tools and Instruments for Live Working

In 1960, the former Liaoji Electric Power Bureau formulated the Specification for Safety Operation for Non-interruption Maintenance of High-voltage Overhead Lines which became the first instructional specification for live working in China. It marked that live working in China had been on the right track. During this period, the nationwide non-interruption maintenance work had shifted from simple technology promotion to the stage of innovation and development combined with local specific conditions and production tasks. In addition to indirect operation and direct equipotential operation, the maintenance method moved forward to water washing, explosion crimping, etc. Maintenance tools changed from original heavy tools such as support rod, pull rod and suspender to light tools and rope tools. The insulated rope ladder and insulated tackle block with oriental characteristics were also been widely used. The operation was marching towards complex projects, such as replacing conductor and overhead ground wire, moving tower and transforming tower head.

In November 1964, live maintenance performance was held in Tianjin, which had a positive impact on the promotion of these new technologies nationwide.

In 1966, the production office of the former Water Conservancy and Power Department held a national live performance conference for live working in Anshan, which marked the live working in China had developed into the stage of popularization, and also promoted the development of live working to a newer and deeper field.

In 1968, the former Anshan Electric Power Bureau successfully experimented the new method of entering 220kV intense electric field along the insulator string. Since it was very convenient to replace single insulator on the double strain insulator string by this method in certain conditions, it was quickly spread to the whole country. single insulator on the double strain insulator string in certain conditions, it was quickly spread to the whole country.

In 1973, the former Water Conservancy and Power Department held the second national experience exchange meeting on live working in Beijing. The technical group of this meeting put forward a discussion draft of related topics on live working safety technology, which laid a technical foundation for the formulation of national regulations for live working.

Fig. 1-5 Operation of Replacing Whole String of Strain Insulators of 220kV line

Fig. 1-6 Equipotential Inspection on 220kV Line, Dealing with
Defects such as Heat of Strain Lead Contact

Chapter 1
Overview of Tools and Instruments for Live Working

Fig. 1-5 shows the operation of replacing the whole string of strain insulators on 220kV line with the third generation of tools and instruments for live working (stringing pull rod and insulator cradle), and Fig. 1-6 shows the equipotential inspection and handling of defects like heat of strain lead contact on 220kV line with insulated rope ladder.

The third generation of tools and instruments for live working have the following characteristics: the main insulating tools are made of epoxy glass cloth laminate, epoxy glass cloth pipe, glass wire drawing rod, nylon rope and silk rope, and the metal parts are made of alloy steel and malleable cast iron. In this stage, the weight of tools and instruments for live working was reduced and the universality was strengthened.

4. Development of Tools and Instruments for Live Working in Later Stages

In 1977, the former Water Conservancy and Power Department brought live working into the safety work regulations issued by the Department, which further affirmed the safety of live working technology. In the same year, China's live working personnel started international exchanges, participated in activities of the live working group of the International Electrotechnical Commission, and established a domestic working group with IEC/TC78 standard to work out relevant standards for live working. In May 1984, China Live Working Standardization Committee was established.

In 1979, the 500kV Yuanbaoshan-Jinzhou-Liaoyang-Haicheng Transmission Line was built in Northeast China. In order to ensure the safe and economic power supply after the 500kV Line was put into operation, the former Anshan Electric Power Bureau carried out the research on new technologies of 500kV live working. In order to solve this problem, retired employees such as Zhang Renjie and Ge Yuanchang were invited back to carry out the research work on live working of 500kV tangent tower and resisting-tensile tower. They fully carried forward the hardworking spirit in previous years, and made tools, conducted field tests and prepared operating procedures in person. In 1984, the research and development supporting work of 500kV partial live working routine items and tools were completed. Soon afterwards, working methods and tools and instruments for 500kV live replacement of linear insulator string and strain insulator string, and live repair of conductor, etc. were successfully developed as scheduled and entered the implementation stage.

In 2000, the former North China Power Grid Co., Ltd. first studied and realized live working of 500kV compact line. Fig. 1-7 shows the operation of Jinzhou Electric Power Bureau using "stringing pull rod-insulated trolley" to replace the whole string of strain insulators on 500kV Line in the 1980s; Fig. 1-8 shows the operation of "entering intense electric field along insulator string" to replace 500kV single insulator in Northeast China.

Fig.1-7　Operation of Replacing Whole String of Strain Insulators of 500kV Line

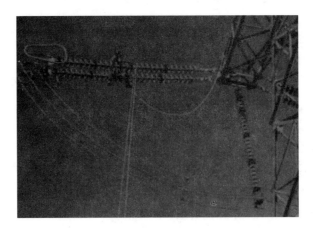

Fig.1-8　Operation of Replacing Single Insulator of 500kV Line

In the 1990s, China introduced aerial devices with insulating booms, which solved many problems that perplexed live working of distribution lines. At present, aerial devices with insulating booms have been widely used in live operation of distribution lines, and the localization of aerial devices with insulating booms with the working voltage of 10-500kV has been realized.

In 2007, North China Power Grid Co., Ltd. realized live working of 500kV line with helicopter after research, which reached the international leading level. Fig. 1-9 shows equipotential operation with helicopter, and Fig. 1-10 shows live washing operation with helicopter.

Chapter 1
Overview of Tools and Instruments for Live Working

Fig. 1-9　Equipotential Live Working with Helicopter

Fig. 1-10　Live Washing Operation with Helicopter

From 2006 to 2009, Shanghai Municipal Electric Power Company of State Grid and Beijing Electric Power Corporation of State Grid researched and implemented non-interruption operation methods of 10kV distribution line.

From 2008 to 2009, Northwest Power Grid researched and implemented live working of UHV AC 750kV line. Central China Power Grid and North China Power Grid began to research and implement live working of UHV 1000kV line, tools and instruments for live working of ±500kV and ±800kV transmission line as well as project implementation.

On April 14, 2009, the first UHV live line defect elimination operation carried out by Hubei UHV Power Transmission and Transformation Company in China was successfully completed on 1000kV Nanjing Line Ⅰ, as shown in Fig. 1-11.

On June 11, 2009, the world's first ±800kV UHV DC transmission live working was successfully completed by 7 professionals (including Hu Guang) of Hubei UHV Transmission and Transformation Company. This live working operation used self-developed tools and instruments for live working, as shown in Fig. 1-12.

Fig. 1-11 1000kV UHV Equipotential Defect Elimination Operation

Fig. 1-12 ±800kV UHV DC Transmission Live Working

1.1.2 Development of Live Working and History of Tools and Instruments in Foreign Countries

1. Live Working of Lines in the United States

The first country in the world to carry out live working is the United States. In 1913, the first live working was carried out in Ohio. At that time, the ground potential method was adopted, and tools for live working were made of wood. Dry wooden rod, due to its good insulation performance, could fully withstand the phase-to-ground voltage. Although tools made at that time were rough and cumbersome, they were the pioneer of live working after all. With the continuous increase of voltage level, it was necessary to have a live operating bar with stronger insulation performance to meet the requirements. The plastic sheathed wooden pole was first introduced and applied in live working by Chance Company in the

United States, and then came the reinforced synthetic resin pipe mixed with glass fiber which had the characteristics of high voltage-withstand strength and uniform electric field distribution. Soon afterwards, it was widely used in live working of transmission line. At present, various power companies in the United States have been engaged in live working overhaul and maintenance projects for transmission lines, and many power companies have also carried out live line patrol and live maintenance projects with helicopters. Fig. 1-13 is the equipotential operation with helicopter carried out in the United States.

Fig. 1-13 Equipotential Operation with Helicopter Carried out in the United States

2. Live Working of Lines in Japan

In Japan, live working on transmission lines is relatively few, and the proportion of live working on distribution lines is large. At the same time, Japan puts forward higher requirements for the reliability of power supply of distribution network, which require to gradually realize complete non-interruption operation. Compared with the past, Japan's live working has gradually moved towards automation, intelligentization and mechanization. It is the first country in the world to carry out research and application of live working robots. In the early 1990s, four Japanese electric power companies developed live working robots with hydraulic mechanical arms and used them in the maintenance of distribution network system. In order to improve the reliability of power supply and support the overall popularization of non-interruption operation, Japan had developed a series of mechanized operation tools and equipment which were continuously promoted and applied, including tools and supporting facilities needed for direct or indirect live working such as high-voltage generator locomotive, live washing transformer trolley, low-voltage non-interruption switching device, and accident point detection vehicle used in emergency accidents. The use of these mechanized tools and instruments for live working reduces labor intensity and improves work efficiency to a certain extent; Fig. 1-14 is the live working carried out by robots in Japan.

Fig. 1-14 Live Working Carried out by Robots in Japan

3. Live Working of Lines in Russia

In the 1930s, Russia (Soviet Union) began to carry out live working of transmission lines. In the 1940s, the new method of equipotential operation was applied and popularized on a large scale; in the mid-1950s, live working had been used in most of line emergency repair work.

A complete and reasonable operation method has been developed for live working of 330-750kV transmission line in Russia. Considering the reliability of operation, transmission line is equipped with tools and instruments for live working suitable for high voltage and intense electric field operation. Its live working mainly includes insulation test, insulator string replacement, conductor repair, anti-corrosion paint coating of overhead line and live equipment, etc., which are in the forefront of the world in live working. In addition, Russia is also actively engaged in research on live working technology of 1150kV line and has in-depth research on safety protection of live working of 1150kV AC transmission line.

4. Live Working of Lines in Britain and France

In the middle of the 20th century, Britain began to use operating poles for live working, typically those for measuring voltage distribution of insulator string, etc. The research on live working of transmission line began till 1965. At that time, France also set up technical committees and experimental research institutes for live working that mainly researched principles of live working, safety regulations of live working, equipotential working methods, indirect working methods with tools, etc.; its operation items included live replacement of insulator strings, replacement of cross arms, replacement of anti-vibration devices. 80% of work on UHV transmission line is done by equipotential operation method, and work on lines with voltage level above 400kV is jointly done by indirect operation method and equipotential operation method with insulating tools. At present, live working in Britain and France is mainly used on distribution line, while that on transmission line is relatively few. This is mainly because transmission line has many backup devices, and Britain and France do not need to build large trans-regional transmission network due to geographical area and energy distribution.

1.2 Materials of Tools and Instruments for Live Working

1.2.1 Common Insulating Materials

For insulating material, under a certain voltage, only a tiny leakage current can pass through, so the insulation resistance value is large. The quality of insulating materials is directly related to the safety of live working, so the insulating materials for making tools for live working must be materials with excellent electrical performance, high mechanical strength, light weight and low water absorption, and must be aging-resistant and easy for processing.

At present, insulating materials used for tools and instruments for live working in China are mainly as follows.

(1) Insulating board. Including hard board and soft board, classified as laminated products, such as 3240 epoxy phenolic glass cloth board, PVC board and polyethylene board in engineering plastics, etc.

(2) Insulating pipe. Including hard pipe and hose, classified as laminated products, such as a 3640 epoxy phenolic glass cloth pipe, and rolled products like belt or silk.

(3) Plastic film. Plastic films such as polypropylene, polyethylene, PVC, polyester, etc.

(4) Rubber. Natural rubber, artificial rubber, silicone rubber, etc.

(5) Insulating rope Rope woven by natural silk and artificial synthetic fiber, such as nylon rope, cotton fiber rope and silk rope (divided into raw silk rope and boiled-off silk rope), including stranded and woven round rope and ribbon woven rope.

(6) Insulating oil, insulating paint, insulating adhesive, etc.

According to the maximum allowable operating temperature of electrical equipment, i.e. heat resistance grade, International Electrotechnical Commission (IEC) classifies insulating materials into seven grades, i.e. Y, A, Z, B, F, H and C, with allowable operating temperatures of 90°C, 105°C, 120°C, 130°C, 155°C, 180°C and above 180°C.

1.2.2 Common Metal Materials

The quality of metal materials is directly related to the safety of live working, so the metal materials for making tools for live working must be good materials with high mechanical strength and light weight. At present, metal materials used in live working are mainly as follows.

(1) Aluminum alloy. Aluminum alloy material used in aerospace field is generally plate which is used to make all kinds of clamps, etc.

(2) Titanium alloy. Titanium alloy used in aerospace field has higher strength and lighter weight than aluminum alloy, but its price is also higher, so it is suitable for making all kinds of UHV clamps.

(3) High-strength alloy steel. High-strength alloy steel is used for tool parts which require high mechanical strength.

1.3 Basic Types of Tools and Instruments for Live Working

China has been engaged in live working for many years, and its tools and instruments for live working have already realized standardization and production specialization. In order to ensure product quality, the state and industry have promulgated more than 40 types of rules and standards for tools and instruments for live working. From the perspective of tools and instruments used in current live working projects, they can be classified into two categories: tools and instruments for live working for transmission line and distribution line. Each category of tools and instruments can be classified into many types.

1.3.1 Basic Types of Tools and Instruments for Live Working for Transmission Line

1. Hard Insulating Tools

Hard insulating tools include insulating bar, insulated insulator cradle, insulated cross arm, insulating supporting and pulling suspender, insulated tackle.

2. Soft Insulating Tools

Soft insulation tools include insulating rope ladder, insulated tackle block, insulating rope and backup protection rope.

3. Metal Tool

Metal tools include aluminum alloy rope ladder head, hanging wheel, wire clamp, all kinds of clamps, metal tackle, somersault tackle.

4. Testing Too

Detection tools include porcelain insulator zero value detector, anemometer, thermohygrometer, insulation resistance tester, tong-type ammeter, phasing tester.

5. Safety and Protective Equipment

Safety and protective equipment include insulating clothes, shielding clothes, electrostatic protection clothes, insulating safety belt, falling protector, etc.

1.3.2 Basic Types of Tools and Instruments for Live Working for Distribution Line

1. Shield Equipment

Shield equipment includes: conductor shield cover, strain device shield cover, pin insulator shield cover, rod insulator shield cover, cross arm shield cover, electric pole shield cover, sleeve shield cover, drop conduit shield cover, partition and insulating blanket, etc.

2. PPE

Personal protective equipment includes insulating sleeves, insulating clothes, insulating shoes (boots), insulating safety helmet, insulating gloves, anti-puncture gloves, etc.

3. Insulating Operating Equipment

Insulating operating equipment includes insulating support bar, pull bar, suspender, etc.

4. Special Operation Vehicles

Special operation vehicles include aerial devices with insulating booms, bypass work vehicle, tool vehicle for live working, mobile box-type substation vehicle, etc.

5. Insulating Hand Tools

Insulating hand tools include insulating screwdriver, insulating wrench, pin puller, wire cutter, peeler, gear bolt clipper, etc.

6. Safety Appliances

Safety appliances include portable ground wire, safety belt, climbers, lifting board, etc.

1.4 Tests of Tools and Instruments for Live Working

Tests of tools and instruments for live working include electrical test and mechanical test. Regular electrical and mechanical tests must be carried out for any type of tools and instruments for live working to check whether they meet specified electrical performance indexes and mechanical strength. Only through the above two tests can we make the conclusion whether they are qualified or not. Defects that can not be observed may be produced or left during production, transportation and storage of tools and instruments, and only through tests can they be exposed, so the test of tools and instruments for live working is the only reliable means to check whether the tools and instruments are qualified.

1.4.1 Test Principles of Tools and Instruments for Live Working

1. Test Types and Cycles of Tools And Instruments for Live Working

Tests of tools and instruments for live working, including electrical test and mechanical test, shall be carried out regularly according to DL/T976-2005 *of Preventive Test Code of Tools, Devices and Equipment for Live Working*, the test items and cycles of tools, devices and equipment for live working are shown in Table 1-1.

Table 1-1 Test Items and Cycles of Tools, Devices and Equipment for Live Working

S/N	Appliance	Test item			
		Electrical test	Test cycle	Mechanical test	Test cycle
1	Insulating support bar, pull bar and suspender	Power frequency withstand voltage test and impulse withstand voltage test	1 year	Static load test and dynamic load test	2 years
2	Insulated insulator cradle	Visual dimension inspection, power frequency withstand voltage test and impulse withstand voltage test	1 year	Bending static load test and bending dynamic load test	2 years

Chapter 1
Overview of Tools and Instruments for Live Working

Table (Cont'd)

S/N	Appliance	Test item			
		Electrical test	Test cycle	Mechanical test	Test cycle
3	Insulated tackle	Power-frequency voltage-withstand test	1 year	Tension test	1 year
4	Insulating bar	Visual dimension inspection, power frequency withstand voltage test and impulse withstand voltage test	1 year	Bending and torsional static load test and bending dynamic load test	2 years
5	Insulating ladder	Power frequency withstand voltage test and impulse withstand voltage test	1 year	Bending static load test and bending dynamic load test	2 years
6	Insulated rope ladder	Power frequency withstand voltage test and impulse withstand voltage test	1 year	Tensile property test, rope ladder head static load test and rope ladder head dynamic load test	2 years
7	Insulating rope tools	Power frequency withstand voltage test and operation impulse withstand voltage test	1 year	Static tensile test	2 years
8	Insulating hand tools	Power-frequency voltage-withstand test	1 year	—	—
9	Insulating (temporary) cross arm and insulating platform	Power-frequency voltage-withstand test	1 year	—	—
10	Insulator clamp	—	—	Static load test and dynamic load test	1 year
11	Stringing wire clamp	—	—	Static load test and dynamic load test	1 year
12	Shielding clothes	Clothes resistance test and shielding efficiency test of whole suit of clothes	Half a year	—	—
13	Electrostatic protective clothes	Shielding efficiency test of whole suit of electrostatic protective clothes	Half a year	—	—
14	Insulating clothes (tippet)	Power frequency withstand voltage test of whole suit of clothes	Half a year	—	—

Table (Cont'd)

S/N	Appliance	Test item			
		Electrical test	Test cycle	Mechanical test	Test cycle
15	Insulating sleeves	Mark inspection, AC withstand voltage and DC withstand voltage test	Half a year	—	—
16	Insulating gloves	AC withstand voltage and DC withstand voltage test	Half a year	—	—
17	Anti-mechanical puncture gloves	AC withstand voltage and DC withstand voltage test	Half a year	—	—
18	Insulating safety helmet	AC voltage withstand test	Half a year	—	—
19	Insulating shoes	AC voltage withstand test	Half a year	—	—
20	Insulating blanket	AC voltage withstand test	Half a year	—	—
21	Insulating mat	AC voltage withstand test	Half a year	—	—
22	Conductor soft shield cover	AC withstand voltage and DC withstand voltage test	Half a year	—	—
23	Protective cover	AC voltage withstand test	Half a year	—	—
24	Aerial device with insulating booms	Leakage current test of AC withstand voltage meter	Half a year	Overall performance test in rated load	Half a year
25	Phasing tester	Power frequency withstand voltage and leakage current test	Half a year	—	—
26	Electricity tester	Power frequency withstand voltage and leakage current test	Half a year	—	—
27	500kV four-bundle conductor hanging wheel	—	—	Static load test and dynamic load test	1 year
28	Spark gap detector	Gap adjustment and discharge test, power frequency withstand voltage test and operation impulse withstand voltage test	1 year	—	—

Chapter 1 Overview of Tools and Instruments for Live Working

2. Electrical Preventive Test Items and Standards of Insulating Tools

The electrical preventive test items and standards of insulating tools are shown in Table 1-2.

Table 1-2 Electrical Preventive Test Items and Standards of Insulating Tools

Rated voltage/ kV	Test length/m	1min high frequency withstand voltage/kV		3min high frequency withstand voltage/kV		15 times of operation impulse withstand voltage/kV	
		Predelivery test and type test	Preventive test	Predelivery test and type test	Preventive test	Predelivery test and type test	Preventive test
10	0.4	100	45	—	—	—	—
35	0.6	150	95	—	—	—	—
63（66）	0.7	175	175	—	—	—	—
110	1.0	250	250	—	—	—	—
220	1.8	450	440	—	—	—	—
330	2.8	—	—	420	380	900	800
500	3.7	—	—	640	580	1175	1050

Note: The test sample shall be in whole piece and shall not be divided into sections.

3. Inspection Test Standard for Insulating Tools

During the power frequency withstand voltage test, divide the insulating tool into several sections and add a voltage of 75kV on each 300mm for 1min, after which if no flashover, breakdown or overheating is found, the insulating tool is qualified.

4. Mechanical Test Standard for Tools for Live Working

(1) When bearing loads of various clamps and insulator link fittings under working load condition, the test shall be conducted according to relevant fitting standards.

(2) When bearing other static loads under working load condition, the test shall be conducted according to the design load and provisions of DL/T976-2005 Preventive Test Code of Tools, Devices and Equipment for Live Working.

(3) Load test shall be conducted regularly for tools and instruments for live working.

1.4.2 Mechanical Tests of Tools for Live Working

Mechanical tests of tools for live working include static load test and dynamic load test under working load. Some tools for live working, such as insulating pulling plate (pole), hanging pole, etc., shall only be subject to static load test; while some tools that may be impacted by loads, such as operating pole, tensioner, etc., shall be subject to dynamic load test in addition to static load test.

1. Static Load Test

Static load test is a test to assess the ability of tools, devices and equipment for live working to withstand mechanical loads (tension, torsion, pressure, and bending force). Static load generally refers to the load applied in the rated test which is 2.5 times of the allowable service load of the tested object for 5min, after which if there is no permanent deformation on each part of the tested object, it shall be deemed as qualified.

The service load can be determined according to the following principles.

(1) For tightening, pulling, hoisting and supporting tools (including tractor and fixator), if they are products manufactured by manufacturers, the allowable working load marked on the nameplate shall be taken as the service load; or the maximum service load can be calculated according to the actual use.

(2) For tools used for loading people (including all kinds of ladders, hanging baskets, hanging wheels, etc. used by single person), the quality of people and tools carried by them shall be taken as the service load.

(3) For tools used for holding, hoisting and hooking insulators, the quality of a string of insulators shall be taken as the service load.

During the static load test, the loading method is: assemble the tool into working state and apply the test load by simulating stress condition on the spot. Fig. 1-15 shows the static load test of tools and instruments for live working.

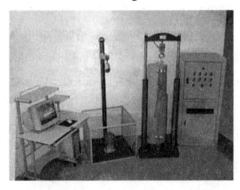

Fig. 1-15 Static Load Test of Tools and Instruments for Live Working

Chapter 1
Overview of Tools and Instruments for Live Working

2. Dynamic Load Test

Dynamic load test is a test item to check whether the mechanism operation is flexible and reliable when the tested object is impacted. Therefore, the load applied shall not be too large. Generally, it is required to apply 1.5 times of service load to the test object assembled into working state, and operate movable parts of the test object (such as leading screw handle, hydraulic tightener handle and unloading valve, etc.) for three times, after which if there is no damage, failure and other abnormal phenomena, it shall be deemed as qualified.

Since the operating bar is often used to pull out cotter pin or spring pin, or wrest screw, it shall be subject to the impact and torsion resistance test with an impact torque of 500N·cm and a torque of 250N·cm. Fig. 1-16 is the arrangement diagram of mechanical tests of insulator clamps.

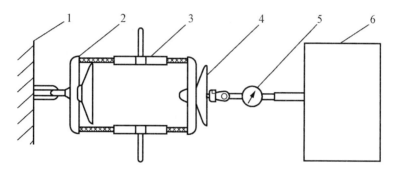

1 - fixed end; 2 - clamp; 3 - screw rod; 4 - insulator; 5 - dynamometer (sensor); 6 - tensile machine diagram

Fig. 1-16　Arrangement Diagram of Mechanical Tests of Insulator Clamps

1.4.3　Electrical Test of Tools for Live Working

Insulating tools and instruments for live working shall be subject to predelivery test before delivery, and the test items and relevant indexes must meet requirements of national standards. Due to long-term overstocking and delivery transportation of products and some manufacturers only conducting random sampling test during predelivery test, acceptance test must be conducted when users receive the products. The test standards shall refer to relevant national standards.

After a period of use and storage, tools for live working may suffer a certain degree of damage or deterioration in electrical performance or mechanical performance, so in addition to the above two tests, regular tests, namely preventive test and inspection test, shall also be conducted. Content and standards of electrical tests of insulating tools and instruments are introduced as follows.

Electrical tests of insulating tools and instruments shall be conducted regularly: preventive test once a year and inspection test once a year, with an interval between the two tests being half a year and test contents including power frequency withstand voltage test and operation impulse test. Since the operation impulse withstand voltage test can only be conducted by a party with the test qualification, so here is only a briefly introduction to test methods and test conditions of the power frequency withstand voltage test.

Fig. 1-17 is the power frequency withstand voltage and operation impulse withstand voltage wiring diagram.

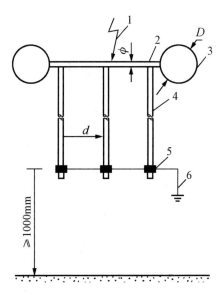

1 - high voltage lead; 2 - simulation conductor (φ>30mm); 3 - uniform ball (D=200-300mm);
4 - test object (spacing between test objects d≥500); 5 - lower test electrode; 6 - grounding lead

Fig. 1-17 Wiring Diagram of Power Frequency Withstand Voltage and Operation Impulse Withstand Voltage Test

1. Test Method

(1) Power frequency withstand voltage test of insulating bar. During the test, the metal head (or metal joint) part of the insulating bar, such as support bar, pull bar, suspender, etc. shall be hung on the high voltage terminal (generally, a metal rod with a length of not less than 2.5m and a diameter of 20mm shall be used as the high voltage terminal to simulate the conductor and hung in the air), and the ground wire shall be connected at the handle part and the boundary of effective insulating length (referring to the operating bar) or the original

Chapter 1
Overview of Tools and Instruments for Live Working

ground terminal (referring to the support bar, pull bar and suspender), and then voltage test shall be conducted for the whole section.

(2) Power frequency withstand voltage test of insulating ladder. Insulating ladders include vertical ladder, herringbone ladder, horizontal ladder and hanging ladder. During the test, a metal rod with a diameter of 20mm and a length of 2.5m shall be used as the high voltage terminal and hung horizontally to simulate the conductor. Then one end of the insulating ladder (generally the end of metal head) shall be hung on the high voltage terminal, the ground wire shall be connected to the shortest effective insulating length (wrapping the surface with tin foil and then bare copper wire for grounding), and finally the voltage test shall be conducted.

(3) Power frequency withstand voltage test of insulating rope and insulating rope ladder. In order to check the voltage withstand level of the whole insulating rope ladder or the whole insulating rope, they shall be wired according to the method shown in Fig. 1-18 during the test. Two metal rods with a diameter of 20mm and a proper length shall be wound with insulating rope ladder or insulating rope. One of the metal rods shall be used as the high voltage terminal, which is horizontally suspended in the air through insulator, while the other metal rod is suspended in the air and grounded at the same time. The distance between the two metal rods shall be equal to the shortest effective insulating length under the highest service voltage.

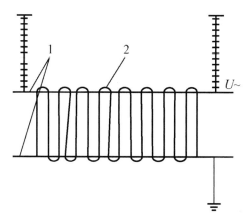

1 - metal rod; 2 - insulating rope

Fig. 1-18 Power Frequency Withstand Voltage Test of Insulating Rope

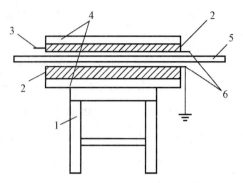

1 - support platform; 2 - foam plastic; 3 - power supply; 4 - connecting piece; 5 - tested item; 6 - tin foil
Fig. 1-19   Power Frequency Withstand Voltage Test of Insulating Shelter

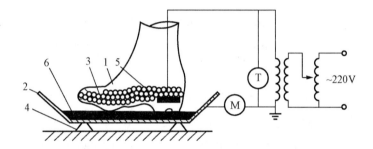

1 - test boots; 2 - metal plate; 3 - metal ball; 4 - metal sheet; 5 - sponge with water; 6 - insulating support
Fig. 1-20   Power Frequency Withstand Voltage Test of Insulating Boots

(4) Power frequency withstand voltage test of insulating shelter. Insulating shelters include all kinds of insulating soft board, hard board, film, etc. During the withstand voltage test, as shown in Fig. 1-19, the shelter shall be horizontally placed on the insulating support platform, and at the same time, both sides of the tested object shall be provided with aluminum foil as electrodes, and pressed with plastic connecting pieces at the upper and lower parts to ensure good contact. Then the upper pole plate shall be connected to the power supply and the lower pole plate to the ground. In addition, the tested object shall be reserved with edge width according to requirements of service voltage.

(5) Power frequency withstand voltage test of insulating boots. During the test, a piece of metal consistent with the size of the sample boots shall be covered with a piece of copper with a diameter of no more than 4mm and a height of no less than 15mm, and a piece of copper with a diameter of more than 4mm shall be connected externally and embedded in the metal ball. The outer electrode shall be a soggy sponge placed in the metal container, and the test circuit is shown in Fig. 1-20. The voltage shall rise from zero to 75% of the specified voltage at a speed of 1kV/s, and then rise to the specified voltage at a speed of 100V/s.

Chapter 1
Overview of Tools and Instruments for Live Working

When the voltage rises to the specified voltage, keep it for 1min, and then record the current value of the milliammeter. If the current value is less than 10mA, the test is passed and the insulating boots are qualified.

(6) Power frequency withstand voltage test of insulating gloves. Insulating gloves are very important personal protective equipment in live working of distribution line, and there are many kinds of them. Its AC withstand voltage test is shown in Fig. 1-21. The length of the exposed part of 10kV insulating gloves is 75mm.

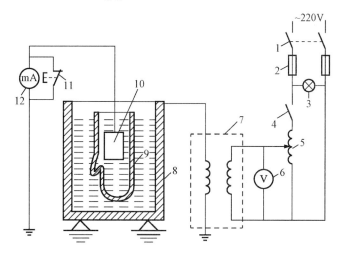

1 - disconnecter; 2 - fusible silk; 3 - power indicator; 4 - overload switch; 5 - voltage regulator; 6 - voltmeter; 7 - transformer; 8 - metal water container; 9 - test sample; 10 - electrode; 11 - milliammeter short circuit switch; 12 - milliammeter

Fig. 1-21 AC Withstand Voltage Test of Insulating Gloves

Electrical tests of insulating gloves include AC withstand voltage test and DC withstand voltage test. When conducting AC withstand voltage test on insulating gloves, the voltage shall rise to the AC withstand voltage test value shown in Table 1-3 for 1min, after which if there's no corona, flashover, breakdown or obvious heating, the insulating gloves shall be deemed as qualified.

Table 1-3 AC Withstand Voltage Value of Insulating Gloves

Model	Rated voltage/V	AC withstand voltage value (effective value)/V
1	3000	10000
2	10000	20000
3	20000	30000

When When DC withstand voltage test is conducted on insulating gloves, the voltage shall rise to the DC withstand voltage test value shown in Table 1-4 for 1min, after which if there's no flashover, breakdown or obvious heating, the insulating gloves shall be deemed as qualified.

Table 1-4 DC Withstand Voltage Value of Insulating Gloves

Model	Rated voltage/V	AC withstand voltage (effective value)/V
1	3000	20000
2	10000	30000
3	20000	40000

(7) Power frequency withstand voltage test of water washing tools. During power frequency withstand voltage test assembly, water washing tools shall be placed at an angle of 30°-45° with the ground. As shown in Fig. 1-22, the simulated conductor shall be a metal rod with a diameter of 20mm and a length of 2.5m suspended in the air through insulator string. The distance between the hydraulic giant nozzle and the simulated conductor shall be determined by the voltage class applied on the hydraulic giant, and microammeters shall be provided for grounding at the handle of the flushing rod and 1.8m from the penstock to the nozzle to measure the leakage current. A grounding switch shall be connected in parallel on the microammeter. When voltage is applied and scale knob is switched, the grounding switch shall always be in the on-position state and shall only be opened in case of reading to prevent the high voltage from acting on the microammeter. In addition, after the voltage of simulated conductor is boosted, the hydraulic giant shall be aimed at the simulated conductor to spray (water resistance 10000 $\Omega \cdot m$) for 1min, and then the microammeter shall be opened and read.

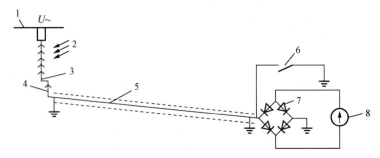

1 - voltage metal bar; 2 - water spray direction; 3 - operating bar in rainy days; 4 - rain cover;
5 - shield wire for measuring; 6 - switch; 7 - diode; 8 - DC microammeter

Fig. 1-22 Power Frequency Withstand Voltage Test of Water Washing Tools

Considering the combined insulation problem (water + insulating bar + aqueduct) of water washing tools, the test voltage shall be calculated and selected according to regulations.

2. Test Conditions

The insulating strength of insulating tools includes two parts: external insulation and internal insulation, and the main factors affecting the external insulation are air pressure, temperature, humidity, rain, pollution and the proximity effects of adjacent objects. Therefore, if the atmospheric state during the test is different from the standard atmospheric state, the discharge voltage shall be corrected to the standard atmospheric state. When the relative humidity is more than 80%, the discharge voltage will change, so in the rain test, only the relative air density shall be corrected, not the humidity.

Chapter 2

Metal Tools and Instruments

The most commonly used metal tools and instruments for transmission and distribution lines are insulator clamp, aluminum alloy clamp device, tackle, hanging wheel, etc., which are mainly used for live replacement of insulators, repair of conductors, overhaul and replacement of metal tools and instruments, etc. Good using habits and safekeeping methods can prevent deformation and damage to metal tools and instruments, extend service life of metal tools and instruments, and effectively ensure the safety of live working.

2.1 Insulator Clamp and Wire Clamp

Insulator clamps can be divided into strain insulator string clamps, linear string clamps and single insulator clamps. They are combined with aluminum alloy wire clamps to form the main tool for live replacement of single or whole string of strain or linear insulators in the power transmission and distribution line.

2.1.1 Insulator Clamp

The main tools used for live replacement of insulators are clamps, leading screws, insulating pulling plates, hanging hooks, insulating bars, etc. Among them, insulator clamps are the most important tools and instruments for replacing insulators in live working of transmission lines. When replacing, first install the clamps at the cross arm position of the insulator string hanging point, tighten the conductor through the leading screw and the insulating pulling plate to loosen the insulator string, and replace the insulator with the insulating bar.

According to different tower types, insulator string types and hardware types, the most commonly used insulator clamps for live replacement of insulator of power transmission line include wing-shaped clamp, broadsword clamp and closed clamp. These three types of insulator clamps will be detailed herein.

1. Basic Structure, Function and Usage of Common Insulator Clamps

(1) Airfoil clamp

① Basic structure

The clamps assembled on the conductor end of the insulator string are called the front clamps, and the clamps assembled on the cross arm end of the insulator string are called the rear clamps, as shown in Fig. 2-1.

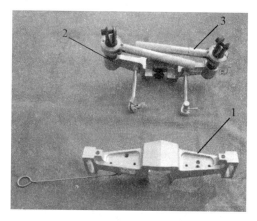

1-front clamp; 2-rear clamp; 3-screw rod
Fig. 2-1 Wing-shaped Clamp Structure Drawing

② Basic function

Mainly used for live replacement of single tension insulator string. During operation, it is necessary to fix the fittings on both sides of the insulator string with clamps and tighten the conductor through lead screw and insulating pulling plate, so that the insulator string can be relaxed on the insulator cradle for replacement (silicone rubber composite insulator is not suitable for the insulator cradle).

③ Usage

Before the electrician climbs up the tower, the ground electrician shall first assemble the insulator clamp, the leading screw and the insulating pulling plate. After the operator on the tower arrives at the designated position, the assembled wing-shaped clamp, the insulator cradle and the operating rod are successively transferred to the tower.

No. 1 electrician holds the clamp, and No. 2 electrician holds the insulating bar. They cooperate with each other to fix the front clamp of the wing-shaped clamp to the lowest position of the bolt-type strain clamp, as shown in Fig. 2-2. Locking device that prevents the clamp from sliding down is locked, as shown in Fig. 2-3. Fix the cross arm side clamp on the cross arm and insulator string insulator set fitting, and confirm that the fitting is properly embedded in the slot, as shown in Fig. 2-4. During the installation process, the body part of the No. 1 electrician shall not be beyond the first insulator.

Fig. 2-2　Laying the Wing-shaped Clamp

Fig. 2-3　Installing the Front Clamp and Locking It up

Fig. 2-4　Installation of the Rear Clamp

The No. 2 electrician holds the insulating bar and cooperates with the No. 1 electrician to insert the insulating tube of the insulated insulator cradle into the conductor side support frame, and the triangular supporting structure of the insulated insulator cradle is connected with the cross arm side clamp and the bolt is tightened. Before installing the insulated insulator cradle, adjust the height of the conductor side support frame so that the insulating tube of the insulator cradle can be parallel and close to the insulator string. After installation, check whether the two insulating tubes of the insulator cradle are inserted into the conductor support frame, as shown in Fig. 2-5.

Fig. 2-5 Installation of Insulator Cradle

The No. 1 electrician performs impact test on the insulator string and the clamp, tightens the leading screw and performs impact test on the stringing tool. When the leading screw is tightened, the stress of the leading screws on both sides shall be balanced and the tightening length shall be about the same. When the leading screw is shaken, the force shall be even, and the twisting amplitude of the insulating pulling plate shall be minimized.

(2) Broadsword clamp

① Basic structure

Broadsword clamp installed at the conductor end of the insulator string is called front clamp whose top is provided with a lifting pulley, and the one assembled at the cross arm end of the insulator string is called rear clamp, as shown in Fig. 2-6. When the broadsword clamp is used, a pulling plate, aleading screw, etc. are also needed, as shown in Fig. 2-7.

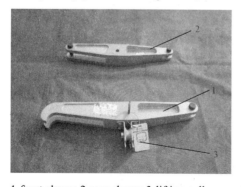

1-front clamp; 2-rear clamp; 3-lifting pulley

Fig. 2-6 Broadsword Clamp

1-insulating pulling plate; 2-leading screw

Fig. 2-7 Assembled Broadsword Clamp

② Basic functions

Method: It is mainly used for replacing tension duplex insulator string with the ground potential operation method.

③ Usage

The ground electrician shall first assemble the broadsword clamp, the leading screw and the insulating pulling plate, and adjust the distance between the leading screw and the insulating pulling plate, and the connection of each part is firm and reliable. After the operator on the tower arrives at the designated position, the assembled broadsword clamp, the insulator cradle and the operating rod are successively transferred to the tower.

The No. 1 electrician holds the insulating pulling plate and the clamp, and the No. 2 electrician holds the insulating bar, and they cooperate with each other to fix the front clamp of the broadsword clamp to the conductor side yoke plate for 2-bundle conductors, and fix the rear clamp of the broadsword clamp on the cross arm side yoke plate for 2-bundle conductors. When installing the front clamp of the broadsword clamp, check that the clamp is inserted in place, as shown in Fig. 2-8. When installing the rear clamp of the broadsword clamp, check that the pin is inserted in place and locked, as shown in Fig. 2-9. During the installation process, the body part of the No. 1 electrician shall not be beyond the first insulator.

Fig. 2-8   Installation of the Front Clamp of the Broadsword Clamp

Fig. 2-9   Installation of the Rear Clamp of the Broadsword Clamp

After adjusting the leading screw and making the stringing device slightly stressed, the No. 2 electrician holds the insulating bar and cooperates with the No. 1 electrician to insert the insulating tube of the insulated insulator cradle into the conductor side support frame, and the triangular supporting structure of the insulated insulator cradle is connected with the rear clamp of the broadsword clamp and the bolt is tightened, as shown in Fig. 2-10. Before installing the insulated insulator cradle, adjust the height of the conductor side support frame so that the insulating tube of the insulator cradle can be parallel and close to the insulator string, and fix the front clam, as shown in Fig. 2-11. After installation, check whether the two insulating tubes of the insulator cradle are inserted into the conductor support frame. When installing the triangular supporting structure of the insulator cradle, the No. 1 electrician shall pay attention to the safe distance between the hand and the foot and the jumper wire under the cross arm for 1.8m and above.

After the tool is assembled, the No. 1 electrician performs an impact test on the insulator string and the clamp. Finally, the No. 1 electrician tightens the leading screw at a constant speed to loosen the insulator string.

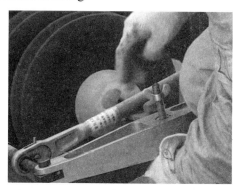

Fig. 2-10   Installation of Insulator Cradle

Fig. 2-11   Fixing the front clamp and installing the insulator cradle

(3) Closed clamp

① Basic structure

Closed clamp includes front clamp movable cover, front clamp bolt, rear clamp movable cover, rear clamp bolt, etc., as shown in Fig. 2-12. When using a closed clamp, you need to tighten the leading screw, as shown in Fig. 2-13.

② Basic functions

It is suitable for replacing conductor at the side of tension string conductor by the equipotential working method.

③ Usage

1-movable cover; 2-bolt
Fig. 2-12　Closed Clamp

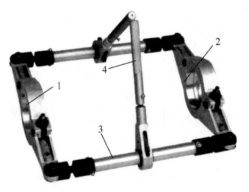

1-front clamp; 2-rear clamp; 3-tightening the leading screw; 4-crankshaft
Fig. 2-13　Assembled Closed Clamp

A. In the case of a porcelain insulator string, the ground electrician transmits the spark gap detector to the equipotential electrician, and the equipotential electrician detects the insulator to determine whether the number of good insulator pieces required for the operation is satisfied.

B. The equipotential electrician fastens a safety belt and an anti-fall rope, carries the insulated transmission rope and enters the operation point along the insulator string. When entering the potential, grasp the work string with the two hands, and step on the non-work string with the two feet. Move by the "two-span and three-short-circuit" method.

C. The ground electrician transfers the assembled closed clamp to the equipotential electrician's work position.

D. The equipotential electrician installs the closed clamp on the adjacent two steel caps of the replaced insulator, as shown in Fig. 2-14.

Fig. 2-14   Installation of Closed Clamp

Fig. 2-15   Removing the Insulator

E. The equipotential electrician tightens the leading screw to make it slightly stressed. After checking and confirming that there is no abnormality in the force, remove the fitting pins on both sides of the replaced insulator. Continue to balance the tightening of the leading screw to remove the insulator, as shown in Fig. 2-15.

2. Classification, Technical Parameters and Precautions for Use of Insulator Clamps

(1) Classification

① Clamps can be classified as the following series: tension string clamps, linear string clamps and monolithic insulator clamps according to the function. The codes are indicated as follows:

"N" - Tension string clamp;

"Z" - Linear string clamp;

"D" - Monolithic insulator clamp.

Tension string clamp is used to replace clamp of tension insulator and hardware, which can be classified as airfoil clamp, broadsword clamp, turning plate clamp, bending plate clamp, oblique card, etc. according to its structure shape. Linear string clamp is used to replace clamp of linear insulator and hardware which can be classified as V-shaped string clamp, bracket clamp, flower-shaped clamp, linear hook clamp, hook plate clamp, etc. according to its structure shape. Monolithic insulator clamp is used to replace clamp of monolithic insulator, which can be classified as closed clamp, end clamp, etc. according to its structure shape.

② Clamp is identified by the first letter of its name in Chinese Pinyin plus "K" according to its structure style. For example: "wing-shaped clamp" is identified as "YK", and "broadsword clamp" is identified as "DK".

(2) Technical Parameters

The general rated load value of clamp is:

$$P=P_0\times 25\%+5$$

Where: P represents the rated load of the clamp, kN; P0 is the applicable insulator or fitting grade, kN.

① Performance parameters of strain insulator string clamps

See Table 2-1 for specifications and technical parameters of tension string clamp.

Table 2-1　Specifications and Technical Parameters of Tension String Clamp

Name	Model	Rated load/kN	Dynamic test load/kN	Static test load/kN	Failure load/kN	Applicable insulator grade/kN
Airfoil clamp	NYK80-300	80	120.0	200.0	240.0	300
	NYK60-210	60	90.0	150.0	180.0	210
	NYK45-160	45	67.5	112.5	135.0	160
	NYK35-120	35	52.5	87.5	105.0	120
	NYK30-100	30	45	75.0	90.0	≤100

Chapter 2 Metal Tools and Instruments

Table (Cont'd)

Name	Model	Rated load/kN	Dynamic test load/kN	Static test load/kN	Failure load/kN	Applicable insulator grade/kN
Broadsword clamp	NKD45-160	45	67.5	112.5	135.0	160
	NKD35-120	35	52.5	87.5	105.0	120
	NKD30-100	30	45.0	75.0	90.0	≤100
Turning plate clamp	NFK60-210	60	90.0	150.0	180.0	210
	NFK45-160	45	67.5	112.5	135.0	160
	NFK35-120	35	52.5	87.5	105.0	≤120
Bending plate clamp	NWK45-160	45	67.5	112.5	135.0	210、160
	NWK35-120	35	52.5	87.5	105.0	≤120
Oblique clamp	NXK60-210	60	90.0	150.0	180.0	210、160
	NXK35-120	35	52.5	87.5	105.0	≤120

② Technical parameters for linear string clamps

See Table 2-2 for specifications and technical parameters of tension string clamp.

Table 2-2 Specifications and Technical Parameters of Linear String Clamp

Name	Model	Rated load/kN	Dynamic test load/kN	Static test load/kN	Failure load/kN	Applicable insulator grade/kN
Linear hook clamp	ZDK60-210	60	90.0	150.0	180.0	210、160
	ZDK35-120	35	52.5	87.5	105.0	120
	ZDK30-100	30	45.0	75.0	90.0	≤100
V-shaped string clamp	ZVK60-210	60	90.0	150.0	180.0	210
	ZVK45-160	45	67.5	112.5	135.0	160
Supporting plate clamp	ZTK60-210	60	90.0	150.0	180.0	210
	ZTK45-160	45	67.5	112.5	135.0	160
Hook plate clamp	ZGK60-210	60	90.0	150.0	180.0	210
	ZGK45-160	45	67.5	112.5	135.0	160
Flower-shaped clamp	ZHK45-160	45	67.5	112.5	135.0	160
	ZHK35-120	35	52.5	87.5	105.0	120
	ZHK30-100	30	45.0	75.0	90.0	≤100

③ Technical parameters for monolithic insulator clamps

See Table 2-3 for specifications and technical parameters of monolithic insulator clamps.

Table 2-3 Specifications and Technical Parameters of Monolithic Insulator Clamp

Name	Model	Rated load /kN	Dynamic test load /kN	Static test load /kN	Failure load/kN	Applicable insulator grade/kN
End clamp	DDK105-400	105	157.5	262.5	315.0	400
	DDK80-300	80	120.0	200.0	240.0	300
	DDK60-210	60	90.0	150.0	180.0	210
	DDK45-160	45	67.5	112.5	135.0	160
	DDK35-120	35	52.5	87.5	105.0	120
	DDK30-100	30	45.0	75.0	90.0	≤100
Closed clamp	DBK105-400	105	157.5	262.5	315.0	400
	DBK80-300	80	120.0	200.0	240.0	300
	DBK60-210	60	90.0	150.0	180.0	210
	DBK45-160	45	67.5	112.5	135.0	160
	DBK35-120	35	52.5	87.5	105.0	120
	DBK30-100	30	45.0	75.0	90.0	100

(3) Precautions for use

① When the clamp is selected, the maximum actual working load shall be checked. If there are special requirements or adverse weather conditions that may cause the working load to exceed the rated load of clamp, the clamp with higher specification shall be selected.

② The appearance of the clamp shall be inspected before use, and the clamp shall not be used if any crack is found. Collision shall be avoided during transportation, and it is not allowed to knock and fall hard in use to avoid affecting its mechanical and working performance.

③ The contact surface between the clamp and the hanging point shall be closely and reliably matched, and the non-contact surface shall have a gap of 1-2mm, so as to facilitate the installation or disassembly of the clamp.

④ All parts of the clamp shall have smooth and flat surface without burr, sharp edge, crack and other defects.

3. Inspection Items of Insulator Clamps

The insulator clamp must be inspected before use as follows.

(1) All parts of the clamp shall have smooth and flat surface without burr, sharp edge, crack and other defects.

(2) Dimensional tolerances, shape tolerances, and overall dimensions of each part of the clamps shall comply with the design drawings.

(3) The raw materials used in the main body of the clamps and other major parts must be re-inspected for chemical composition and mechanical properties before use. The aluminum alloy material shall also be re-inspected for macrostructure.

(4) The clamp shall be fitted with the hanging point to ensure that it is reliable and easy to handle.

4. Precautions for Storage of Insulator Clamps

(1) The insulator clamps shall be kept in a dry and well-ventilated tools and instruments warehouse, stored in a special tools and instruments box or placed at the fixed point of the shelf. The tool cabinet for storing the insulator clamps is shown in Fig. 2-16.

(2) The clamp label shall be provided on an easily identifiable part by imprinting or other methods, and the indentation depth shall not exceed 0.1mm.

(3) The label includes the model specification of the clamps, the name of the manufacturer or the code name, trademark, and factory number.

Fig. 2-16   Storage Cabinet for Insulator Clamps

5. Preventive Test for Insulator Clamps

For preventive tests, static load test and dynamic load test shall be carried out on the insulator clamps as required, and the test cycle is 1 year.

2.1.2 Aluminium Alloy Stringing Wire Clamp

1. Basic Structure and Function of Aluminium Alloy Stringing Wire Clamps

(1) Basic structure

The wire clamp comprises an upper clamp plate, a lower clamp plate, a wing-shaped pulling plate, a pulling plate and a pressure plate, and the structure and main parts thereof are as shown in Fig. 2-17.

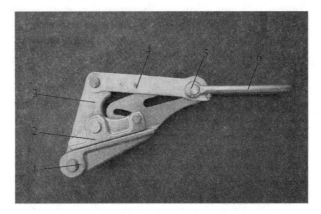

1-lower clamp plate; 2-upper clamp plate; 3-pressure plate; 4-pulling plate;
5-wing-shaped pulling plate; 6-pulling ring

Fig. 2-17　Structure and Main Functions of Aluminium Alloy Stringing Wire Clamps

(2) Basic function

When the conductor is released and tightened on the overhead power line, it can automatically clamp the conductor, and connect the intermediate connecting tool for the conductor and the hauling unit conveniently, and is suitable for adjusting the conductor or ground wire sag and tightening the conductor.

2. Classification and Technical Parameters of Aluminium Alloy Stringing Wire Clamps

(1) Classification

The aluminium alloy stringing wire clamp is divided into a single-hauling type (U-shaped pulling ring type) and a double-hauling type (wing-shaped pulling plate type) according to the hauling mode.

Fig. 2-18 shows the stringing wire clamp in use. Fig. 2-19 shows the installed stringing wire clamp.

Chapter 2
Metal Tools and Instruments

Fig. 2-18 Stringing Wire Clamp in Use   Fig. 2-19 Installed Stringing Wire Clamp

(2) Technical Parameters

See Table 2-4 for the model, specifications and dimensions of the stringing wire clamp.

See Table 2-5 for technical performance indicators of the stringing wire clamp.

Table 2-4 Models, Specifications and Dimensions of the Stringing Wire Clamp

Name	Type A	Type B	Type C	Type D	Type E	Type F	Type G	Type H
Model	LJKa25-70	LJKb95-120	LJKc150-240	LJKd300	LJKe400	LJKf500	LJKg630	LJKh720
Specification	φ12/14	φ16/20	φ22/24	φ26/28	φ30/32	φ31/33	φ35/37	φ37/39
Jaw clamping length (mm)	112	136	167	167	187	187	208	280
Clamping camber diameter (mm)	12	16	22	26	30	31	34	37
Maximum opening mm	14	20	24	28	32	33	37	39

Table 2-5 Technical Performance Indicators of the Stringing Wire Clamp

Model	Maximum mass/kg	Rated load/ kN	Maximum dynamic test load/kN	Maximum static test load/ kN	Minimum failure load/kN	Allowable maximum relative slippage of the conductor under the maximum dynamic test load/mm
LJKa25-70	1.0	8.0	12.0	20.0	24.0	5
LJKb95-120	1.5	15.0	22.5	37.5	45.0	5
LJKc150-240	3.0	24.0	36.0	60.0	72.0	5

Table (Cont'd)

Model	Maximum mass/kg	Rated load/ kN	Maximum dynamic test load/kN	Maximum static test load/ kN	Minimum failure load/kN	Allowable maximum relative slippage of the conductor under the maximum dynamic test load/mm
LJKd300	4.0	30.0	45.0	75.0	90.0	5
LJKe400	4.5	35.0	52.5	87.5	105.0	5
LJKf500	6.5	42.0	63.0	105.0	126.0	5
LJKg630	7.0	47.0	70.5	117.5	141.0	5
LJKh720	10.0	49.0	73.5	122.0	147.0	5

3. Inspection Items of aluminium Alloy Stringing Wire Clamps

The aluminium alloy stringing wire clamp must be inspected before use as follows.

(1) Dimensional tolerances and shape tolerances of each part of the aluminium alloy stringing wire clamps shall comply with the design requirements.

(2) The upper clamp plate, the lower clamp plate, the wing-shaped pulling plate, the pulling plate and the pressure plate of the aluminum alloy wire clamp shall be die forgings of super-strength aluminum alloy, and the pulling ring of the aluminum alloy wire clamp shall be die forgings made of high-quality structural alloy steel.

(3) The surface of each component after processing shall be smooth and free of defects such as sharp edges, burrs, cracks and metal inclusions.

4. Precautions for Use of Aluminium Alloy Stringing Wire Clamps

The precautions for the use of aluminium alloy stringing wire clamps are as follows.

(1) Various aluminium alloy stringing wire clamps shall not be relatively slipped under the rated load with the clamped conductor, and the surface of the conductor is not allowed to be damaged.

(2) At the maximum test load, a certain amount of slip is allowed, but the maximum slip amount shall not exceed 5mm. The parts of the wire clamp are not deformed, and the average diameter of the conductor shall not be less than 97% of the diameter before clamping, and the surface of the conductor shall be free of obvious indentation.

(3) After the maximum static test load is removed, the parts of various aluminium alloy stringing wire clamps shall not be permanently deformed.

(4) The connection of each component shall be tight and reliable, and the opening and

closing of the clamp is convenient and flexible with good integrity.

5. Precautions for storage of aluminium alloy stringing wire clamps

(1) The stringing wire clamps shall be kept in a dry and well-ventilated tools and instruments warehouse, stored in a special tools and instruments box or placed at the fixed point of the shelf.

(2) The label shall include the symbol, manufacturer's trademark, model and factory number, rated load and date of manufacture.

(3) There shall be a rectangular mark on the wire clamp, and the inspection cycle and the test date are marked in the rectangular mark.

(4) The label is stamped or riveted with signs.

6. Preventive Test for Aluminium Alloy Stringing Wire Clamps

For preventive tests, the appearance and main dimensions shall be checked for the aluminium alloy stringing wire clamps and the tensile test shall be carried out according to the requirements. The test cycle is 1 year.

2.2 Tackle and Hanging Wheel

The tackles and hanging wheels are common metal tools for live working on transmission lines. According to different operation conditions, the tackles are divided into iron tackles and aluminum tackles. The hanging wheels are divided into single-conductor hanging wheels, 220kV double-bundle hanging wheels and 500kV four-bundle hanging wheels, all of which are the most commonly used tool for equipotential live working on transmission lines.

2.2.1 Tackle

1. Basic Structure and Function of Tackles

(1) Basic structure

The tackle is mainly composed of hooks (links), pulleys, shafts, bushings (or bearings) and clamp plates, as shown in Fig. 2-20.

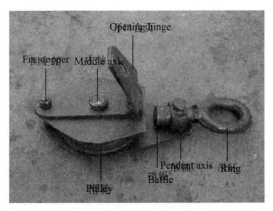

Fig. 2-20 Sample of Tackle Structure

(2) Basic function

The tackle is one of the commonly used tools for power line maintenance and construction and installation. It can generate a rotary motion by means of a hoisting rope, thereby changing the direction of the force. Because the tackle is easy to use and carry, it is widely used in line maintenance and construction and installation.

2. Classification and Technical Parameters of Tackles

(1) Classification

① According to the material, it can be divided into iron tackle, aluminum tackle, stainless steel tackle and insulated tackle.

② According to the use, it can be divided into lifting tackle, tackle for paying-off, and somersault tackle. The lifting tackle is divided into a ring type lifting tackle, a hook type lifting tackle, a hanger type lifting tackle, and a chain ring lifting tackle; the tackle for paying-off is divided into a ground cable pulley, a cable pulley, a skyward pulley, a nylon pulley, a cable conductor paying-off pulley, a large diameter tackle for paying-off, and a lifting pulley.

③ According to the number of sheaves of the tackle, it can be divided into a single-sheave tackle, a double-sheave tackle and a multi-sheave tackle.

(2) Technical Parameters

① Naming method for the model of the tackle

Model specifications of H series tackles are represented by a set of text codes, and the code consists of 4 parts.

② Model code of the tackle

The type code of the tackle is shown in Table 2-6.

Table 2-6 Type Code of the Tackle

Type	Opening	Closing	Hook	Chain ring	Ring	Lifting beam	Picking opening
Code	K	No K	G	L	D	W	KB

③ Selection of tackles

When selecting the tackles, according to the lifting mass and the number of pulleys required, check the diameter of the bottom of the sheave groove of the tackle and the diameter of the wire rope used in accordance with Table 2-7 to check whether the selected wire rope meets the requirements.

Table 2-7 H-Type Tackle Selection Table

Sheave groove bottom diameter (mm)	Lifting mass (t)														Used wire rope φ (mm)	
	0.5	1	2	3	5	8	10	15	20	32	50	80	100	140	Applicable	Maximum
	Number of pulleys															
70	1	2													5.7	7.7
85		1	2	3											7.7	11
115			1	2	3	4									11	14
135				1	2	3	4								12.5	15.5

Table (Cont'd)

Sheave groove bottom diameter (mm)	Lifting mass (t)														Used wire rope φ (mm)	
	0.5	1	2	3	5	8	10	15	20	32	50	80	100	140	Applicable	Maximum
	Number of pulleys															
165					1	2	3	4	5						15.5	18.5
185						2	3	4	6						17	20
210									5						20	23.5
245							1	2	4	6					23.5	25
280									2	3	5	7			26.5	28
320									1		4	6	8		30.5	32.5
360									1	2	3	5	6	8	32.5	35

3. Inspection Items of Tackles

The tackle must be inspected before use as follows.

(1) Visual inspection must be carried out before use. Where any hooks, rings, guard plates, partitions, etc. have severe deformation and cracks, rims are broken, bearings are deformed, bearing bushes are worn and rotation is inflexible, parts are incomplete, and tonnage is unclear, they shall not be used.

(2) Check if the torsional deformation of the lifting ring, the misalignment of the hinge plate, and the radial wear of the sheave groove and the uneven wear of the sheave groove affect the use.

(3) Check the appearance of the tackle for defects such as cracks and breakage.

(4) The test certificate of tackle shall be checked for its validity.

(5) Mainly check whether the pulley of the tackle and the lifting ring are flexible, whether there is jamming phenomenon, whether the opening and closing hinges and the clamp plate are flexible.

4. Usage of Tackles

(1) It shall be safely used according to the factory nameplate and shall not be overloaded.

(2) To use the open-door tackle, the door holder must be locked.

(3) During the loading and unloading process, the tackle must not be thrown to prevent deformation of the components.

Chapter 2
Metal Tools and Instruments

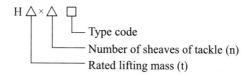

(4) The commonly used multi-sheave tackles are all of the opening type, and the connecting part includes a rotatable lifting ring type and a fixed lifting ring type, and the tackle of the fixed lifting ring has a directionality when it is suspended. The tackle is not allowed to bear the torsion when force is applied, otherwise it shall be adjusted by adding a U-ring.

(5) If the direction of the force changes greatly or when the tackle is used in a high place, a ring-type tackle shall be used. If a lifting hook-type tackle is used, the lifting hook opening must be sealed. The hook-type hook is easily decoupled from the binding rope after repeated stress.

(6) When the hook or the tackle hook is used, the sealing plate shall be installed. If the sealing plate is damaged or lost, use the #10 iron wire to seal the hook. Do not tie the binding rope to the hook with the iron wire, as shown in Fig. 2-21.

(7) In order to prevent the hook or the lifting ring of the single-sheave tackle from rotating with the tackle block during the stress process, so that the hauling rope is twisted, the hook or the lifting ring is fastened to the guard plate by using the #8 iron wire, as shown in Fig. 2-22.

(8) The tackle block is composed of single or multi-sheave single-whip tackle and movable tackle. When using, the size and model of the two tackles shall be basically the same. It is not suitable to use the tackles of different sizes to form the tackle block.

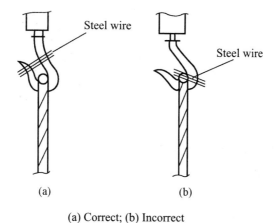

(a) Correct; (b) Incorrect

Fig. 2-21 Specifications for Use of Iron Tackle (I)

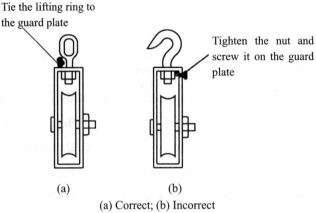

(a) Correct; (b) Incorrect

Fig. 2-22　Specifications for Use of Iron Tackle (II)

(9) The method of stringing the tackle block includes two types: the ordinary winding method and the flower winding method. Generally, the ordinary winding method is adopted. It is to fix two tackles (single or multiple sheaves) on the ground at an appropriate distance, and wind the wire rope ends from the sheave groove on one side of the tackle, and then wind it into another corresponding sheave groove. In the same direction of winding, it is wound up to the last sheave groove. After the wire rope is inserted and when it is tightened, the strands shall be parallel to each other and must not be twisted and displaced.

5. Precautions for Storage of Tackles

(1) The tackles shall be stored in a well-ventilated and dry environment, and put on shelves. Fig. 2-23 shows the placement of the iron tackle on the shelf.

(2) When the tackle is not used for a long time, it shall be cleaned and coated with anti-rust oil to avoid rust. Each group of tackles shall be numbered.

Fig. 2-23　Placement of Iron Tackle on Shelf

6. Requirements for Inspection Cycle of Tackles

The visual inspection and static load test for the tackle shall be carried out on a quarterly basis. The dismantling inspection shall be carried out semi-annually based on the visual inspection and the static load test. The new tackle shall be inspected according to the above requirements after half a year of use.

2.2.2 Hanging Wheel

1. Basic Structure and Function of Hanging Wheels

(1) Basic structure

The hanging wheel is a special tool for driving on overhead conductors, and Fig. 2-24 shows the structure of the manual hanging wheel for double bundle conductors. It consists of swing frame, front road wheel, auxiliary wheel, front road wheel handle, auxiliary wheel handle, chain, chain cover, cushion, backrest, frame, rear road wheel handle, rear road wheel, bumper joint, bumper, chain gear, pedal and brake. Fig. 2-25 shows a single-conductor hanging wheel.

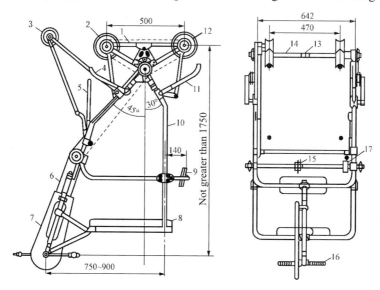

Fig. 2-24　Structure of Manual Hanging Wheel for Double Bundle Conductors

1-swing frame; 2-front road wheel; 3-auxiliary wheel; 4-front road wheel handle; 5-auxiliary wheel handle; 6-chain; 7-chain cover; 8-cushion; 9-backrest; 10-framework; 11-rear road wheel handle; 12-rear road wheel; 13-bumper joint; 14-bumper; 15-chain gear 16-pedal; 17-brake

Fig. 2-25　Single-conductor Hanging Wheel

(2) Main functions

The hanging wheel is used as a vehicle when the spacer, anti-vibration hammer and other accessories of the transmission line are installed or repaired, or the line crimping pipe is inspected, replaced, and maintained.

2. Basic Type and Technical Parameters of Hanging Wheels

(1) Basic type

① According to the number of conductors on which the wheel walks: There are single-conductor, double-bundle conductor, three-bundle conductor and four-bundle conductor hanging wheels. A single-conductor hanging wheel generally uses two wheels; the other multi-bundle conductor hanging wheels have 4 to 10 wheels, respectively.

② According to the driving mode of the hanging wheels: There are man-powered and motor-powered hanging wheels. Most of the man-powered hanging wheels use a pedal-like driving method like the bicycle; the motor-powered hanging wheels use a small gasoline engine as the power and are driven by hydraulic transmission or chain transmission.

③ According to the overall structure of the hanging wheel: It can be divided into two types: basket type and frame type hanging wheels. The transmission of the basket type is installed in an easy-to-purchase metal basket, and the operator stands in the basket while working. The shape of the frame-type hanging wheel is similar to that of a bicycle or a scooter, and the operator sits on the cushion to work.

Fig. 2-26 shows the pedal type hanging wheel of double-bundle conductor, and Fig. 2-27 shows several common hanging wheels.

Fig. 2-26　Pedal Type Hanging Wheel of Bundled Conductor

Chapter 2
Metal Tools and Instruments

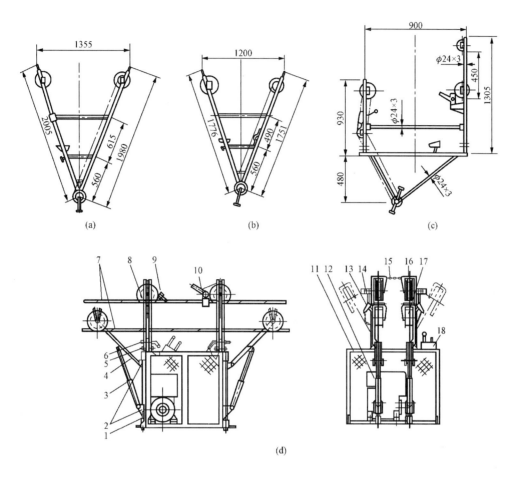

(a) SFS-1×150 single-conductor hanging wheel; (b) SFC-2×150 double-bundle conductor hanging wheel;
(c) man-powered pedal type four-bundle conductor hanging wheel;
(d) four-conductor eight-sheave frame type motor-powered hanging wheel

1-long pin axle; 2-hinged connecting plate; 3-hydraulic cylinder; 4-driven wheel arm; 5-pin shaft; 6-plug pin; 7-conductor; 8-drive wheel; 9-distance measuring device; 10-cable clamp brake; 11-frame; 12-hydraulic pump; 13-plug pin for preventing falling; 14-hydraulic motor; 15-anti-falling chain; 16-drive wheel arm; 17-driven wheel; 18-operation box

Fig. 2-27 Tackle Diagram

(2) Performance data

① Product code

The product code consists of voltage class, hanging wheel symbol, characteristics, and applicable range, such as F500-SB-ZA (B).

Wherein: F indicates the hanging wheel;

500 indicates 500kV;

SB indicates double drive roll type;

Z indicates linear type;

A indicates a maximum slope of 14°;

B-maximum slope of 20°.

② Technical requirements

a. Material requirements: The hanging wheel shall be made from high strength, light weight and non-brittle materials in accordance with national standards.

b. Rated load: The rated load of the hanging wheel shall be no less than 900N.

c. Safety: The hanging wheel shall have double safety devices.

③ Driving performance requirements

a. The hanging wheel shall be able to pass the spacer, the anti-vibration hammer and the suspension insulator string.

b. The active road wheel sheave groove shall be inlaid with conductive abrasion resistant rubber.

c. There shall be no charging or discharging when driving on an alive line;

d. The hanging wheel distance shall be able to vary within 450 (-30 to +50) mm.

e. The dimensions of the hanging wheel shall be compact and not greater than the data shown in Fig. 2-2-8.

f. The climbing ability of the A-type hanging wheel shall reach 14°, and the climbing ability of the B-type hanging wheel shall reach 20°.

g. The hanging wheel shall be equipped with the forward gear, the reverse gear and the neutral gear. When it is reversed, it can also pass the spacer, the anti-vibration hammer and the suspension insulator string. The B-type hanging wheel shall be equipped with high speed and low speed gears.

④ Structural requirements: The structure of the hanging wheel shall be light, easy to operate, transport and store. The weight of the A-type hanging wheel is below 40kg, and the weight of the B-type hanging wheel is below 50kg.

⑤ Process requirements: The surface of parts of the hanging wheel shall be free of sharp edges and corners, and the interfaces shall be rounded; the processing, heat treatment and assembly shall be carried out in accordance with the provisions of the national standard, and the surface of the material shall be subjected to anti-corrosion treatment.

Chapter 2
Metal Tools and Instruments

3. Inspection Items Before Use of the Hanging Wheel

(1) Visual inspection shall be carried out before use, including inspection of the overall shape, materials, signs, and technology.

(2) Before use, the hanging wheel shall be inspected to see if the structure of each part is firm and reliable, if the transmission structure is flexible, and if the safety devices are intact and reliable.

(3) Before use, inspect if the hanging wheel is subjected to mechanical load test and if the test cycle is within the validity period.

4. Precautions for Use of Hanging Wheel

(1) When the operator is working on..., the cross-sectional area of the conductor and ground wire shall satisfy the requirement that it shall not be less than 120mm^2 for steel-cored aluminum stranded conductors, not be less than 50mm^2 for steel strands, and not be less than 95mm^2 for copper strands.

(2) In any of the following cases, the work shall be carried out after verification and approval by the chief engineer of the unit, including: work on the conductor and ground wire in the isolated span; work on the conductor and ground wire with broken strands; work on the ground wire corroded; work on conductors and ground wires with small cross sections or other models; work on conductors or ground wires by more than two people.

(3) Before hanging the ladder (or hanging wheel) on the conductor and ground wire, the operator must check the fastening of the conductor and ground wire at the pole and tower at both ends of the span. After the hanging ladder (or the hanging wheel) is loaded, the minimum distance between the ground wire and the human body to the electrified body is 0.5m greater than the minimum safe distance between the human body and the electrified body, and the minimum distance between the conductor and the human body to the crossed power line is 1.0m greater than the minimum safe distance between the human body and the electrified body.

(4) It is strictly forbidden to hang ladders (or hanging wheels) on the porcelain cross arm line. Before the ladders (or hanging wheels) hung on the line are rotated, the cross arm shall be fixed.

5. Precautions for Storage of Hanging Wheel

(1) The hanging wheels shall be stored in a well-ventilated and dry environment, and put on shelves.

(2) When the hanging wheel is not used for a long time, it shall be cleaned and coated with anti-rust oil to avoid rust.

6. Preventive Test of the Hanging Wheel

The preventive test of the hanging wheel includes visual inspection, stepping test, static load test, dynamic load test, bumper test and brake test, and the contents are as follows.

(1) Mainly inspect the shape, materials, technologies and signs of the hanging wheel, and the frequency is once a year.

(2) Stepping test. Under the rated load of 900N, the stepping of the hanging wheel on the analog line shall be normal, and it can smoothly pass the spacer, the anti-vibration hammer and the suspension insulator string in the analog line. The frequency is once a year.

(3) Static load test. A load of 2250N is applied to the cushion of the hanging wheel and it lasts 5 minutes on the analog line. The cushion of the hanging wheel shall be complete, good, and free of permanent deformation. The test cycle is 1 year.

(4) Dynamic load test. A load of 900N (including the operator's load) is applied to the cushion of the hanging wheel, and it can step on the analog line back and forth three times normally, which is equipped with a spacer, an anti-vibration hammer and a suspension insulator string. The test cycle is 1 year.

(5) Bumper test. The front and rear active road wheels on the side of the hanging wheel are disengaged from the conductors, so that one side of the two bumpers is directly placed on the conductors, and it is normal when the load of 1350N is applied, without permanent deformation. The test cycle is 1 year.

(6) Brake test. The braking performance of the hanging wheel shall be good, and the test carried out in the following ways shall meet the requirements. For the type A hanging wheel, when it bears a person on a slope of $\alpha=14°$ (total load 900N), it glides at the speed of 3km/h, and the brake displacement shall be ≤ 0.2m; for the type B hanging wheel, when it bears a person on the slope of $\alpha=20°$ (total load 900N), it glides at the speed of 3km/h, and the brake displacement shall be ≤ 0.3m. The test cycle is 1 year.

Chapter 3

Insulating Tools and Instruments

Insulating tools for live working shall have good electrical insulation properties and high mechanical strength. At the same time, they shall also have the characteristics of low hygroscopicity, aging resistance, light weight, convenient operation and damage prevention. At present, insulating tools for live working can be roughly classified into three types: hard insulating tools, soft insulating tools and other insulating tools. Hard insulating tools mainly refer to tools with insulating tubes, rods and plates as the main body. Soft insulating tools mainly refer to tools with insulating rope as the main body. Other insulating tools refer to insulated tackles and insulated stringing tools, which have the characteristics of both hard and soft insulating tools.

3.1 Hard Insulating Tools

Among the hard insulating tools, the most widely used are insulating rods, as well as insulating platforms and insulating hard ladders made of insulating tubes or plates. Insulating rods are divided into insulating bars, insulating struts and pull rods (hangers) according to the application and operation method. In addition, insulates cross arms and insulated insulator cradles are also frequently used as hard insulating tools. Hard insulating tools are basically made of glass fiber boards or fiber glass epoxy plastics. In particular, fiber glass epoxy plastics (referred to as FRP) are made from glass fiber and epoxy resin. Because of the excellent electrical insulation properties of glass fiber and epoxy resin, the glass fiber reinforced plastics have excellent mechanical and electrical properties.

3.1.1 Insulating Bar

1. Basic Structure and Function of Insulating Bar

(1) Basic structure

① Insulating bar is composed of two parts: joint and insulating rod, and some insulating bars are equipped with grab handle at the end. The joint of can be fixed type or disassembled type, and the joint fixed on the bar is made by high strength material. Fig. 3-1 shows the structure of the insulating bar.

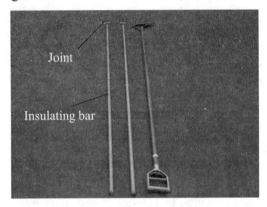

Fig. 3-1　The Structure of the Insulating Bar.

② The length of each part of the insulating bar shall meet the requirements of Table 3-1.

Table 3-1　Requirements for Length of Each Part of the Insulating Bar

Rated voltage (kV)	Minimum effective insulation length (m)	End metal joint length (m)	Handheld length (m)
10	0.70	≤0.10	≥0.60
35	0.90	≤0.10	≥0.60
66	1.00	≤0.10	≥0.60
110	1.30	≤0.10	≥0.70
220	2.10	≤0.10	≥0.90
330	3.10	≤0.10	≥1.00
500	4.00	≤0.10	≥1.00
750	5.00	≤0.10	≥1.00
±500	3.50	≤0.10	≥1.00

Chapter 3 Insulating Tools and Instruments

(2) Function

In live working, operators shall install corresponding accessories on the insulating bar, so that they can use the insulating bar to complete multiple operations such as lead disconnection and connection while maintaining a safe distance from the electrified body.

2. Classification and Technical Specifications of Insulating Bars

(1) Classification

① The insulating bars can be divided into 3m, 4m, 5m, 6m, 8m, and 10m insulating bars by length.

② Insulating bars can be divided into 10kV, 35kV, 110kV, 220kV, 330kV, and 500kV insulating bars according to voltage class.

③ Insulating bars can be classified into two types according to the material: manual roll forming bar with FRP epoxy resin bar and mechanical pultrusion bar. The two types of insulating bars have their own advantages: The advantage of the manual roll forming bars is that the tension is large, but the longitudinal strength is relatively small compared with the mechanical pultrusion bars; the advantage of the mechanical pultrusion bars is that the strength is large, but the transverse tension is relatively small compared with the manual roll forming bars.

④ Insulating bars can be classified as interface insulating bar, retractable insulating bar and maneuver insulating bar according to the structure. The interface insulating bar is a commonly used insulating bar. The spiral joint is used at the section separation part, which can reach 10m at most. It can be conveniently carried as the sections can be bagged. The retractable insulating bar is designed with 3 sections that are retractable, which can generally reach 6m at most. It is light in weight, small in size, and easy to carry and use. It can be telescopically positioned to any length according to the used space, effectively overcoming the shortcomings of the inconvenience to use of interface insulating bars due to the fixed length. The maneuver design is adopted for the interface of the maneuver insulating bar, so it does not overturn when it is tightened.

(2) Technical specifications

① The insulation length of the insulating bar shall not be less than 0.7m.

② The material of the insulating bar has the advantages of high compressive strength, corrosion resistance, moisture resistance, good mechanical property, light weight and portability.

③ The connection between the three sections shall be firm and reliable and they must not fall off during operation.

3. Inspection Items of Insulating Bars

The insulating bar must be inspected before use as follows.

(1) First check whether the specifications of the insulating bar are the same as the working line voltage class. It is forbidden to use the insulating bar that does not meet the specifications.

(2) Check whether the test certificate of the insulating bar is within the validity period.

(3) Check the appearance of the insulating bar for external damages such as cracks and scratches.

(4) The connection between the insulating bar sections shall be firm and reliable.

4. Usage of Insulating Bars

(1) When it is necessary to operate outdoors in rainy and snowy days, special insulating bar with rain and snow cover shall be used.

(2) When connecting the insulating bar, screw thread connecting section and section shall be off the ground, so as to avoid weeds or soil from entering the screw thread or adhering to the surface of the rod. The screw thread shall be tightly fastened before use.

(3) When in use, the bending force of the rod shall be minimized, so as to avoid damaging the rod body. The usage of the insulating bar is as shown in Fig. 3-2.

Fig. 3-2　Usage of Insulating Bar

5. Precautions for Storage of Insulating Bars

(1) After the insulating bar is used, the surface shall be wiped clean in time.

(2) The insulating bar shall be stored in a well ventilated, dry environment and stored in a well ventilated, clean and dry rack or suspended.

(3) The insulating bar must be kept by a specially assigned person.

Fig. 3-3 shows the tool cabinet for storing insulating bars.

Fig. 3-3   Tool Cabinet for Storing Insulating Bars

6. Preventive Test of Insulating Bars

The routine test of the insulating bar shall include the following items.

(1) Electrical test. For the electrical test of the insulating bars, the samples with voltage class of 220kV and below shall be able to pass the short-duration power-frequency withstand voltage test, and the samples with voltage class of 330kV and above shall be able to pass the long-duration power-frequency withstand voltage test and operating impulse withstand voltage test. The test cycle is 1 year.

(2) Mechanical test. The mechanical test items of the insulating bar include a bending static load test and a bending dynamic load test. The test cycle is 2 years.

3.1.2   Insulating Support Bar, Pull Bar and Suspender

1. Basic Structure and Function

(1) Basic structure. The insulating support rod, pull rod and suspender are generally formed by a combination of metal accessories and hollow tubes, filling tubes and insulating plates.

(2) Functions of insulating support rod and pull rods. It is an insulating load-bearing tool supported by the electric pole body, which can make the conductor horizontally and vertically displaced at the same time. This kind of tool can not only be used to replace the insulator, but also be used to replace the cross arm and the electric pole.

(3) Functions of insulating stringing pull rod (plate). It is an insulating component that bears horizontal loads (tightening force of conductors). To replace a single string of strain insulators, two pull rods are required. To replace the double strings of insulators, only one pull rod is used in most cases and can only be used with a fixed clamp (such as a broadsword clamp).

(4) Functions of insulating hanging pole. Two insulating components that are subjected to vertical loads generally form one group (also used as a single piece) and are used in applications where vertical loads are large.

2. Type and Performance Parameters of Insulating Support Rod, Pull Rod and Suspender

(1) Insulating hanging pole can be classified as common type, fixed type, compression bar type and insulator bracket type according to the structure, among which the first two types shall be used together with fixators and traction machines.

(2) Performance parameters: The shortest effective insulation length, total length and length of each part of the insulating support rod, pull rod and suspender shall comply with the requirements of Table 3-2. It shall be noted that the total length of the support rod is determined by the sum of the shortest effective insulation length of the support rod, the length of the fixed part and the length of the movable part. The total length of the support rod and the pull rod is determined by the sum of the shortest effective insulation length and the length of the fixed part.

Table 3-2 Length Requirements for Insulating Suspender, Pull Rod and Support Rod

Rated voltage/ kV	Minimum effective insulation length/m	Length of fixed part/m			Length of movable part of the support rod/m
		Support rod	Suspender and pull rod		
10	0.4	0.6	0.2		0.5
35	0.6	0.6	0.2		0.6
63	0.7	0.7	0.2		0.6
110	1.0	0.7	0.2		0.6
220	1.8	0.8	0.2		0.6

3. Usage of Insulating Support Rod, Pull Rod and Suspender

The insulating support rod, pull rod and suspender are mainly used for transfer of conductors, fittings, insulators and so on. Fig. 3-4 shows the usage of insulated suspender.

Fig. 3-4 Usage of Insulated Hanging Pole

4. Inspection Items of Insulating Support Rod, Pull Rod and Suspender

(1) First check whether the specifications of the insulating support rod, pull rod and suspender are the same as the working line voltage class.

(2) Check whether the test certificate of insulating support rod, pull rod and suspender is within the validity period. It is strictly forbidden to use the insulating support rod, pull rod and suspender that do not meet the specifications.

(3) Check whether the surface is smooth and there are external damages such as cracks and scratches on the surface of the insulating support rod, pull rod and suspender; the connection between bar sections shall be firm.

(4) The connection between the metal accessories, the hollow tube, the foam filling tube, the solid rod and the insulating plate on the insulating support rod, pull rod and suspender shall be firm and flexible and convenient to use.

5. Precautions for Storage of the Insulating Support Rod, Pull Rod and Suspender

(1) The insulating support rod, pull rod and suspender shall be stored in a well-ventilated and dry environment.

(2) The insulating support rod, pull rod and suspender shall be stored vertically in a special tool cabinet and shall not be placed on the shelves transversely.

6. Preventive Test of the Insulating Support Rod, Pull Rod and Suspender

The routine test of the insulating support rod, pull rod and suspender shall include the following items.

(1) Electrical test. The electrical test items of the insulating support rod, pull rod and suspender include the power frequency withstand voltage test and the operating impulse withstand voltage test. The test cycle is 1 year.

(2) Mechanical test. The mechanical test items of the insulating support rod, pull rod

and suspender include a bending static load test and a bending dynamic load test. The test cycle is 2 years.

3.1.3 Insulated Insulator Cradle

1. Basic Structure and Function of the Insulated Insulator Cradle

(1) Basic structure

Insulated insulator cradle is an important tool for live replacement of tension and linear insulator strings on power transmission line. Both of its sides are composed of oval hollow pipe or rectangular insulating plate, and its main materials are connected by several curved plates, on which 2-3 insulating pipes are fixed to facilitate string movement of insulator. Fig. 3-5 shows the insulated insulator cradle.

Fig. 3-5 Insulated Insulator Cradle

(2) Basic function

The insulator string is held in the operation of replacing the strain insulator string to keep it in a relaxed state, so as to facilitate the replacement of the insulator string.

2. Classification and Technical Specifications of the Insulated Insulator Cradle

(1) Classification

Insulated insulator cradle can be classified as linear insulator cradle and tension insulator cradle according to the function. Fig. 3-5 shows the insulated linear and strain insulator cradle.

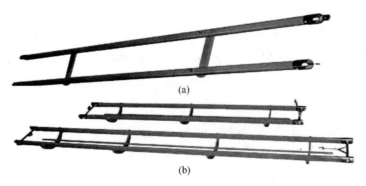

a-insulated linear insulator cradle; b-insulated strain insulator cradle

Fig. 3-6 Insulated Linear and Strain Insulator Cradle

(2) Technical specifications

① The metal accessories at both ends of the insulated insulator cradle shall be treated with chrome plating and other surface anti-corrosion. After the parts of the insulating laminated materials are formed, the machined surface shall be insulated, that is, the joints at the inner holes of the interface shall be filled with high-strength insulating adhesive. The surface shall be coated with the insulating paint.

② After the components of the insulated insulator cradle are formed, the machined surfaces shall be regularly flat.

③ The spacing between insulating supports in insulated insulator cradle (insulator cradle with voltage level of 220kV and below) is generally about 500mm (the height of 3 pieces of insulators), and that of the insulator cradle with the voltage class of 330kV and above should generally be about 600~800mm (height of 3 to 4 insulators).

④ The width of the two upper connecting rods in the insulated insulator cradle shall be compatible with the disk diameter of the insulator. The distance (depth) of the upper and lower connecting rods shall be more than half of the disk diameter of the insulator.

3. Inspection Items of the Insulated Insulator Cradle

(1) The surface of the insulated insulator cradle shall be smooth, free from air bubbles, wrinkles and cracks. The glass fiber cloth and the resin are well bonded, and the rods, segments and plates are firmly connected.

(2) The insulated insulator cradle shall be inspected for insulation before use, as shown in Fig. 3-7.

(3) When the insulated insulator cradle is used, the electrical and mechanical test certificates shall be checked for validity period. The insulated insulator cradles beyond the validity period cannot be used.

Fig. 3-7　Insulated Insulator Cradle

4. Usage of the Insulated Insulator Cradle

(1) Visual inspection must be carried out before use of the insulated insulator cradle to check whether its main materials, curved plate and insulating pipe are deformed or slackly connected.

(2) When replacing insulator strings, it is necessary to pay attention to the match of clamps to ensure that the connection is firm and reliable, and the impact test shall be carried out to verify the connection.

(3) In the course of delivery, attention shall be paid that the insulator cradle shall not collide with the tower and fittings, which may lead to damage. Fig. 3-8 shows the replacement of the insulator string with an insulated insulator cradle.

Fig. 3-8 Replacement of the Insulator String with the Insulator Cradle

5. Storage of the Insulated Insulator Cradle

The insulated insulator cradle shall be stored in a well-ventilated room for special tools for live working. If the storage period exceeds the production date by 12 months, the electrical performance test shall be carried out according to relevant standards. The insulated insulator cradles shall be transported in special tool bags, tool cases or special tool cars to protect them from moisture and damage. Fig. 3-9 shows the storage of insulated insulator cradles on shelves.

Fig. 3-9 Storage of Insulated Insulator Cradles on Shelves

6. Preventive Test of the Insulated Insulator Cradle

The preventive test of the insulated insulator cradle includes the following items.

(1) Electrical test. Power frequency withstand voltage test and operating impulse withstand voltage test: Short-duration power frequency withstand voltage test is carried out for test samples with voltage class of 220kV and below, and long-duration power frequency withstand test and operating impulse withstand voltage test are carried out for test samples with voltage class of 330kV and above. The test cycle is 1 year.

(2) Mechanical test. The mechanical test items include a bending static load test and a bending dynamic load test. The test cycle is 2 years.

3.1.4 Insulated Cross Arm

1. Basic Structure and Function of Insulated Cross Arm

(1) Basic structure

Insulated cross arms are generally composed of glass fiber boards and metal accessories. In recent years, insulated cross arms made of FRP and composite materials have appeared. They all have the characteristics of light weight, strong insulation, corrosion resistance and small volume.

The composite insulated cross arm consists of a glass fiber epoxy core rod, a silicone rubber skirt, a front end link fitting and a connecting flange, as shown in Fig. 3-10. The core rod and the fittings are connected by a cementing process, and the core rod and the skirt are integrally molded by one-time forming injection process. The core rod has high mechanical strength and high tensile strength compared to ordinary steel.

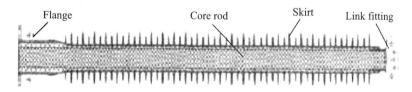

Fig. 3-10   Composite Insulated Cross Arm

(2) Basic function

The main role of the insulated cross arm is to support the conductors and keep the conductors insulated from the pole and tower. It is widely used in the straight line cross arm operation of the live replacement of 10 to 35kV distribution line pole and tower, as well as the steel cross arm of replacement of 110kV-220kV transmission line steel tube pole.

2. Classification and Technical Specifications of Insulated Cross Arms

(1) Classification

① According to the composition of materials, it can be divided into glass fiber board insulated cross arm, FRP insulated cross arm, composite insulated cross arm and so on.

Fig. 3-11 shows the glass fiber board insulated cross arm, Fig. 3-12 shows the used FRP cross arm, and Fig. 3-13 shows the composite insulated cross arm.

Fig. 3-11　Glass Fiber Board Insulated Cross Arm

Fig. 3-12　FRP Cross Arm

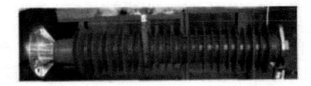

Fig. 3-13　Composite Insulated Cross Arm

② According to the shape and use, it can be divided into insulated branch cross arm, temporary pole cross arm, rubber rod auxiliary cross arm, and three-phase insulated branch cross arm for pole, as shown in Fig. 3-14.

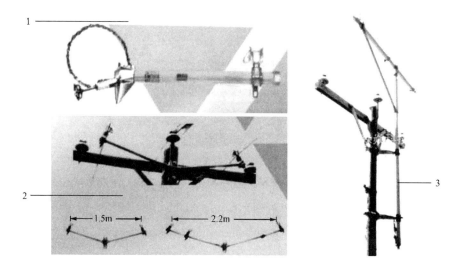

1-insulated branch cross arm; 2-temporary pole cross arm; 3-three-phase branch cross arm for pole

Fig. 3-14 Insulated Cross Arm for Distribution Lines

(2) Performance specification

See Table 3-3 for the technical specifications for the most commonly used FRP insulated cross arms for live replacement of linear cross arms.

Table 3-3 Specifications for FRP Insulated Cross Arms and Insulated Pedals

Angle FRP specifications (mm)	Thickness of arm of angle (mm)	Angle FRP length (mm)	Scope of application
∠50 × ∠50 (±2)	6 ~ 8 (±0.5)	1200 ~ 2000 (±10)	Pedals
∠50 × ∠50 (±2)	6 ~ 8 (±0.5)	1200 ~ 2000 (±10)	Cross arm
∠50 × ∠50 (±2)	6 ~ 8 (±0.5)	1200 ~ 2000 (±10)	Cross arm
∠50 × ∠50 (±2)	6 ~ 8 (±0.5)	1200 ~ 2000 (±10)	Cross arm
∠50 × ∠50 (±2)	6 ~ 8 (±0.5)	1200 ~ 2000 (±10)	Cross arm

3. Inspection Items of the Insulated Cross Arm

(1) Insulated cross arms must be visually inspected before use. Mainly inspect the insulating main material (or insulating rod) and metal parts; inspect whether the insulation part is reliable (with relevant instruments), and whether the metal parts are deformed or damaged; whether the connection of metal parts is firm and reliable.

(2) In addition to checking whether the surface of the FRP insulated cross arm is broken or dirty, it must be checked before use for stress, and it can be used without any problem.

Short circuit of the FRP cross arm shall be strictly prevented.

(3) The composite insulated cross arm must be checked before use to see if the skirt, the core rod and the front and rear end link fittings are broken and damaged. The load test and power frequency withstand voltage test must be carried out.

4. Usage of Insulated Cross Arms

(1) The insulated branch cross arm used in the distribution line is mainly used for rewiring or maintenance of single-phase and three-phase conductor support frames. The temporary pole cross arm is mainly used as the temporary cross bar frame to be connected with live cables when replacing the electric pole cross arm. The three-phase branch cross arm for poles is mainly used for replacing electric poles, brackets and insulated porcelain insulators, etc.

(2) FRP insulated cross arm is used for live replacement of the cross arm for the 10 to 35kV line. It has the characteristics of having no need to move the conductor, keeping the conductor at the original position, and operating more simply and reliably.

(3) Composite insulated cross arm is a cross arm made of a new type of material. It is small in size, light in weight, excellent in electrical and mechanical properties, and is used as the cross arm on 110kV and 220kV double-circuit steel tube poles, which can further reduce the line corridor.

In addition, the insulated cross arm can also be used in the straight rod and strain rod operations in the charged state.

5. Precautions for Storage of Insulated Cross Arms

The insulated cross arm shall be placed on the shelf in a room for special tools for live working, which is dry and well ventilated with suitable temperature. Fig. 3-15 shows the storage of the insulated cross arm.

Fig. 3-15 Storage of Insulated Cross Arm

6. Preventive Test of Insulated Cross Arms

The preventive test of the insulated cross arm includes the following items.

(1) Electrical test. It shall pass the corresponding 45kV or 95kV short-duration power-frequency withstand voltage test at the voltage class of 10kV and 35kV (qualified if there is no breakdown, flashover and obvious heat), and the test cycle is 1 year.

(2) Mechanical test. Including static and dynamic load tests. The test cycle is 2 years.

3.2 Soft Insulating Tools

Among the soft insulating tools, the most widely used is insulating rope. The insulating rope is one of the most important insulating materials commonly used in live working, which can be used as a carrier, climbing tool, lifting rope, connecting sleeve and security rope, etc. The tool with insulating rope as the main insulating component is flexible, light, portable and suitable for on-site operation. In addition, insulating rope or insulating belt can be used to make insulating rope ladder, waist belt, etc. Soft insulating tools are mainly made from silk or synthetic fiber, among which silk ropes are most commonly used.

3.2.1 Insulating Rope

1. Basic Structure and Function of Insulating Rope

(1) Basic structure

Insulating ropes are mainly made from silk or synthetic fiber, among which silk ropes are most commonly used. Silk is a good electrical insulating material in the dry state. However, silk has a strong hygroscopicity because of the hydrophilicity and fiber porosity of the glue. When silk is used as an insulating material, special care shall be taken to avoid moisture. The tool with insulating rope as the main insulating component is flexible, light, portable and suitable for on-site operation. Fig. 3-16 shows the insulating rope.

Fig. 3-16　Insulating Rope

(2) Basic function

The insulating rope is widely used in live working, and is one of the most important insulating materials in live working, which can be used as a carrier, climbing tool, lifting rope, connecting sleeve and security rope, etc. In addition, insulating rope or insulating belt can be used to make insulating rope ladder, waist belt, etc.

Insulating ropes commonly used in live working are silk ropes (including raw silk ropes and boiled-off silk ropes) and nylon ropes (including nylon wire ropes and nylon thread ropes). Insulating ropes are used for traction, lifting of objects, temporary pulling of wires, making of ladders and tackle ropes, etc. during live working. Insulating ropes shall not only have high mechanical strength, but also have wear resistance.

2. Classification and Model Specifications of Insulating Ropes

(1) Classification

① According to the material, the insulating rope can be divided into natural fiber insulating rope and synthetic insulating rope.

② According to the electrical properties in the wet state, the insulating rope can be divided into conventional insulating rope and moisture-proof insulating rope.

③ According to the mechanical strength, the insulating rope can be divided into conventional strength insulating rope and high strength insulating rope.

④ According to the weaving process, the insulating rope can be divided into braided insulating rope, stranded insulating rope and woven insulating rope.

⑤ According to the use, the insulating rope can be divided into arc suppression rope, insulating rope noose, insulating safety rope, insulating distance measuring rope and lifting rope.

(2) Model and specification

① Arc suppression rope

The model of the arc suppression rope consists of materials, code numbers and specifications. Example: SCXHS—10×20m

Wherein: SC - silkworm; XHS - arc suppression rope; 10 - rope diameter 10mm; 20m - length 20m. The diameter of the arc suppression rope is 12mm, and the length can be divided into 15m, 20m and 30m.

② Insulating rope noose

a. Electrodeless insulating rope noose. The model of the electrodeless insulating rope noose consists of materials, code numbers and specifications. The specifications of the electrodeless insulating rope noose can be made into various rope diameters and lengths as needed.

Example: JCSTW—16×400m

Wherein: JC-nylon filament; ST-noose; W-electrodeless noose; 16-rope diameter 16mm; 400-length 400m.

b. Two-eye insulating rope noose. The model of the two-eye insulating rope noose consists of materials, code numbers and specifications. The specifications of the two-eye insulating rope noose can be made into various rope diameters and lengths as needed.

Example: JCSTL—16×400m

Wherein: JC-nylon filament; ST-noose; L-two-eye noose; 16-rope diameter 16mm; 400-length 400m.

③ Insulating safety rope

a. Personal insulating safety rope. The model of the personal insulating safety rope consists of materials, code numbers and specifications. The rope diameter of the personal insulating safety rope shall not be less than 14mm, and the length may be various specifications of 2~7m.

Example: JCSTL—14×4m

Wherein: SC - silkworm; RBS - personal safety rope; 14 - rope diameter 14mm; 4 - length 4m.

b. Conduct insulating safety rope. The model of the conductor insulating safety rope consists of materials, code numbers and specifications. Example: SCDBS—30×4.5m

Wherein: SC - silkworm; DBS - conductor safety rope; 30 - rope diameter 30mm; 4.5 - length 4.5m.

See Table 3-4 for the classification and specifications of conductor insulating safety ropes.

Table 3-4 Classification and Specifications of Conductor Insulating Safety Rope

Model	Rope diameter (mm)	Rope length (m)	Rated load (kN)	Voltage level (kV)	Remarks
SCDBS—18×2.5m	18	2.5	8	35 ~ 110	
SCDBS—22×3.5m	22	3.5	12	220	
SCDBS—34×3.5m	34	3.5	24	220	Suitable for 2-bundle conductors
SCDBS—34×4.5m	34	4.5	24	330	Suitable for 2-bundle conductors
SCDBS—34×5.5m	2×34	5.5	2×24	500	Suitable for 4-bundle conductors

Note: Insulating safety ropes for 500kV four-bundle conductors shall be 2 SCDBS-34×5.5m ropes

Chapter 3
Insulating Tools and Instruments

④ Insulating distance measuring rope

The model of the insulating distance measuring rope consists of materials, code numbers and specifications. Example: SCCS—4×50m

Wherein: SC - silkworm; CS - distance measuring rope; 4 - rope diameter 4mm; 50m - length 50m. The diameter of the insulating distance measuring rope is generally 4-5mm, and the length is 50m.

3. Inspection Items of Insulating Ropes

(1) Insulating rope must be visually checked before use. It is strictly prohibited to wrap insulating rope with metal wire. The twisted synthetic rope strands of all insulating rope tools shall be tightly twisted without looseness or separation. There shall be no fold marks, bulges, reverse strands, cramps and other defects on each strand of the rope and silk threads of each strand.

(2) Before the insulating rope is used, the test certificate shall be checked for validity period. Insulating ropes beyond the validity period cannot be used.

(3) Before the insulating rope is used, the type and diameter of the insulating rope shall be selected according to the working load, and the 2,500V and above tramegger or insulating detector shall be used for segmented insulating detection (electrode width 2cm, interelectrode width 2cm), and the resistance value shall not be less than 700mΩ. Fig. 3-17 shows the inspection of insulation performance of the insulating rope.

Fig. 3-17　Insulation Performance Inspection of Insulating Ropes

4. Usage of Insulating Ropes

(1) Insulating rope can only be used after passing the test and it is strictly prohibited to use unqualified insulating rope; moisture-proof insulating rope must be selected according to

the environmental humidity; dirty insulating rope is strictly prohibited to be used in live working.

(2) Operators of insulating rope shall wear clean and dry gloves.

(3) Insulating rope must be placed on moisture-proof cloth during use to prevent moisture and dirt.

(4) During the use of insulating rope, the safety distance of corresponding voltage level shall be met. Fig. 3-18 shows the usage of insulating ropes.

Fig. 3-18 Usage of Insulating Ropes

5. Precautions for Storage of Insulating Ropes

Insulating ropes shall be placed in special tool bags, tool cars or special tool cars during transportation to prevent moisture and damage. The insulated tackles shall be stored in a clean and dry special room with good ventilation, infrared light bulbs or dehumidifying equipment. If the storage period exceeds the production date by 12 months, the electrical performance test shall be carried out according to relevant standards. Fig. 3-19 shows the storage of the insulating rope.

Chapter 3
Insulating Tools and Instruments

Fig. 3-19 Storage of Insulating Rope

6. Preventive Test of Insulating Rope

The preventive test of the insulating rope includes the following items.

(1) Electrical test. Mainly including the power frequency withstand voltage test and the operating impulse withstand voltage test. The test cycle is 1 year.

(2) Mechanical test. Mainly including the static tensile test. The test cycle is 2 years.

3.2.2 Insulated Rope Ladder

1. Basic Structure and Function of Insulating Rope Ladder

(1) Basic structure

The side rope and the endless rope of the insulating rope ladder shall be made of mulberry silk or flame-retardant insulating fiber raw materials not lower than the performance of mulberry silk. The cross pedal shall be made of epoxy phenolic laminated glass cloth tube raw materials.

Fig. 3-20 shows the insulating rope ladder.

Fig. 3-20 Insulating Rope Ladder

(2) Basic function

In the process of live working, the insulating rope ladder is used for climbing at high places, working at high places and equipotential working.

2. Classification and Technical Specifications of Insulating Rope Ladders

(1) Classification

① According to the step distance of insulating rope ladder, there are three kinds of insulating rope ladders: 300mm, 400mm and 500mm.

② According to the electrical performance in damp conditions, insulating rope ladder can be classified as conventional type and moisture-proof type insulating rope ladder.

(2) Technical specifications

① It is used as a cross pedal's laminated epoxy glass cloth tube, its outer diameter is 22mm, the wall thickness is 3mm, the length is 300mm, the nozzles at both ends are arc-shaped R1.5, It shall be flat and smooth, with the outer surface coated with insulating paint.

② The rope diameter of the endless rope and the side rope is 10mm, and the laying length of the strand is 32±0.3mm.

③ All parts of the rope ladder head shall shall be subject to anti-corrosion treatment.

3. Inspection items of insulating rope ladders

(1) The connection of each component shall be tight and firm, and the overall structure is good.

(2) The main components of the rope ladder head shall have a smooth surface and no defects such as sharp edges, brushes, notches and cracks.

(3) The connection between the endless rope and the side rope shall be firm and flat, and the twisted synthetic rope strands shall be tightly twisted without looseness or separation. There shall be no fold marks, bulges, crushing, reverse strands, cramps and other de-

fects on each strand of the rope and silk threads of each strand, and there shall be no disorganized or intersecting filaments, wires or strands.

(4) Before the insulating rope ladder is used, the 2500V and above tramegger or insulating detector shall be used for segmented insulating detection (electrode width 2cm, interelectrode width 2cm), and the resistance value shall not be less than 700mΩ.

(5) After the installation of the insulating rope ladder, the impact test of the force must first be carried out to check the installation and bearing capacity of the insulating rope ladder. Fig. 3-21 shows the force check before the insulating rope ladder is used.

Fig. 3-21　Stress Check before Using the Insulating Rope Ladder

(6) Before the insulating rope ladder is used, the test certificate shall be checked for validity period. Insulating rope ladders beyond the validity period cannot be used.

4. Precautions for Use Insulating Rope Ladders

(1) The length of insulating ladder shall be selected according to the height of the operation point.

(2) When the operator enters the equipotential electric field, the somersault tackle and lifting rope shall be hung first; then the insulating rope ladder and backup protection rope shall be lifted through lifting rope.

(3) A backup protection rope must be used to climb the rope ladder. There are two ways to climb, namely side climbing and front climbing.

(4) When the potential is transferred between the climbing rope ladder and the ladder head, attention must be paid to the distance between the exposed part of human body and the electrified body.

(5) For operation with rope ladder or mobile operation with ladder head on the overhead line, only 1 person is allowed to work on the rope ladder or ladder head. The seal of the

ladder head shall be reliably closed before the staff reaches the ladder head and the ladder head begins to move. Otherwise, the protection rope shall be used to prevent the ladder from being unhooked. Fig. 3-22 shows the use of an insulating ladder in equipotential operation.

Fig. 3-22   Use of Insulating Rope Ladder in Equipotential Operation

5. Precautions for Storage of Insulating Rope Ladders

Insulating rope ladders shall be placed in special tool bags, tool cars or special tool cars during transportation to prevent moisture. injury. The insulating rope ladders shall be stored in a clean and dry special room with good ventilation, infrared light bulbs or dehumidifying equipment. If the storage period exceeds the production date by 12 months, the electrical performance test shall be carried out according to relevant standards. Fig. 3-23 shows the storage of the insulating ladders.

Fig. 3-23   Example of Storage of Insulating Rope Ladder

6. Preventive Test of Insulating Rope Ladders

The preventive test of the insulating rope ladders includes the following items.

(1) Electrical test. Mainly including the power frequency withstand voltage test and the operating impulse withstand voltage test. The test cycle is 1 year.

(2) Mechanical test. Mainly including the tensile test of the insulating rope ladder and the dynamic and static load tests of the rope ladder head. The test cycle is 2 years.

3.3 Other Insulating Tools

Other insulating tools include insulated tackles and insulated stringing tools, which are also widely used in live working operations of transmission and distribution lines.

3.3.1 Insulated Tackle

1. Basic Structure and Function of Insulated Tackles

(1) Basic structure

Insulated tackle is composed of baffle, pulley, bearing, central shaft, pulling plate, lifting shaft and hook, as shown in Fig. 3-24. The pulleys are generally made of materials such as nylon, polymethyl methacrylate or engineering plastics. They are formed by car or compression molding methods. The bearings are installed inside, the number of which is determined according to the load; the partition and the reinforcing plate are made of 3240 epoxy glass cloth. The hook is the same as the general tackle, and is forged from high-quality structural steel. The hooks of a few key insulation parts can also be made of 3240 epoxy glass cloth board.

(2) Basic function

It is used as a tool for guiding or carrying a load during hoisting of the live working. In the line operation, the operators shall use the insulated tackle to lift the insulator string, live working tools and instruments and the equipotential operators to enter and exit the electric field, and the operating force steering and the inverting need the insulated tackle. Depending on the work position, hooks of different shapes can be selected; depending on bearing capacity, the tackle blocks bearing corresponding tension can also be selected to save labor.

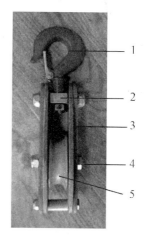

1-hook; 2-hanger shaft; 3-pulling plate; 4-tackle; 5-central shaft

Fig. 3-24 Structure of Insulated Tackle

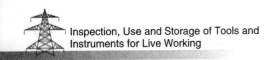
Inspection, Use and Storage of Tools and Instruments for Live Working

2. Classification and Technical Requirements for Insulated Tackles

(1) Classification

Insulated tackles are divided into 15 models, and their model specifications are shown in Table 3-5. The models are expressed by the first letter of pinyin and Arabic numerals: JH indicates insulated tackles, the number after JH indicates the rated load, the number after the dash indicates the number of pulleys, and the last letter indicates the structural characteristics. For example, B - side panel closed type, K - side panel open type, D - short hook type, C - long hook type, J - insulated hook type, X - conductor hook type.

Table 3-5 Model Specifications of Insulated Tackle

Model	Name	Rated load	Number of pulley	Remarks
JH5-1B	Single-wheel closed type insulating tackle	5	1	
JH5-1K	Single-wheel opening type insulating tackle	5	1	
JH5-1DY	Single-wheel multi-purpose hook type insulating tackle	5	1	
JH5-2D	Two-wheel short hook type insulating tackle	5	2	
JH5-2X	Two-wheel conductor hook type insulating tackle	5	2	
JH5-2J	Two-wheel insulating hook type insulating tackle	5	2	
JH5-3D	Three-wheel short hook type insulating tackle	5	3	
JH5-3X	Three-wheel conductor hook type insulating tackle	5	3	
JH10-2D	Two-wheel short hook type insulating tackle	10	2	
JH10-2C	Two-wheel long hook type insulating tackle	10	2	
JH10-3D	Three-wheel short hook type insulating tackle	10	3	
JH10-3C	Three-wheel long hook type insulating tackle	10	3	
JH15-4D	Four-wheel short hook type insulating tackle	15	4	
JH15-4C	Four-wheel long hook type insulating tackle	15	4	
JH20-4D	Four-wheel short hook type insulating tackle	20	4	
JH20-4C	Four-wheel long hook type insulating tackle	20	4	

(2) Technical requirements

① Parts and assemblies can only be used and assembled after inspection according to the drawings.

② After assembly, the pulley shall be flexible on the middle shaft, without jamming and friction against the rim.

③ Hooks and lifting rings shall be flexible on the lifting beam.

④ The cotter pins shall not be bent outward and the excess shall be cut off.

⑤ The lateral bolt is not more than 2mm above the nut.

⑥ The side plate opening has no jamming in the range of opening and closing of 90°.

⑦ All types of insulated tackles shall be tested for electrical performance in accordance with the relevant regulations. After the AC power frequency withstand voltage is conducted for 1 min,

there shall be no heat or breakdown.

⑧ The mechanical performance test shall be based on the load 1.6 times the rated load for 5 minutes without permanent deformation or cracking. The breaking force of the tackle shall not be less than 3 times the rated load.

3. Inspection Items of the Insulated Tackles

(1) The insulation part of the insulated tackle shall be smooth and free from air bubbles, wrinkles and cracks.

(2) The pulley shall be flexible on the middle shaft, without jamming and friction against the rim.

(3) The side plate opening has no jamming in the range of opening and closing of 90°.

(4) Before the insulated tackles are used, the test certificate shall be checked for validity period. Insulated tackles beyond the validity period cannot be used.

(5) Check if the insulated tackle is in compliance with the requirements.

4. Usage of Insulated Tackles

(1) According to the needs of use, select the tonnage of the insulated tackle.

(2) Short hook type or long hook type can be selected depending on the characteristics of the place of use.

(3) The pulley block is labor-saving when replacing the linear insulator, and the distance control is easier. Fig. 3-25 shows the insulated tackle lifting tool.

Fig. 3-25 Insulated Tackle Lifting Tool

5. Precautions for the Storage of Insulated Tackles

Fig. 3-26 Example of Storage of Insulated Tackles

Insulated tackles shall be placed in a well-ventilated, clean and dry special room. If the storage period exceeds the production date by 12 months, the electrical performance test shall be carried out according to relevant standards. Fig. 3-26 shows the storage of the insulated tackles.

6. Preventive Test of Insulated Tackles

The preventive test of the insulated tackles includes the following items.

(1) Electrical test. Mainly including the power frequency withstand voltage test. The test cycle is 1 year.

(2) Mechanical test. Mainly including the tensile test. The test cycle is 1 year.

3.3.2 Insulating Stringing Tool

1. Basic Structure and Function of Insulating Stringing Tool

(1) Basic structure

The insulating ratchet tightener consists of the operating rod, the ratchet mechanism, the insulating tape and the stringing hook, as shown in Fig. 3-27.

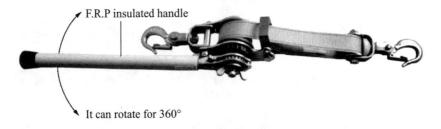

Fig. 3-27 Insulating Ratchet Tightener

(2) Basic function

It is often used to tighten steel-cored aluminum stranded conductor, steel strands and

Chapter 3
Insulating Tools and Instruments

lines.

2. Classification and Technical Specifications of Insulating Stringing Tools

(1) Classification

There are two main types of insulating stringing tools: insulating double groove tighteners and insulating ratchet tighteners. The insulating double hook tightener is shown in Fig. 3-28.

Fig. 3-28 Insulating Double Hook Tightener

(2) Technical specifications

① The insulating rod of the insulating ratchet tightener is made of fiberglass material. It is safe and reliable. The hook can be rotated 360°. It is equipped with a safety lock device. The stainless steel spring can prevent corrosion. The copper connector can prevent moisture.

② The insulating double hook tightener consists of operating rod and outer rubber. The insulating double hook can be operated with the operating rod. The conductor hook has a small hole connecting the operating rod. A spring lock is mounted to the operating rod for safety. The metal part consists of heat-treated aluminum alloy and copper alloy, which is firm and reliable.

In the United States, only the insulating ratchet tightener is produced. The technical parameters are shown in Table 3-6. The parameters of the double-groove insulating tightener and the insulating ratchet tightener produced in Japan are shown in Tables 3-7 and 3-8.

Table 3-6 Technical Parameters of American Insulating Ratchet Tightener

Model	Single tape			Double tape			Handle length (mm)	Mass (kg)
	Load (kg)	Lifting length (m)	Minimum head distance (mm)	Load (kg)	Lifting length (m)	Minimum head distance (mm)		
E153-10	1500	10	22	3000	5	27	20	12
E153-10	1500	10	22	3000	5	27	20	12
EH24-12	2000	15	17	4000	7.5	30	30	12

Table 3-7 Technical Parameters of Japanese Insulating Double Hook Tightener

Model	Service voltage (kV)	Maximum load (kg)	Maximum stringing distance (mm)	Lengthening (mm)	Elongation (mm)	Insulation length (mm)	Mass (kg)	Remarks
3464	5/35	1814	305	1499	1804	660~914	5.4	Double tape
3465	46/69	1814	305	1804	2108	965~1219	5.6	Double tape

Table 3-8 Technical Parameters of Japanese Insulating Ratchet Tightener

Model	Load (t)	Braid size (mm)	Lengthening (mm)	Head (mm)	Handle length (mm)	Mass (kg)	Remarks
N-1000	1	2	420	850	400	3.2	Double tape
N-1500R	1.5	2	450	850	460	4.5	Double tape

3. Inspection Items of Insulating Stringing Tools

(1) Check that the mechanical part is flexible and has no jamming.

(2) When using, please keep the insulation ribbon away from sharp edges and corners to avoid wear or cutting.

(3) Do not use the stringing tool as a load lifting adjustment tool.

(4) Insulating straps or insulating tubes shall not be dirty, burned, etc.

(5) Check the nameplate of the insulating stringing tool and do not overload it.

(6) 为 Before the insulating stringing tools are used, the test certificate shall be checked for validity period. Insulating stringing tools beyond the validity period cannot be used.

Fig. 3-29 shows the insulation inspection of the Japanese insulating ratchet tightener.

Fig. 3-29 Japanese Insulating Ratchet Tightener Inspection

4. Usage of the Insulating Stringing Tools

The insulating stringing tool is used to tighten the conductor during the live mainte-

Chapter 3
Insulating Tools and Instruments

nance of the distribution and overhead line. Select the appropriate insulating stringing tool according to the load of the line. The method of using the insulating ratchet tightener is described as an example. When using, first loosen the insulating ribbon on the tightener, fix the handle end on the cross arm, and clamp the conductor with the wire clamp at the other end. When the insulating ribbon passes through the insulator part, the special wrench is pulled. Due to the anti-reverse action of the ratchet, gradually wind the insulating ribbon around the ratchet roller to tighten the conductor and fix the tightened conductor on the insulator. Then loosen the pawl, loosen the insulating ribbon, release the wire clamp, and finally wrap the insulating ribbon around the roller of the ratchet. The usage of the insulating tightener is shown in Fig. 3-30.

Fig. 3-30   Usage of Insulating Tightener

5. Precautions for Storage of Insulating Stringing Tools

Insulating stringing tools shall be kept on shelves in a well-ventilated, clean and dry special room. If the storage period exceeds the production date by 1 year, the electrical performance test shall be carried out according to relevant standards. Fig. 3-31 shows the storage of the insulating ratchet tightener and insulating double hook tightener.

1-insulating ratchet tightener; 2-insulating double hook tightener

Fig. 3-31   Storage of the Insulating Ratchet Tightener and Insulating Double Hook Tightener.

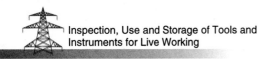

6. Test of Insulating Stringing Tools

The preventive test of the insulating stringing tools includes the following items.

(1) Electrical test. Including the power frequency withstand voltage test. The test cycle is 1 year.

(2) Mechanical test. Including the tensile test. The test cycle is 1 year.

Chapter 4

Special Vehicles for Live Working

In the live working of transmission and distribution lines, especially in the live working of distribution lines, with the introduction of foreign live working equipment and the advancement of domestic live working technology, the manufacturers of live working tools and instruments have gradually developed the compound tools combining live working tools and automobiles, such as aerial devices with insulating booms, live working tool cars, bypass cable aerial devices with insulating booms, mobile box-type substation vehicles and other special vehicles for live working, which have initially realized the mechanization of live working, effectively reduced the labor intensity of live working, and improved the level of safe operation and work efficiency. This chapter will focus on aerial devices with insulating booms, bypass work vehicles, load switch vehicles, mobile box-type substation vehicles, and live working tool cars.

4.1 Aerial Devicc with Insulating Booms

Aerial devices with insulating booms are a special vehicle that performs equipotential operation with convenient transportation in complicated wiring conditions. They can realize the main insulation necessary for live working through its insulating boom and work bucket, and can perform live working at heights on the power line of 10kV and above. High-altitude aerial devices with insulating booms with only insulating work buckets are generally not included in the range of aerial devices with insulating booms. In actual work, due to various factors such as line and pole and tower positions, traffic conditions, supporting chassis, use efficiency, product price and so on, the live working on the transmission line is rarely carried out with aerial devices with insulating booms, while they are usually used in the maintenance work of 10kV, 35kV and 66kV urban power distribution lines, so the aerial devices with insulating booms play a very important role in the live working of the distribution lines.

4.1.1 Basic Structure and Function of The Aerial Devices with Insulating Booms

Aerial devices with insulating booms are generally modified from automobile engine and chassis. The rear part of the vehicle is equipped with a moving turntable and a hydraulic bucket arm device. The arm adopts a folding or telescopic structure, and the front end is provided with a square or circular insulating bucket. The entire arm is mounted on a turntable that can be rotated 360°. In order to increase the stability of the vehicle, thereby maximizing the length of the arm, it is generally equipped with hydraulic legs, the structure of which is shown in Fig. 4-1.

The work bucket, work arm, control oil line and arm joint of the aerial devices with insulating booms can meet certain insulation performance indicators, and are equipped with grounding device. The insulating arm is generally made from glass fiber reinforced epoxy resin material, which is generally formed into a cylindrical or rectangular cross-section structure, and has the advantages of light weight, high mechanical strength, good electrical insulation performance, strong water repellency, etc. It provides insulation protection relative to ground for the human body during live working. The insulating bucket in divided into single layer bucket and double layer bucket. The outer layer bucket is generally made from fiber glass epoxy plastics, and the inner layer bucket is made from PTFE material.

1-insulating work bucket; 2-moving turntable; 3-hydraulic legs; 4-insulating arm; 5-grounding wire

Fig. 4-1 Structure of Aerial Devices with Insulating Booms

4.1.2 Classification of Aerial Devices with Insulating Booms

(1) According to the structure of the work arm, the aerial devices with insulating booms can be divided into folding arm type, straight arm type, multi-joint arm type, vertical lifting type and hybrid type. The most commonly used are straight arm type and folding type, as shown in Fig. 4-2.

(2) According to the lifting height, the aerial devices with insulating booms can be divided into 6m, 8m, 10m, 12m, 16m, 20m, 25m, 30m, 35m, 40m, 50m, 60m, and 70m devices, etc.

(3) According to the operating line voltage class, the aerial devices with insulating booms can be divided into 10kV, 35kV, 46kV, 63 (66) kV, 110kV, 220kV, 330kV, 345kV, 500kV, and 765kV devices, etc. China's aerial devices with insulating booms are typically used on 10kV, 35kV and 66kV lines.

1-straight arm type aerial devices with insulating booms;
2-folding arm type aerial devices with insulating booms
Fig. 4-2   Aerial Devices with Insulating Booms

4.1.3 Technical Requirements for Aerial Devices with Insulating Booms

1. Working Conditions

(1) The wind speed does not exceed 10.8 m/s.

(2) The ambient temperature is -25 to 40°C.

(3) The relative humidity does not exceed 90%.

(4) Requirements for areas above 1,000m sea level: The selected chassis power shall be adapted to the plateau driving and operation. For every 100m increase in altitude, the insulation level of the insulator shall be increased by 1%.

(5) The ground is firm and flat, and the legs do not sink during the operation.

(6) The turntable plane is level.

(7) The electrical insulation performance of the insulating work bucket must meet the corresponding requirements.

(8) The surface of the insulating work bucket is smooth, without pits or pitted surface, and with strong water repellency.

(9) The height of the insulating work bucket should be between 0.9 and 1.2m.

(10) The rated load capacity of the work bucket shall be clearly indicated on the insulating work bucket.

2. Insulating Arm Requirements

(1) The electrical performance test of the insulating arm must meet the corresponding requirements.

(2) The surface of the insulating arm is smooth, without pits or pitted surface, and with strong water repellency.

(3) The minimum insulation length of the insulating arm of aerial devices with insulating booms of each voltage class should not be less than the corresponding requirements.

3. Vehicle Requirements

(1) For the aerial devices with insulating booms with only the insulating arm insulated between the grounding part and the work bucket, the electrical insulation performance of the whole vehicle shall meet the corresponding requirements.

(2) For the aerial devices with insulating booms with up and down operation function and automatic balance function, the electrical insulation performance of the whole vehicle also needs to meet the requirements of partial routine test leakage current of less than 200μA.

4. Insulating Hydraulic Oil Requirements

The hydraulic oil used to withstand the corresponding voltage of the live working line shall be tested for the breakdown strength.

5. Insulating Property Requirements

The rated voltage of the aerial devices with insulating booms shall be clearly indicated on the instructions and nameplate. Each vehicle shall be subjected to routine insulation performance tests before leaving the factory.

4.1.4　Usage of Aerial Devices with Insulating Booms

Proper use and operation of the aerial devices with insulating booms is the basis for personal safety and vehicle safety.

1. Engine Start and Operation of the Power Takeoff and Legs

(1) Apply the hand brakes and lay the triangular blocks.

(2) Make sure the transmission lever is in the correct position.

(3) Depress the clutch pedal to the bottom and start the engine.

(4) Depress the clutch pedal and move the power takeoff switch to the "ON" position.

(5) Slowly release the clutch pedal.

(6) Throttle high and low speed operation.

(7) Horizontal leg operation. In the change-over levers, select the change-over lever of the horizontal leg to be operated, and switch it to the "horizontal" position; when the "retraction" lever is pulled to the "extending" position, the horizontal leg will extend, as shown in the Fig. 4-3.

Fig. 4-3　Operating Rod of Legs of Aerial Devices with Insulating Booms

(8) Vertical leg operation. Operate the legs correctly to prevent the legs from running out or contracting, causing damage to the vehicle. The legs of the aerial devices with insulating booms during working are shown in Fig. 4-4.

Fig. 4-4　Legs of Aerial Devices with Insulating Booms During Working

(9) The retraction operation method is: retract the legs to the original state in the order of "vertical legs - horizontal legs"; retract them in the reverse order of items (8) and (9); after retraction, return the operating rods to the middle position.

2. Connect Grounding Wire

The grounding wire on the outgoing line coil is retracted and the grounding wire of the aerial devices with insulating booms is reliably connected to the grounding lead of the pole and tower. Fig. 4-5 shows the grounding wire installed for the aerial devices with insulating booms.

Fig. 4-5　Installation of Grounding Wire

3. Upper Operation (Work Bucket Operation)

(1) Work arm operation

① Lower arm operation (arm lifting operation), Fig. 4-6 shows the operating panel inside the work bucket.

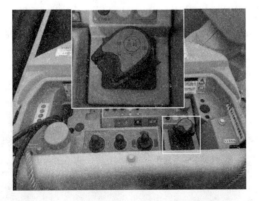

Fig. 4-6　Operating Panel in the Work Bucket

② Rotary operation.

③ Upper arm operation (flexible operation).

(2) Work bucket swing operation

Move the bucket swing lever in the direction of the sign arrow to swing the work bucket to the right or left.

(3) Emergency stop operation

When the emergency stop lever is turned on, the upper operations will stop, but the engine will not stop. Emergency stop operation is required in the following cases:

① The operator on the work bucket needs to stop the action of the work arm in order to avoid dangerous situations;

② The operation control is out of control.

(4) Small boom operation

As shown in Fig. 4-7 and Fig. 4-8, some aerial devices with insulating booms are equipped with small boom for the work bucket. For the operation, refer to the relevant manual.

Fig. 4-7　Operation of Small Boom Inside the Bucket

Fig. 4-8　Small Boom Device Inside the Bucket

(5) Auxiliary device operation

The oil inlet, return oil and drain connections are reliably connected to the corresponding joints of the hydraulic tool, and the auxiliary lever is switched to the "tool" position.

4. Lower Operation (Operation of the Operating Panel at the Turntable)

(1) Work arm operation

In the state where the work arm is retracted to the bracket, the swing operation of the work arm is not allowed, and the operating panel of the work arm at the turntable is as shown in Fig. 4-9.

(2) Emergency stop operation

Use the emergency stop lever for emergency stop operation. When the emergency stop lever is turned on, all operations of the upper and lower parts are stopped.

(3) Emergency pump operation

The operation button of the emergency pump at the turntable is shown in Fig. 4-10.

Fig. 4-9 Work Arm Operating Panel at the Turntable

Fig. 4-10 Operating Button of the Emergency Pump at the Turntable

Chapter 4
Special Vehicles for Live Working

4.1.5 Maintenance, Repair and Inspection of Aerial Devices with Insulating Booms

1. Maintenance of the Aerial Devices with Insulating Booms

Maintenance during transportation. The work bucket must be returned to the driving position. For the aerial devices with insulating booms with boom, the boom shall be removed or fully retracted, the upper and lower arms shall be restored to their respective support frames, and must be fixed firmly to prevent impact and damage due to shaking during transportation. During the travel of the aerial devices with insulating booms, the hydraulic operating system of the two arms must be shut off to prevent the hydraulic balance device of the working bucket from swinging back and forth.

When the aerial devices with insulating booms is exposed to the open air environment, the insulation tolerance level will be reduced for the insulation properties of the insulating working bucket and boom affected by rain, road dust, corrosion and other air pollution. Long-term UV exposure also affects its insulation properties. Therefore, a moisture-proof protective cover shall be used for protection during transportation.

2. Maintenance of Aerial Devices with Insulating Booms

(1) General requirements. A set of periodic inspection procedures for aerial devices with insulating booms must be established, including detailed visual inspection and insulation strength tests.

(2) Cleaning. Where the surface of the insulating part is contaminated with small dirt, the dirt can be wiped with a lint-free cloth. Where the insulating part is very dirty, high-pressure hot water (with the temperature not exceeding 50°C and the pressure not exceeding 690 kPa) can be used to rinse.

(3) Silicon coating. The insulating surface should be cleaned and dried before applying silicon.

3. Check

The inspection of the aerial devices with insulating booms includes visual inspection and function test.

The visual inspection of the insulating parts shall be carried out after the outer surface is cleaned; the upper and lower booms, working bucket, and crane jib must be wiped with a lint-free cloth, and can be gently wiped with a clean cloth dampened with a little isopropyl alcohol or other suitable solvent if necessary. The visual inspection is mainly for structural damage. Structural damage includes insulation breakdown due to impact, bare fiber, holes and grooves, crack marks caused by collisions with sharp objects such as branches and poles, and cracks, bulges, etc. that occur at the upper and lower arm joints or near the steel joint portion due to overloading.

(1) Inspection before work

The purpose of the inspection is to eliminate defects and malfunctions that may have been left in the pre-work of the aerial devices with insulating booms and during inventory. The visual inspection is mainly carried out by the operator, and supplemented with function test, aiming to listening to whether the components have abnormal noise.

Inspect and confirm that the insulation test of the insulation parts of the aerial devices with insulating booms is within the validity period.

① Entire inspection

The visual inspection and function test must be completed before official start of work, and the results of the inspection shall be recorded in the form of a chart.

Visual inspection: Inspect the damage of the surface of the insulation parts, such as cracks, insulation peeling, and deep scratches.

Function test: After starting the aerial devices with insulating booms, the lower control system should be used to operate the expansion arm expansion, rotation and work bucket lifting cycle in the case where the working bucket is unmanned, to check for fluid seepage, hydraulic cylinder leaks, abnormal noise, failure, oil leakage, unstable motion or other faults.

In order to ensure the safety, the flexibility and reliability of each power supply and emergency braking system should be checked and operated, and visual and audible alarms should be verified.

For aerial devices with insulating booms with ultra high voltage class and above, the grading ring must be inspected when it is used in the equipotential working part to conform the equipotential point and perform leakage current test.

② Working bucket inspection

Inspect the bottom plate of the working bucket for dirt or other objects that may damage the work bucket, or objects that interfere with the bottom plate and the conductive shoes in the case of equipotential operation.

Inspect the mechanical damage of working bucket, whether there are holes, cracks or delamination, and wipe the working bucket with a soft, lint-free cloth with suitable solvent, and remove the debris and peeling material from the working bucket.

(2) Weekly inspection

The aerial devices with insulating booms should be inspected weekly.

① Entire inspection. Including the visual inspection and function test.

② Inspection of working bucket. The inner working bucket must be taken out from the outer working bucket to clean the dirt up. In case of damage or peeling, it is necessary to identify that the damage or peeling is caused by mechanical damage or chemical factors. Any mechanical damage will reduce the wall thickness of the inner working bucket. Where the wall thickness is less than the minimum wall thickness recommended by the manufacturer, the inner and outer working buckets must be electrically tested before reuse.

(3) Regular inspection. The regular inspection cycle can be determined based on manufacturer's recommendations and other influencing factors such as operating conditions, maintenance levels, and environmental conditions. In general, the maximum period of normal periodic inspection should be 1 year.

4.1.6 Preventive Test for Aerial Devices with Insulating Booms

Preventive test for aerial devices with insulating booms includes the power-frequency voltage-withstand test and leakage current test for insulating working bucket, power frequency voltage withstand test and leakage current test for insulation arm, the power-frequency voltage-withstand test and leakage current test for the entire aerial devices with insulating booms, and the breakdown strength test for insulating hydraulic oil, and the test period should be half a year.

4.2　Bypass Work Vehicle and Load Switch Vehicle

The method for implementing distribution network without power failure operation by using bypass equipment has been widely used in distribution lines at home and abroad and achieved good results in improving the reliability of power supply. A set of bypass equipment has a large number of bypass cables, switches, pulleys, link fittings, etc., so all parts of the bypass operation could not be customized, and some equipment were easily damaged during transportation in the past. In addition, the application of bypass equipment in emergency repair and distribution line uninterrupted work is greatly limited due to the road traffic restrictions. The successful development and application of the bypass work vehicles have solved many difficult problems at the work site. In addition, the cooperation of the bypass work load switch vehicles, the preparation time in the application process of the bypass operation equipment is greatly shortened, providing favorable conditions for vigorously carrying out the power distribution line without power failure.

The bypass work vehicle is shown in Fig. 4-11 and the load switch vehicle is shown in Fig. 4-12.

Fig. 4-11　Bypass Work Vehicle

Fig. 4-12　Load Switch Vehicle

Chapter 4
Special Vehicles for Live Working

4.2.1 Basic structure of Bypass Work Vehicle and Load Switch Vehicle

1. Basic Structure of Bypass Work Vehicle

The bypass work vehicle mainly includes a vehicle platform, a cable retracting device, a component storage box, etc., as shown in Fig. 4-13. The vehicle platform includes a vehicle chassis, a vehicle body (compartment) structure, etc., which is a transport carrier of the bypass work vehicle. The cable retracting device is mainly composed of a circular track, a triplex overhead cable reel, a reel driving mechanism, a lifting device, etc. The component storage box is used for stationary storage (excluding the bypass flexible cable) bypass work equipment components, including bypass load switch, adapter cable, cable connector, etc. The bypass work vehicle adopts a subdivision design with an independent cab, component storage box, cable retracting device, and tool kit.

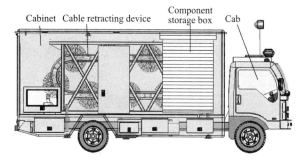

Fig. 4-13　Overall Structure of the Live Working Bypass Work Vehicle

2. Basis Structure of Load Switch Vehicle

A chassis of the engineering vehicle is adopted, and a set of electro-hydraulic drive cable reel and a bypass load switch are arranged in the vehicle, with the structure shown as Fig. 4-14.

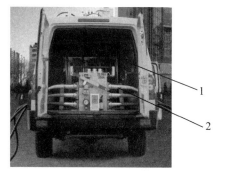

1 - Cable reel device; 2 - Bypass load switch
Fig. 4-14　Structure of Load Switch Vehicle

4.2.3 Main Technical Requirements for Bypass Work Vehicle

1. Working Conditions

The bypass work vehicle shall be in normal working under the following condition:

(1) The altitude is not more than 1,000m;

(2) Ambient temperature: −40 to 40°C;

(3) Relative air humidity is not more than 95% (+25°C);

Where the vehicle is used in special conditions, it is necessary to design according to the requirements of use.

2. Functional Requirements

(1) Bypass flexible cable should be loaded in a fixed manner. The whole vehicle is a van-type engineering vehicle, and a cable retracting device is arranged in the vehicle compartment. The cable retracting device is mainly composed of a circular track, a triplex overhead cable reel, a reel driving mechanism, a lifting device, etc. The cable retracting device should be loaded with no less than 18 sets of bypass flexible cables in a fixed manner, as shown in Fig. 4-15.

Fig. 4-15 Cable Retracting Device

(2) Component storage box The component storage box is used for stationary storage bypass work equipment components, including bypass load switch, adapter cable, cable connector, etc. Various bypass working parts are classified and designed, and the special tooling fixtures are designed for reliable fixing. The molding die for storing small parts should be marked with quantity to realize the fixed management of each part to prevent the tools from colliding and bumping during transportation.

(3) Function of manually receiving and releasing the bypass flexible cable. The triplex

overhead cable reels are installed in a parallel manner in the circular track. By driving each set of reels with an electric or hydraulic mechanism, the reels can be moved one by one to the position specified for receiving and releasing bypass flexible cables at the tail of the compartment. Each set of reels should be automatically locked to prevent it from moving during driving. A set of reel devices shall be provided with continuous, jog, three-phase simultaneous and single-phase retracting functions according to the working needs at the position for receiving and releasing bypass flexible cables. The cable reel should be provided with positioning lock function to prevent the cable reel from moving and rotating during driving. The receiving and releasing of cable should be realized by the configured wired remote control device, and the control cable length should be not less than 3m. The mechanism should be designed with sufficient space for maintenance.

(4) The function of quickly disassembling the cable reel on site. The equipped on board hoisting device can hang the cable reel outside the compartment. A set of cable reels can be quickly split into three single reels for easy transportation and maintenance of reel. The rated lifting capacity of the boom should not be less than 500kg.

(5) The function of lighting for work site at night. The outer top of the cab is provided with a on-board lift lighting device, equipped with omni-direction steering pan-tilt for lighting the work site at night. The vehicle chassis battery DC 24V power supply should be used for the lighting power supply, LED lights and other energy-saving lamps should be used for lighting fixtures, and the illumination shall meet the site work requirements.

(6) The function of vehicle storage support. The bottom of the compartment should be equipped with 4 hydraulic vertical telescopic legs, and the tires should not be carried and can withstand the total mass of the vehicle and the cargo after the legs are extended. The control system of the hydraulic telescopic legs should be installed in a position easy to operate.

(7) Extended function. As technology advances and matures, new features can be added.

(8) Modification. The bypass work vehicle should be modified with a finalized vehicle.

3. Auxiliary System

(1) Electrical system. A power master switch shall be provided in the circuit system and shall be arranged in a position convenient for the operator to operate.

(2) Illumination system. The illumination of the bypass work vehicle includes vehicle body lighting, work lighting, and emergency lighting.

4.2.3 Main Technical Requirements for Bypass Work Load Switch Vehicle

The bypass cable and connector are rated at 200A, so the rated current of the bypass load switch shall be not less than 200A, and the bypass load switch is shown in Fig. 4-16. According to the rated current of 200A, combining the technical requirements of 10kV three-phase load switch, the performance parameters are shown in Table 4-1.

Fig. 4-16 Bypass Load Switch

Table 4-1 Performance Parameters of Bypass Load Switch

S/N	Parameter		Standards	S/N	Parameter		Standards
1	Rated voltage (kV)		12	11	Impulse withstand voltage (kV)	To ground	75
2	Rated current (A)		200			Interphase	75
3	Rated frequency (Hz)		50			Between open contacts	85
4	Rated breaking load current (A)		200	12	3s thermal stability current (kA)		16
5	Number of rated drop-out load current		20	13	Dynamic stability current (kA)		40
6	Rated disconnect charging current (A)		20	14	Drop-out contact resistance of the switch (μΩ)		<200
7	Number of rated disconnect charging current		20	15	Three-phase segmentation on-synchronism (ms)		<5
8	Breaking period (ms)		20	16	Mechanical life		3000
9	Closing short circuit current capability (kA)		40	17	Annual leak rate		<0.5%
10	Power frequency withstand voltage (kV)	To ground	42	18	Protection grade		IP68
		Interphase	42				
		Between open contacts	48				

4.2.4 Precautions for Use of Bypass Work Vehicle and Load Switch Vehicle

1. Bypass Work Vehicle

(1) Bypass work vehicle is a special transport vehicle for the transfer bypass equipment, with high cost, and it is necessary to strictly prevent the vehicle from being crushes and bumped during transportation and use.

(2) Special personnel shall be designated for the bypass work vehicle, and shall not be operated by the untrained or licensed personnel.

(3) During parking and using, the hydraulic auxiliary legs should be supported in time to avoid deformation and damage of the wheel load.

2. Load Switch Vehicle

(1) In order to make the bypass load switch and the bypass flexible cable easy to connect, the external interface shall be plug-in interface, which is easy and reliable to connect. The indicative marks for closing and opening of the switch shall be clear and obvious. No matter the manner by which the switch is closed, it is necessary to lock after closing. It is strictly forbidden to open the switch without operation and without any measures. It also has a pressure display and a decompression locking device.

(2) The bypass load switch has a nuclear phase function, so that the voltage on both sides of the switch can be phased in the opening position to avoid phase-to-phase short circuit caused by incorrect phase sequence closing during bypass circuit assembly. Otherwise, the bypass load switch shall have a terminal for checking the nuclear phase function, and the terminal voltage shall be no more than 800V. The phasing device or phasing detector must be provided with obvious in-phase or out-of-phase indication signals, audible alarm signals, etc.

(3) When the power side bypass load switch is pulled off, the bypass load switch is disconnected by the capacitor current of the no-load bypass cable, that is, the charging current. The charging current of the no-load cable is related to the length, wire diameter, voltage, etc. of the cable. According to the testing analysis of the China Electric Power Research Institute, when the single-core (phase) cross-sectional area of the cable line is not more than 300mm^2 and the length is not more than 3km, the maximum single-phase steady-state capacitive current that is disconnected should be 5A. In general, the length of the overhead bypass work line is relatively short, and the cross-sectional area of the bypass flexible cable is also small,

generally 35 to 50mm^2. In this case, when the power side bypass load switch is opened, the disconnected charging current should be much less than 20A. Therefore, the life of the bypass switch cannot be simply measured by the number of times the load current or the charge current is disconnected. Of course, it is not equivalent to the mechanical life and its life is also related to the use, preservation and maintenance.

4.2.5 Test and Storage of Bypass Work Vehicle and Load Switch Vehicle

1. Test For Vehicle

(1) Tests of live working bypass vehicle include type test, predelivery test and acceptance test. The type test and predelivery test should be carried out by the manufacturer.

(2) The acceptance test should be carried out jointly by the user and the manufacturer, including visual inspection, power switching test, cable retraction, switch switching, lighting system and vehicle storage support functions.

(3) In addition, the power frequency withstand voltage test between the fractures in the open state and the phase and phase power frequency withstand voltage test in the closed state should be performed periodically on the bypass load switch of the load switch vehicle. Before use, the insulation resistance detector shall be used to check the insulation resistance between the phase-earth, phase-phase and open contacts of the switch to ensure that the insulation resistance are greater than 700MΩ.

2. Storage of Vehicles

(1) The bypass work vehicle and the load switch vehicle shall be stored in a dedicated warehouse. The storage conditions of the warehouse should meet the conditions of anti-theft, moisture and ventilation, and equipped with corresponding fire-fighting facilities.

(2) For vehicles stored in the warehouse, such moving parts as doors and windows and drawers shall be closed, and all devices shall be securely fastened or tied.

(3) Vehicles shall be regularly maintained in accordance with the motor vehicle manual.

4.3 Mobile Box-type Substation Vehicle

The mobile box-type substation vehicle is also called the load transfer vehicle, which is a mobile power supply equipped with a box-type substation. A set of high-voltage load switches and low-pressure air switches are installed on the high and low pressure sides of the box-type substation. Through the load transfer, non-interruption maintenance of the distribution transformer on the pole can be achieved and the power can be temporarily taken from the high voltage line to supply power to the low voltage user.

4.3.1 Classification, Structure and Function of Mobile Box-type Substation Vehicle

1. Classification of Mobile Box-type Substation Vehicle

(1) Classified by automobile product. Classified according to Type Terms and Definitions forMotor Vehicles and Trailers (GB/T3730.1—2001), the mobile box-type substation vehicle belongs to the special work vehicle. Classified according to Terms Marks and Designation for Special Purpose Vehicles and Special Trailers (GB/T17350—2009), and the mobile box-type substation vehicle belongs to the box-type automobile. The mobile box-type substation vehicle is shown in Fig. 4-17.

(2) Classified by vehicle configuration equipment. The mobile box-type substation vehicle is divided into basic type and extended type according to the vehicle configuration equipment.

① Basic type. The vehicle is used to carry out simpler power distribution lines and temporary power supply projects for cables.

② Extended type. The vehicle is used to carry out more complicated distribution lines and temporary power supply projects for cables.

Fig. 4-17 Example of Mobile Box-Type Substation Vehicle

2. Basic Structure of Mobile Box-type Substation Vehicle

A mobile box-type substation vehicle is mainly composed of a vehicle platform, an on-board device, an auxiliary system, etc.

(1) Vehicle platform. The vehicle platform includes a vehicle chassis, a vehicle body (compartment) structure, etc., which is a transport carrier of the mobile box-type substation vehicle.

(2) On-board device. The on-board device is mainly composed of transformers, bypass load switches, bypass flexible cables, and low-voltage power distribution panels.

(3) Auxiliary systems. The auxiliary systems of the mobile box-type substation vehicle mainly include electrical, lighting, grounding, hydraulic, and safety protection systems. The internal structure of the mobile box-type substation vehicle is shown as Fig. 4-18. The working diagram for bypass work of mobile box-type substation vehicle is shown in Fig. 4-19.

High voltage cable reel and high voltage quick plug

HV ring main unit

Rear view of the compartment

Installation method and protection of transformer

Fig. 4-18　Internal Structure of Mobile Box-Type Substation Vehicle

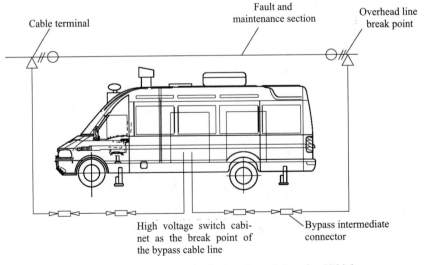

Fig. 4-19　Bypass Work of Mobile Box-Type Substation Vehicle

Chapter 4
Special Vehicles for Live Working

3. Basic Function of Mobile Box–type Substation Vehicle

The mobile box-type substation vehicle should have the uninterrupted power supply capability for conveying and converting electric energy. The main functions are shown in Table 4-2. As technology advances and matures, new features can be added.

Table 4-2 Main functions of Mobile Box-Type Substation Vehicle

Equipment name	S/N	Function/Item	Basic	Extended type
Bypass flexible cable reel	1	Manual reeling	●	●
	2	Mechanical or hydraulic reeling	○	○
LV cable reel	1	Manual reeling	○	●
	2	Mechanical or hydraulic reeling	○	●
Phase detection	1	HV side phase detection	●	●
	2	LV side phase detection	●	●
	3	Automatic phase detection	○	○
LV side reverse phase sequence	1	Manual reverse phase sequence	●	●
	2	Automatic reverse phase sequence	○	○
HV and LV sides outlet	1	HV side outlet quick interface	○	●
	2	LV side outlet quick interface	○	○
Bypass load switch and ring main unit	1	The bypass load switch shall be provided with a reliable safety locking mechanism	●	●
	2	Be equipped with a ring main unit with at least one inlet and two outlets	○	○
HV and LV protection	1	HV protection	○	●
	2	LV protection switch rating > 2/3 of transformer capacity	○	●
Auxiliary equipment	1	Hydraulic vertical telescopic hydraulic support	●	●
	2	Emergency lighting	○	○

Note: ● indicates the function that should be available; ○ indicates the function that should be available as needed.

Fig. 4-20 shows temporary power taken from overhead lines to power the mobile box-type substation vehicle, and Fig.4-21 shows temporary power taken from the ring main unit to the mobile box-type substation vehicle.

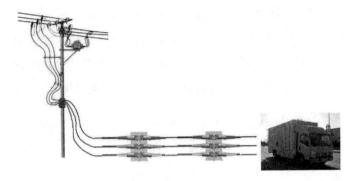

Fig. 4-20 Temporary Power Taken from Overhead Lines to Power the Mobile Box-Type Substation Vehicle

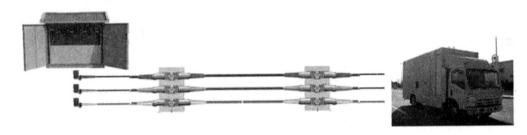

Fig. 4-21 Temporary Power Taken from Ring Main Unit to Power the Mobile Box-Type Substation Vehicle

4.3.2 Technical Requirements of Mobile Box-type Substation Vehicle

1. Working Conditions

The mobile box-type substation vehicle shall be in normal working under the following condition:

① Altitude: ≤ 1000m;

② Ambient temperature: –40 ~ 40°C;

③ Relative air humidity: ≤95% (+25°C).

Where the vehicle is used in special conditions, it is necessary to design according to the requirements of use.

2. Functional Requirements

(1) Overall requirements

① Transportation. Mobile box-type substation vehicle shall be of good maneuverability, anti-vibration, anti-shock, dust-proof and other performance, and shall meet the requirements of reliable transportation vehicle equipment.

② Modification. The mobile box-type substation vehicle should be modified with a finalized vehicle. The modification of mobile box-type substation vehicle shall meet the Trucks - Engineering Approval Evaluation Program (GB/T1332—1991), Bus and Coach - Engineering Approval Evaluation Process (GB/T13043—2006), Light Buses—Engineering Approval Evaluation Program (GB/T13044—1991), The Approval Test Procedure for Special Purpose Vehicles (QC/T252—1998) and other requirements for automotive modification technical standards.

③ Production. In addition to complying with the provisions of this document, the production of mobile box-type substation vehicle shall also meet the relevant laws promulgated by the state.

(2) Vehicle equipment

① General requirements

a. Maintenance inspection On-board devices shall be regularly calibrated, maintained or inspected in accordance with relevant regulatory requirements or its instructions.

b. Performance and parameters. In addition to meeting the requirements, performance and parameters of on-board devices shall also comply with the relevant technical standards or regulations.

c. Wiring method. The high-voltage side wiring is a set of incoming and outgoing lines, of which one set is used to connect the transformer, and the other set can be used to transfer the load. The low-voltage side outlet is the output of two groups of loads (one master and one standby).

d. Earthquake resistance. On-board elements or components shall be securely mounted and have good shock resistance. The earthquake resistance of on-board devices shall meet the relevant provisions of Environmental Conditions Existing in the Application of Electric and Electronic Products - Section 5: Ground Vehicle Installations (GB4798.5—2007).

② Distribution transformer. The distribution transformer shall comply with the provisions of Standard for Hand-over Test of Electric Equipment Electric Equipment Installation Engineering (GB50150—2006), and the capacity can be three-phase oil-immersed direct-cooling coil non-excitation voltage-regulating distribution transformer or dry-type transformer with specifications of 250-630kVA.

③ Bypass load switch. The bypass load switch shall comply with the 10kV Bypass

Equipment Conditions (Q/GDW249—2009), fully insulating and fully sealed and interconnected with ring main units and branch boxes, with good operational performance (mechanical life ≥ 3,000 cycles) and arc-extinguishing, with a reliable safety locking mechanism.

④ Bypass flexible cable. The bypass flexible cable shall comply with the 10kV Bypass Equipment Conditions (Q/GDW249—2009), which can be bent and reused.

⑤ Bypass connector. The bypass connector includes the incoming connector device, terminal connector, intermediate connector, T-connector, etc., which shall comply with the provisions of 10kV Bypass Equipment Conditions (Q/GDW249—2009). The connection joint requires compact structure, convenient docking, and has a firm and reliable function to prevent automatic falling off of the lock, so that it is convenient to change the separation state in the docking state.

⑥ Bypass cable connection accessory. Bypass cable connection accessory includes touchable terminal elbow cable plugs, detachable cable connectors, auxiliary cables, down cables, etc., and shall comply with the provisions of 10kV Bypass Equipment Conditions (Q/GDW249—2009). The model shall be matched with flexible cable, live working arc extinguishing switch, box transformer, ring network cabinet, branch box and high and low voltage inlet cabinet.

⑦ Low voltage power distribution panel. The low voltage power distribution panel shall comply with the provisions of Low-voltage Switchgear and Controlgear Assemblies—Part 1: Type-tested and Partially Type-tested Assemblies (GB7251.1—2005), and the switchgear, measuring instrument, protection device and auxiliary equipment required for the low-voltage circuit shall be arranged and installed in the metal cabinet according to a certain wiring manner.

⑧ Low voltage flexible cable. The low voltage flexible cable shall comply with the provisions of Wire and Cable Rubber Insulation and Rubber Sleeves Part 1: General (GB75941—1987), which can be bent and reused.

⑨ Ring main unit. The ring main unit shall comply with the provision of Common Technical Requirements of High Voltage Switchgear and Controlgear Standards (GB11022—2011), and shall be divided into separate isolation chambers with a metal closure, including load switch room (circuit breaker), busbar room, cable room and control instrument room, of which the load switch room (circuit breaker), busbar room and cable room are provided with independent pressure relief channels.

⑩ Configuration requirements. The configuration requirements include the typical equipment configuration for basic mobile box-type substation vehicle and the extended mobile box to change the typical equipment configuration, according to the relevant national standards.

(3) Auxiliary system

① Circuit and control. A power master switch shall be provided in the circuit system and shall be arranged in a position convenient for the operator to operate.

② Lighting system. The illumination of the mobile box-type substation vehicle includes vehicle body lighting, work lighting, and emergency lighting.

③ Grounding system. The mobile box-type substation vehicle shall be provided with a dedicated centralized grounding point with an obvious grounding mark. The grounding resistance shall be greater than 4Ω, and the distance between the protective grounding and working grounding shall be 5m or above.

The grounding wire shall have sufficient cross-section and length. The cross section of the main ground return ground wire shall meet the requirements of heat capacity and wire voltage drop.

④ Hydraulic system. Where the mobile box-type substation vehicle is parked in the garage or the unit is working, the hydraulic system can provide support for protecting the tires and the axle. The four hydraulic legs are provided with locking devices, and each leg can operate independently.

⑤ Safety protection, warning, and protection.

a. Safety protection: Where the moving parts such as hydraulic, mechanical, electric and other moving parts of the mobile box-type substation vehicle have obvious influence on the safety of load bearing and transmission, locking protection equipment shall be provided. The locking protection equipment shall be flexible and reliable.

Where the movable parts that can be moved manually have obvious safety impact on transportation, fixing, etc., the limit locking device shall be provided. The locking device shall be convenient for workers to operate, with flexible action and reliable limit.

b. Warning: Mobile box-type substation vehicle shall be provided with audible and visual alarm and can be controlled by the operator on the vehicle. The equipment area can be equipped with smoke alarm and toxic gas alarm according to the need for electrification detection.

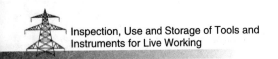

c. Protection: Mobile box-type substation vehicle should be provided with common safety tools and protective equipment. Fire-fighting equipment should be equipped in different functional areas such as the cab and equipment area. Fire-fighting equipment should be installed firmly and easily accessed.

4.3.3 Test and Storage of Mobile Box-type Substation Vehicle

1. Test of Mobile Box-type Substation Vehicle

(1) Tests of live working special vehicle include type test, predelivery test and acceptance test. The type test and predelivery test should be carried out by the manufacturer.

(2) The acceptance test should be carried out jointly by the user and the manufacturer, including visual inspection, driving test, vehicle insulation performance test, cable retraction test, vehicle equipment test, etc.

2. Storage of Mobile Box-type Substation Vehicle

The mobile box-type substation vehicle belongs to the live working special vehicle, and the following requirements shall be met when storing.

(1) The live working special vehicle should be stored in the garage to reduce direct sun exposure or rain, away from high temperature heat sources.

(2) Where the vehicle is parked for a long time, the main power switch shall be turned off to cut off the circuit system of the entire vehicle.

(3) Before starting the vehicle, it is necessary to turn on the main power switch to restore the circuitry of the entire vehicle.

4.4 Live Working Tool Car

With the continuous development of power technology and the increasing demands of the society for power supply safety and stability, live working plays an increasingly important role in power supply companies. In the task of live working on distribution lines and electric emergency rescue, the special tool car for live working effectively solves the problem of transportation and storage of live working tools, so that the live working tool can be kept in a normal state before use, ensuring the reliability and safety of the live working, and improving the work response speed, work efficiency and standardization level.

4.4.1 Basic Types, Structures and Functions of Live Working Tool Special Vehicle

1. Basic Types of Live Working Tool Special Vehicle

The live working tool special vehicle can be divided into the following three types according to the characteristics and scope of use.

(1) Type I - Transmission lines (including substations) live working tool special vehicle, suitable for storing insulation tools, metal tools and PPE used in transmission lines (including substations). The temperature range of the space and tool storage compartment can be adjusted according to the type, size and quantity of live working tools stored in the tool.

(2) Type II - Distribution lines live working tool special vehicle, suitable for storing insulation tools, insulation shielding and PPE used in distribution lines. The temperature range of the space and tool storage compartment can be adjusted according to the type, size and quantity of live working tools stored in the tool.

(3) Type III - Hybrid live working tool special vehicle, suitable for storage of insulating tools, metal tools and PPE used for transmission lines (including substations), and insulation tools, insulation shields and PPE suitable for storage and distribution work. Partition isolation can be adapted to different temperature range requirements.

At present, distribution live working (type II) special vehicle has been widely used in distribution lines (including 10kV cable lines without power outages) live operation, and the transmission line live working vehicle (type I) shall also has been used.

The distribution live working special vehicle is shown in the Fig. 4-22, and the distribution line emergency bypass tool car is shown i Fig. 4-23.

Fig. 4-22 Distribution Live Work Special Vehicle

Fig. 4-23 Distribution Line Emergency Bypass Tool Car

In recent years, the distribution live working special vehicle has been updated very quickly, has been developed into a multi-functional and intelligent way, and has fully utilized modern mechatronics and measurement and control technology. Integrated with insulation tool storage environment monitoring, on-site repair process and on-site environmental monitoring, on-site lighting, dedicated power supply, 3G wireless remote data transmission, alarm function, vehicle positioning and navigation, etc., the needs of the live working of the distribution line and the on-site repair work can be satisfied to the maximum. Fig. 4-24 shows the intelligent live working tool special vehicle developed and produced by Shaanxi Ha Electrical Power Technology Co., Ltd.

Fig. 4-24 Intelligent Live Working Tool Special Vehicle

2. Basic Structure of Live Working Tool Special Vehicle

The live working tool special vehicle is mainly composed of a vehicle platform, an on-board device, an auxiliary system and a tool storage compartment. Fig. 4-25 shows the structure of the live working tool special vehicle.

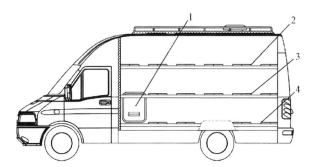

1 - generator; 2 - hard insulating tool storage area; 3 - insulation protective equipment, soft insulating tool storage area; 4 - meter, instrument, power tool, and metal tool storage areas

Fig. 4-25　Structure of Live Working Tool Special Vehicle

① Vehicle platform. It is modified on the basis of engineering vehicles or passenger cars, including the vehicle chassis and the box body. It is the transportation carrier for the live working tool special vehicle.

② On-board device. It mainly includes on-board generator and power conversion device, and can ensure the switching between the vehicle power supply and the mains, followed by dehumidifiers, baking and heating equipment and exhaust equipment.

Fig. 4-26 shows the portable generator adopted by the live working tool special vehicle.

Fig. 4-26　On-board Portable Generator

③ Auxiliary system. It includes control system, car GPS navigation system, fire smoke alarm detection and SMS alarm device. The car roof is provided with high-beam lighting and field video surveillance system, and equipped with micro weather system, car radio, GPS positioning system, on-board computer, hot meal device, etc. Fig. 4-27 shows video surveillance terminal equipment for live working tool special vehicle. Fig. 4-28 shows the lift vehicle lighting system.

Fig. 4-27　On-board Video Surveillance Terminal Equipment

Fig. 4-28　Lift Vehicle Lighting System

④ Tool storage compartment. It is a compartment for storing tools, shown as Fig. 4-29. With temperature adjustment and humidity control function, the compartment can meet the moisture-proof and storage temperature requirements of insulation tools and insulation protection appliances.

Fig. 4-29　Compartment for Storing Tools

3. Basic Functions of Live Working Tool Special Vehicle

It can meet the task requirements of independently completing the maintenance and emergency treatment of general live working and transmission operations of transmission and distribution lines, can regulate the management of distribution live working, and use the special tools for power distribution and working tools in live work, and can ensure the reliability of live working tools, thus ensuring the safety of live working, and providing auxiliary work on the live working site through the special tools for live working tools.

4.4.2 Main Technical Requirements for Live Working Tool Special Vehicle

1. General Requirements

① General requirements for tool storage compartment.

a. Space. The minimum dimensions of the tool storage compartment are shown in Table 4-3.

Table 4-3   Minimum Dimensions of Tool Storage Compartment   mm

Type	Long	Width	High
Type I	2400	1850	1750
Type II	1600	1850	1750
Type III	3200	1850	1750

b. The tool storage compartment is enclosed and shall have a thermal insulation layer with a sealing strip attached to the door. The tool storage compartment shall be provided with grounding device.

c. The fire protection, lighting and decoration materials of the storage compartment for tools and instruments shall meet the relevant requirements of the Live Working Tool Special Vehicle (GB25725-2010).

② Storage rack for tools and instruments

It is necessary to meet the relevant requirements of the Live Working Tool Special Vehicle (GB2725—2010). The internal structure of the live working tool special vehicle is shown in Fig. 4-30.

Fig. 4-30　Internal Structure of Live Working Tool Special Vehicle

2. Temperature and Humidity Requirements

① Temperature

a. The live working tools shall be stored according to the type of tool, and each storage area can have different temperature requirements. Metal There is no requirement on the tool storage.

b. The temperature in the storage areas of hard insulating tools, soft insulating tools, testing tools, and shielding tools shall be controlled within 5 - 40°C.

c. The temperature of the storage areas of the insulating shielding device and the insulating protective device shall be controlled within 10 - 28°C.

d. The temperature of Type I vehicle shall be controlled with 5 -40°C.

e. The temperature of Type II & III vehicles shall be controlled within 10 ~ 28°C.

② Humidity

The humidity in the tool storage compartment shall be not more than 60%.

3. Equipment Requirements

Dehumidification equipment, drying equipment, ventilation equipment, temperature and humidity control equipment, alarm equipment, video monitoring equipment and other equipment shall meet the relevant requirements of Live Working Tool Special Vehicle (GB25725—2010).

4.4.3　Test and Storage of Live Working Tool Special Vehicle

1. Test of Live Working Tool Special Vehicle

(1) Tests of live working tool special vehicle include type test, predelivery test and acceptance test. The type test and predelivery test should be carried out by the manufacturer.

(2) The acceptance test should be carried out jointly by the user and the manufacturer, including visual inspection, power switching test, vehicle insulation performance test, tool compartment sealing test, tool compartment storage temperature and humidity test, tool compartment temperature and humidity control test, drying equipment power cutoff and reset test and manual control test.

2. Storage of Live Working Tool Special Vehicle

The live working special vehicle should be stored in the garage to reduce direct sun exposure or rain, away from high temperature heat sources.

Chapter 5

Safety Protection Devices

The safety protection device for transmission and distribution line work mainly includes insulation protection, insulation shielding appliance and shielding appliance. The insulation protection and insulation shielding appliances are mainly used for live working on distribution lines, of which the insulation protection appliance is mainly used for operator safety protection, and the insulation shielding appliance is mainly used for limited shielding of the charged body during live working to prevent the operator from getting an electric shock; the shielding appliance is mainly used in the live working of the transmission line and shielding the impact of the high-voltage electromagnetic field on the human body during the live working of the transmission line, and dividing the power frequency current through the human body to protect the human body from the high voltage electric field and electromagnetic wave.

5.1 Insulation Protective Equipment

Insulation protective equipment commonly used in live working of distribution lines mainly includes insulating gloves, insulating sleeves, insulation suits, insulating shoes, puncture proof gloves, and insulating helmets, which are the most commonly used insulating protective tools for live working on distribution lines.

5.1.1 Insulating Gloves for Live Working

1. Basic Structure and Function of Insulating Gloves for Live Working

(1) Basic structure

① The gloves are made of synthetic rubber and the gloves can be lined to prevent chemical corrosion and reduce the aging effects of ozone on the gloves.

② According to the structure, the gloves can be divided into the baseline of the thumb, the wrist, the flat cuff, the curling cuff, and the midpoint height of the middle finger.

(2) Basic function

The insulating gloves for live working are gloves that are electrically assisted and insulating during live working on high-voltage electrical equipment or devices, which are made of synthetic rubber or natural rubber and are mainly used for insulation protection of the hands of live workers. At present, the maximum insulation level is up to 2, that is, the gloves can be used at 10kV or below.

2. Classification and Technical Specifications of Insulating Gloves for Live Working

(1) Classification

① According to the usage, the gloves can be divided into conventional insulating gloves and composite insulating gloves. This chapter focuses on conventional insulating gloves, shown as Fig. 5-1.

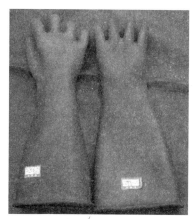

1 - baseline of thumb; 2 - wrist; 3 - flat cuff; 4 - curling cuff; 5 - midpoint height of the middle finger

Fig. 5-1　Conventional Insulating Gloves

② According to the shape, the gloves can be divided into single insulating gloves, mitten insulating gloves, long-sleeved composite insulating gloves and arc-shaped cuff insulating gloves.

③ According to the electrical performance, the gloves can be divided into five grades: 0, 1, 2, 3 and 4. Gloves with different nominal voltages are shown in Table 5-1.

Table 5-1 Gloves with Different Nominal Voltages

No.	Grade	AC^a/V
1	0	380
2	1	3000
3	2	10000
4	3	20000
5	4	35000

Note: AC^a refers to line voltage in a three-phase system.

④ Gloves with special properties are divided into 5 types: A, H, Z, R and C5, shown in Table 5-2.

Table 5-2 Gloves with Special Properties

No.	Model	Special properties
1	A	Acid resistant
2	H	Oil resistant
3	Z	Ozone resistance
4	R	Acid, oil, ozone resistance
5	C	Ultra-low temperature resistant

Note: Class R has the property of A, H and Z.

(2) Technical parameters for conventional insulating gloves

① Length of conventional insulating gloves. The length of conventional insulting gloves of different grades is shown in Table 5-3.

Table 5-3 Length of Conventional Insulting Gloves of Different Grades

No.	Model	L/mm			
1	0	280	360	410	460
2	1	—	360	410	460
3	2	—	360	410	460
4	3	—	360	410	460
5	4	—	—	410	460

② Thickness of conventional insulating gloves. In order to maintain proper flexibility, the maximum thickness of the glove plane (when the surface is not ribbed) is divided into five levels: 0, 1, 2, 3, and 4, and the thicknesses are 1.00mm, 1.50mm, 2.30mm, 2.90mm, and 3.60mm.

3. Inspection Content of Conventional Insulating Gloves

Before use, the conventional insulating gloves must be inspected, including the following:

(1) Identify no harmful and tangible surface defects by inspecting the inner and outer surfaces of the gloves.

(2) Where one of the gloves may not be safe, the gloves cannot be used and should be returned for test.

(3) Before use, air leak test shall be carried out: first, insulate the insulating gloves and squeeze them on the face to check for air leaks. The air leak test is shown in Fig. 5-2.

Fig. 5-2　Air Leak Test of Insulting Gloves

4. Usage of Conventional Insulating Gloves

(1) When using, mechanical protective gloves (such as sheepskin gloves) should be used on the outermost layer of the insulating gloves. The use of insulating gloves is shown in Fig. 5-3.

Fig. 5-3　Usage Example of Insulating Gloves

(2) Avoid using the insulating gloves to directly press a sharp object.

5. Precautions for Storage of Conventional Insulating Gloves

(1) Insulating gloves shall be packaged in separate parts. The gloves shall be stored in a special box to avoid direct sunlight or stored near artificial heat sources, to prevent rain and snow from immersing, and prevent extrusion and sharp objects from colliding, otherwise the gloves may be punctured or scratched.

(2) It is prohibited to contact the gloves with oil, acid, alkali or other harmful substances, and it shall be kept more than 1m away from the heat source. The storage environment temperature should be 10 - 21°C.

(3) In case that the gloves are dirty, it is necessary to wash with soap and water, and apply talcum powder after thorough drying. In case that a mixture of tar and paint adheres to the gloves, it is necessary to wipe it off with suitable solvent.

6. Test for Conventional Insulating Gloves

The conventional test contents of conventional insulating gloves are as follows.

(1) Appearance inspection includes visual inspection, dimensional inspection, thickness inspection, process and molding inspection, marking inspection, and packaging inspection.

(2) Mechanical test includes tensile strength and elongation at break test, tensile permanent deformation test, mechanical puncture test, abrasion test, cut resistance test, and tear resistance test.

(3) Electrical test includes AC verification tests, AC withstand tests, DC verification tests, and DC voltage tests.

(4) The electrical test cycle: the preventive test is conducted once every six months, the inspection test is once a year, and the interval between the two tests is half a year. The test of insulating gloves is shown in Fig. 5-4.

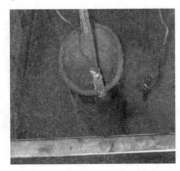

Fig. 5-4　Test for Insulating Gloves

5.1.2 Insulating Sleeves

1. Basic Structure and Function of Insulating Sleeves

(1) Basic structure

Insulating sleeves include impregnated and molded types. Impregnated rubber insulating sleeves have the same process with the insulating gloves, but the molded insulating sleeves are formed by a mold press type.

The insulating sleeve buckles and sleeves are shown in Fig. 5-5 and 5-6.

Fig. 5-5　Insulating Sleeve Buckles

Fig. 5-6　Insulating Sleeves

(2) Basic function

Insulating sleeves for live working are used in contracted grade insulating gloves to protect the shoulder arm; the sleeve buckles are used with sleeves and require good electrical properties, high mechanical properties, and good insulation properties.

2. Classification and Technical Specifications of Insulating Sleeves

(1) Classification

① Insulating sleeves are divided into two types: straight and elbow, according to the shape.

② Insulating sleeves with special properties can be divided into five types: A, H, Z, S and C, which are resistant to acid, oil, ozone and low temperature.

③ The sleeves are divided into 0, 1, 2 and 3 levels according to the pressure level.

(2) Technical specifications

① Thickness: Insulating sleeves shall be sufficiently flexible and flat, and the maximum thickness of the synthetic rubber must meet the requirements of Table 5-4.

Table 5-4 Maximum Thickness of Synthetic Rubber

No.	Grade	Thickness (mm)
1	0	1.00
2	1	1.50
3	2	2.50
4	3	2.90

② Insulating sleeves shall be seamlessly made, and the small holes left by the sleeves and the cuffs must be made of non-metallic reinforced edges, typically 8 mm in diameter.

③ The inner and outer surfaces of the sleeves shall be even, smooth, regular, without small holes, cracks, local bulges, no inclusions of conductive foreign matter at the incision, no creases, irregular corrugations and cast marks.

3. Inspection Items for Insulating Sleeves

Before use, insulating sleeves must be inspected, including visual inspection (visual inspection of the inner and outer surfaces of the cuff, requiring uniform surface), dimensional inspection, thickness inspection, process and molding inspection, marking inspection, ad packaging inspection.

4. Usage of Insulating Sleeves

(1) Before use, it is necessary to carry out appearance inspection.

(2) Connect the insulating sleeve and the insulation buckle securely.

(3) Avoid being scratched by sharp objects or sharp objects.

5. Precautions for Storage of Insulating Sleeves

(1) The insulating sleeves cannot be folded, for crease will cause the rubber to be oxidized and reduce the insulation performance. The storage of the insulating sleeves is shown in Fig. 5-7.

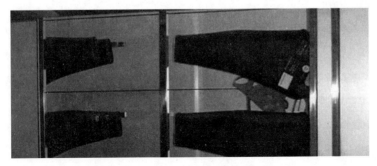

Fig. 5-7 Storage of Insulating Sleeves

(2) The sleeves should be stored in the correct size storage, placed flat, and a storage bag placed in a pair of cuffs.

(3) Insulating sleeves shall be packed by pieces. Attention shall be paid to prevent direct sunlight or being stored near artificial heat sources. In particular, it is necessary to avoid direct contact with sharp objects, which may result in punctures or scratches.

(4) It is not allowed to contact with oil, acid, alkali or other harmful substances, and be more than 1m away from the heat source. The storage environment temperature should be 10 - 21°C.

(5) In case that the sleeves are dirty, it is necessary to wash with soap and water, and apply talcum powder after thorough drying. In case that a mixture of tar and paint adheres to the gloves, it is necessary to wipe it off with suitable solvent.

(6) In case that the sleeve is wet during use or dried thoroughly after washing, the drying temperature shall not exceed 65 °C.

6. Test Items for Insulating Sleeves

The conventional test contents of conventional insulating sleeves are as follows.

(1) Mechanical test. The mechanical test includes tensile strength and elongation test, mechanical puncture resistance test, and tensile permanent deformation test.

(2) Electrical withstand voltage test. Including type test, sampling test and factory routine test. The voltage duration for the type test and sampling test is 3min and that for the pre-delivery test is 1min.

(3) The preventive test of the insulating sleeve shall be conducted once every six months.

5.1.3 Insulation Suit

Insulation suit are made of clothing made of insulating materials and are a personal safety protection device that protects live workers from electric shock when they come into contact with live parts.

1. Basic Structure and Function of Insulation Suit

(1) Basic structure

① The insulation suit is made of ERV resin material or synthetic rubber, which has the function of preventing mechanical abrasion, chemical corrosion and ozone. The insulation suit made of ERV resin material will be introduced, as shown in Fig. 5-8.

② The structure of the insulating cloth is divided into sleeves, cuffs, belts, body and buttons.

③ The structure of the insulating pants is divided into a belt and a waist elastic band.

(2) Basic functions of insulation suit for live working Insulating the human body in addition to the head, hands and feet is achieved to protect the operators of live working from electric shock when they are in contact with the electrified body.

Fig. 5-8 Example of Insulation Suit

2. Classification and Technical Specifications of Insulation Suit

(1) Classification

The complete set of insulation suit includes insulating cloth and insulating pants, of which the insulating cloth is divided into ordinary insulating tops, mesh insulating tops and insulating shawls.

(2) Technical specifications

① The models of insulating cloth and insulating pants are divided into small, medium, large and extra large models.

② The surface of the insulation suit shall be flat, uniform and smooth, without small holes, local bulges, inclusions of foreign matter, creases, voids, etc., and the joint part shall be made in a seamless manner.

③ The average tensile strength of the surface should not be less than 9MPa, and the minimum value should be no less than 90% of the average value.

④ The average surface mechanical anti-piercing force should be no less than 15N, and the minimum value should be no less than 90% of the average value.

⑤ The average value of the surface tearing breaking force should be no less than 150N, and the minimum value should be no less than 90% of the average value.

⑥ Electrical performance should meet the requirements of Table 5-5.

Table 5-5 Electrical Performance of Insulation Suit

AC voltage tests	Verification voltage of whole suit of clothes (kV)	20
	Withstand voltage of whole suit of clothes (kV)	30
	Power frequency withstand voltage along the (kV)	100
Resistivity measurement when drilling services	Inner layer material volume resistivity ($\Omega \cdot cm$)	$\geqslant 1 \times 10^{15}$

3. Inspection Items of Insulation Suit

Insulation suit must be inspected before use, including visual inspection, focusing on the process and molding inspection. The inner and outer surfaces should be visually inspected. The surface should be flat, uniform and smooth, and free of small holes, local bulges, inclusions, creases, voids, etc. The joint part should be made in a seamless manner; mark inspection; and packaging inspection.

4. Usage of Insulation Suit

(1) Visual inspection of the inner and outer surfaces of the insulation suit before each use.

(2) It is forbidden to use the insulation suit when it has defects that may affect the safety performance, and the insulation suit should be tested.

(3) Avoid being scratched by sharp objects or sharp objects. Wearing method of insulation suit is shown in Fig. 5-9.

5. Precautions for Storage of Insulation Suit

(1) The insulation suit cannot be folded, because crease will cause the rubber to be oxidized, reducing the insulation performance.

Fig. 5-9　Wearing of Insulation Suit

(2) Insulation suits should be hung one by one on a special stainless steel metal frame in a dry, well-ventilated live working tool warehouse, as shown in Fig. 5-10.

Fig. 5-10　Storage of Insulation Suit

(3) Insulation suits are prohibited from being stored near steam pipes, radiators or other artificial heat sources; do not store in direct sunlight; in particular, avoid touching sharp objects directly, causing punctures or scratches.

(4) It is prohibited to contact the insulation suit with oil, acid, alkali or other harmful substances, and it shall be kept more than 1m away from the heat source. The storage environment temperature should be within 10-21°C.

(5) In case that the insulation suit are dirty, it is necessary to wash with soap and water, and apply talcum powder after thorough drying. In case that a mixture of tar and paint adheres to the clothing, it is necessary to wipe it off with suitable solvent.

(6) In case that the insulation suit is wet during use or dried thoroughly after washing, the drying temperature shall not exceed 65°C.

6. Test Items of Insulation Suit

The conventional test items of conventional insulation suit are as follows.

(1) Appearance inspection includes process and molding inspection, marking inspection, and packaging inspection.

(2) The mechanical test includes tensile strength and elongation test, mechanical puncture resistance test, and surface tear resistance test.

(3) The preventive test items for the insulation suit include mark inspection, AC withstand voltage test or DC withstand voltage test, and the test period is once every six months. The preventive test of insulation suit is shown in Fig. 5-11.

Fig. 5-11　Preventive Test of Insulation Suit

5.1.4　Insulating Shoes (Boots) for Live Working

1. Basic Structure and Function of Insulating Shoes (Boots)

(1) Basic structure

The basic insulating shoes (boots) are composed of soles, upper, heel, boot leg, etc. Insulating shoes (boots) are shown in Fig. 5-12.

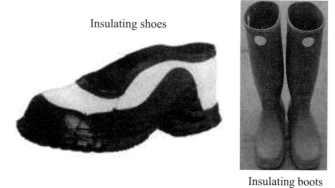

Insulating shoes

Insulating boots

Fig. 5-12 Insulating Shoes (Boots)

(2) Basic function

① Insulating shoes (boots) are auxiliary safety tools used during live working on distribution lines.

② The insulating shoes (boots) for live working are electrically assisted and insulating during live working The insulating shoes (boots) are made of synthetic rubber or natural rubber, mainly protecting the feet of live working operators from electrical insulation. The insulating shoes (boots) require good electrical performance and good insulation properties. At present, the maximum insulation level is up to 2, that is, the shoes (boots) can be used at 10kV and below.

2. Classification and Technical Specifications of Insulating Shoes (Boots) for Live Working

(1) Classification

① According to the system voltage, it can be divided into 3~10kV (power frequency) insulating shoes (boots) and insulating shoes (boots) below 0.4kV.

② According to material, it can be divided into cloth insulating shoes, leather insulating shoes (boots), and rubber insulating shoes (boots).

(2) Technical specifications

① Insulating shoes (boots) should be flat, and the outsole should have a non-slip pattern.

② Insulating shoes (boots) can only be used as an auxiliary safety device within the specified range.

3. Inspection of Insulating Shoes (Boots) for Live Working

Before use, the insulating shoes (boots) must be inspected, including the following.

(1) Appearance inspection, the inner and outer surfaces of the insulating shoes (boots) should be flat, without surface defects such as cracks and holes.

(2) In case that one of the pair of insulating shoes (boots) may not be safe, the shoes cannot be used and should be returned for testing.

(3) For the mark inspection, the test certificate shall be complete and within the validity period.

4. Usage of Insulating Shoes (Boots) for Live Working

(1) Wear correctly. Insulating shoes (boots) should be placed before entering the insulating stand or the insulation bucket. After wearing, it is not allowed to walk and step on the ground or other sharp objects.

(2) The model of the insulating shoes (boots) should be adapted to the operator's foot code and should not be too large or too small.

(3) Insulating shoes (boots) shall not be used in case that the shoes are damaged, the soles of the soles are flattened, and the outer soles are exposed to the insulation layer or the preventive inspection is unqualified.

(4) In case that the insulating shoes (boots) are wet during use or dried thoroughly after washing, the drying temperature shall not exceed 65 °C.

5. Precautions for Storage of Insulating Shoes (Boots) for Live Working

(1) The storage place should be dry and ventilated. When it is stacked, it should be more than 20cm away from the ground and the wall. Leave all heating elements more than 1m. It is strictly forbidden to store with oil, acid, alkali or other corrosive materials.

(2) In case that the shoes are dirty, it is necessary to wash with soap and water, and apply talcum powder after thorough drying. In case that a mixture of tar and paint adheres to the shoes, it is necessary to wipe it off with suitable solvent.

6. Test for Insulating Shoes (Boots) for Live Working

(1) Mechanical performance test. Including tensile strength and elongation at break test, wear resistance test, Shore A hardness test, strip and upper adhesion strength test, upper and sole peel test, folding performance test.

(2) Electrical test. Including AC verification voltage test and leakage current test.

(3) The preventive inspection period for various insulating shoes (boots) should not exceed 6 months.

Chapter 5
Safety Protection Devices

5.1.5 Anti-mechanical Piercing Insulating Gloves for Live Working

1. Basic Structure and Function of Anti-mechanical Piercing Insulating Gloves for Live Working

(1) Basic structure

① The gloves are made of synthetic rubber and the gloves can be lined to prevent mechanical, chemical and ozone oxidation.

② The main part consists of the baseline of the thumb, the wrist, the flat cuff, the curled cuff, and the middle point of the middle finger bending, as shown in Fig. 5-13.

Fig. 5-13 Puncture Proof Gloves

(2) Basic function

Puncture proof insulating gloves for live working are gloves for electrical auxiliary insulation during live working on high-voltage electrical equipment or devices., mainly providing insulation protection for the hands of live working operators, requiring good electrical performance and high performance. Mechanical behavior.

2. Classification and Technical Specifications oOf Anti-mechanical Piercing Insulating Gloves for Live Working

(1) Classification

① According to the method of use, the gloves can be divided into composite insulating gloves and long-sleeved composite insulating gloves.

② According to the shape, the gloves can be divided into single insulating gloves and mitten insulating gloves.

③ Three grades of gloves are specified according to electrical properties: 00, 0 and 1. Anti-mechanical piercing gloves with different nominal voltages are shown in Table 5-6.

Table 5-6 Anti-mechanical Piercing Gloves with Different Nominal Voltages

No.	Grade	AC RMS (V)	DC (V)
1	00	500	750
2	0	1000	1500
3	1	3000	11250

Note: it refers to line voltage in a three-phase system.

③ Gloves with special properties are divided into five types: A, H, Z, P and C5, shown in Table 5-7.

Table 5-7   Gloves with Special Properties

No.	Model	Special properties
1	A	Acid resistant
2	H	Oil resistant
3	Z	Ozone resistance
4	R	Acid, oil, ozone resistance
5	C	Ultra-low temperature resistant

Note: Class P combines the performance of A, H, and Z.

(2) Technical specifications

① The puncture proof insulating gloves for live working have mechanical protection properties, can be used without mechanical protective gloves, and have good insulation properties.

② Glove cuffs can be made with or without crimping.

③ The length of the insulating gloves. The length standards for different grades of gloves are shown in Table 5-8.

Table 5-8   Length of Conventional Insulating Gloves of Different Grades

No.	Grade	Length (mm)				
1	00	270	360	—	—	800
2	0	270	360	410	460	800
3	1	—	—	410	460	800

Note: The composite insulating gloves are allowed to have a length deviation of ±20 mm.

④ Thickness of puncture proof insulating gloves for live working. To maintain proper softness, the maximum thickness of the glove plane (when the surface is not ribbed) is shown in Table 5-9.

Table 5-9   Maximum Thickness of Synthetic Rubber

No.	Grade	Thickness (mm)
1	00	1.80
2	0	2.30
3	1	

Note: The thickness value corresponding to the level "1" is not determined.

Chapter 5
Safety Protection Devices

3. Inspection Items of Mechanical Puncture Resistance Insulating Gloves for Live Working

(1) Identify no harmful and tangible surface defects by inspecting the inner and outer surfaces of the gloves.

(2) The palm and finger surfaces designed to improve grip performance should not be considered surface defects.

(3) Where one of the gloves may not be safe, the gloves cannot be used and should be returned for test.

4. Usage of Mechanical Puncture Resistance Insulating Gloves for Live Working

(1) Before use, air leak test shall be carried out: first, insulate the insulating gloves and squeeze them on the face to check for air leaks.

(2) Avoid using the insulating gloves to directly press a sharp object.

5. Precautions for Storage of Mechanical Puncture Resistance Insulating Gloves for Live Working

(1) Insulating gloves shall be packaged in separate parts. The gloves shall be stored in a special box to avoid direct sunlight or stored near artificial heat sources, to prevent rain and snow from immersing, and prevent extrusion and sharp objects from colliding, otherwise the gloves may be punctured or scratched.

(2) It is prohibited to contact the gloves with oil, acid, alkali or other harmful substances, and it shall be kept more than 1m away from the heat source. The storage environment temperature should be 10 - 21°C.

(3) In case that the gloves are dirty, it is necessary to wash with soap and water, and apply talcum powder after thorough drying. In case that a mixture of tar and paint adheres to the gloves, it is necessary to wipe it off with suitable solvent.

(4) In case that the gloves are wet during use or dried thoroughly after washing, the drying temperature shall not exceed 65°C。

(5) Do not unnecessarily expose the gloves to heat, light, or contact with oil, grease, turpentine or weak acids.

6. Test items of Mechanical Puncture Resistance Insulating Gloves

The conventional test contents of conventional insulating gloves are as follows.

(1) Mechanical test. Including wear test, anti-mechanical puncture test, tear test, tensile strength and elongation at break test, anti-cut test.

(2) Electrical performance test. Including AC verification test, AC tolerance test, and

leakage current test after damp.

(3) The preventive test of the insulating gloves shall be conducted once every six months.

5.1.6 Insulating Safety Helmet

Insulating safety helmet is a protective device used to protect the head of an electrician from electric shock, bruises or falling objects during live working.

1. Basic Structure and Function of Insulating Safety Helmet for Live Working

(1) Basic structure

Insulating safety helmet is composed of shell, lining, chin strap, rear band, etc., without air hole. The inner and outer structures of the insulating safety helmet are shown as Figs. 5-14 and 5-15.

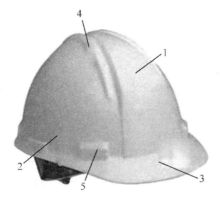

1 - lining; 2 - connecting hole; 3 - lining joint; 4 - girdle/rear band; 6 - rear band regulator;
7- chin strap; 8 - sweatband; 9 - backing; 10 - headband; 11 - locking; and 12 - protective band

Fig. 5-14　Inner Structure of Insulating Safety Helmet

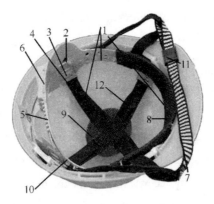

1 - shell; 2 - brim; 3 - visor; 4 - top support; and 5 - plug

Fig. 5-15　Outer Structure of Insulating Safety Helmet

(2) Basic function

The helmet is worn on the head to protect the operator's head during live working, can shield the arc, cushions shock absorption and disperses stress, and protect against electric shock or mechanical damage.

2. Classification and Technical Specifications of Insulating Safety Helmet for Live Working

(1) Classification

The insulating safety helmet for live working mainly includes American, European, Japanese and domestic, and the colors are mainly white and yellow.

(2) Technical Parameters

The insulating safety helmet for live working is made of high-density composite polyester material. In addition to the mechanical strength that meets the safety testing standards of the helmet, it shall also meet the electrical testing standards for the relevant distribution and live working. The strength of the dielectric must meet the test requirements of 20kV/3min.

3. Inspection Items of Insulating Safety Helmet

(1) For new insulting safety helmet, it is necessary to check whether there is a certificate of production and product certification issued by the labor department; then check if it is damaged, whether the thickness is uniform, whether the buffer layer and the adjustment belt and the elastic belt are complete and effective; and finally check the trademark, model, manufacturer name, production date and production license number on the helmet.

(2) The appearance inspection must be carried out on the insulated safety helmet before its use. The components of the safety helmet, such as the shell, the headband, the top lining, the chin strap, the rear buckle (or the headband buckle), etc. shall be intact. The cushion space between the shell and the top lining shall be between 25mm and 50mm. The test certificate is in good condition and within the validity period of the test.

(3) Periodical inspection shall be done for checking, depressions, cracks and wear and tear, it shall be replaced immediately if any abnormal phenomena are found, and it shall not be used any more; any insulated safety helmet subject to thump, electric shock or crack shall be scrapped, regardless of whether it is damaged or not; its insulation performance shall be tested.

The insulated safety helmet shell shall not have breathable holes and shall not be confused with ordinary ones.

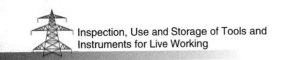
Inspection, Use and Storage of Tools and Instruments for Live Working

4. Usage for Insulated Safety Helmet

(1) When the operator wears the insulated safety helmet, the rear adjusting belt of the helmet is adjusted to a suitable position according to the head shape (the head is slightly restrained, but not uncomfortable to the extent that the safety helmet will not fall off when the chin strap is not fastened). The chin band must be fastened when the operator wears the insulated safety helmet. the chin strap shall be close to the jaw so that the jaw is restrained but not uncomfortable. Then elastic strap in the helmet is fastened. The tightness of the buffering cushion is regulated by the strap, and the vertical distance between the top of the head and the top of the helmet body is generally between 25mm and 50mm, and at least not less than 32mm.

(2) After the safety helmet is worn, the rear buckle shall be screwed to an appropriate place (or the headband buckle is adjusted to an appropriate place), and the chin strap is locked to prevent slipping due to tilting forward and backward or other reasons during work. It is prohibited to put the insulated safety helmet askew, or put your hat on the back of the brain. Otherwise, it will reduce the protective effect of the safety helmet.

(3) The lower jaw strap of the insulated safety helmet must be buckled and fastened under the jaw, with moderate tightness. This prevents the safety helmet from being blown off by strong winds, knocked off by other obstacles, or falling off due to back-and-forth swings of the head.

(4) It is strictly prohibited to use the insulated safety helmet with only the chin strap connected to the shell, that is, the insulated safety helmet without a cushion layer therein.

(5) It is strictly prohibited to use the safety helmet in a non-standard manner. During the field operation, the operators shall not take off the safety helmets and put them aside or use them as cushions, shall not unfasten the buckle or not tighten it, and shall not put the buckle strap into the lining.

(6) When the insulated safety helmet is used usually, it shall be kept clean, can not contact the source of fire, and shall not applied with paint at will.

5. Precautions for Storage of Insulated Safety Helmet

(1) The insulated safety helmet shall not be stored in acid, alkali, high-temperature, sunlight and humid places, and even shall not be placed with hard objects together.

(2) The insulated safety helmet shall be kept in the warehouse of special live working tools with constant temperature and humidity, and shall be placed separately to avoid mixing

with ordinary safety helmets.

(3) The suitable temperature of the safety helmet is −10~+50°C. It shall be placed in a dry place with breathable air, and avoid contact with fire (heat) source and erosive substances, so as not to affect its normal service life. The storage period of the safety helmet shall be shortened as far as possible, and it is better to use it immediately after being purchased. The products with short time of delivery shall be purchased, so as not to affect the actual effective use time due to the excessive storage time.

6. Test Contents of the Insulated Safety Helmet

(1) The delivery test of insulated safety helmet includes impact absorption property test (the impulse test shall be done after low temperature, high temperature and water spraying pretreatment, and the force transmitted to the head model shall not exceed 4,900N), puncture resistance property test, electrical insulation property test, flame retardant property test, lateral rigidity test and antistatic property test.

(2) The insulation safety helmet shall be tested periodically, and the mechanical test and electrical test shall be conducted once a year. Only after passing the test can be used continuously.

5.2 Insulated Shielding Appliance

The insulated shielding appliance made of insulating materials is the insulated protection tool mainly used to shield live conductors or non-live conductors in live working of distribution lines. The common insulated shielding appliance has insulating shield cover, insulating barrier, insulating sleeve, insulating blanket, insulating mat and so on.

5.2.1 Insulating Shield Cover

The insulating shield cover is arranged between the operator and the shielded object to prevent the operator from coming into direct contact with the electrified body and play the protective role in shielding or isolation.

1. Basic Structure and Function of Insulating Shield Cover

The following is a brief introduction of the insulating shield cover for live working of 10kV distribution line (hereinafter referred to as the shield cover).

(1) Basic structure

① The shield cover is made of insulating materials such as epoxy resin material, rubber material, plastic material and polymeric material.

② The shield cover may be hard-shelled or soft-sheathed, shall be suitable for the shielded object, and shall have the function of preventing the human body from directly contacting the electrified body or the grounding body, with its length no more than 1.5m generally. In addition to meeting the necessary electrical characteristics, the dimensions shall be minimized.

③ The protected area of the shield cover shall be clearly, obviously and firmly marked.

④ All shield covers shall be provided with insulating rods, and shall be provided with lifting rings, holes, hooks and other components. The insulating shield cover is shown in Fig. 5-16.

Chapter 5
Safety Protection Devices

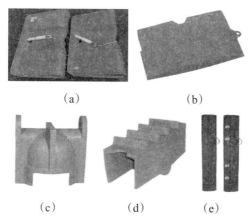

(a)   (b)

(c)   (d)   (e)

(a) Hard falling fuse insulating shield cover; (b) Soft falling fuse insulating shield cover; (c) Pin insulator insulating shield cover; (d) Cross arm insulating shield cover; (e) Pole insulating shield cover

Fig. 5-16 Sample Drawing of Insulating Shield Cover

(2) Basic function

In live working of 10kV distribution line, the shield cover does not play the main insulation role, and it is only suitable for insulation and shielding or isolation protection when the live working personnel are accidentally and briefly collided with each other, that is, when they are in contact with each other.

2. Classification and Technical Specifications of Insulating Shield Cover

(1) Classification

① According to the different shielded objects, it can be divided into hard shell type, soft type or deformed type, and can also be divided into fixed type or flat type.

② According to the use of the shield cover, it can be divided into conductor shield cover, pin insulator shield cover, strain device shield cover, hanging device shield cover, cable clamp shield cover, bar insulator shield cover, cross arm shield cover, pole shield cover, sleeve shield cover, falling fuse shield cover, arrester shield cover, etc. It can also be specially designed according to the shielded object.

③ There are five kinds of shield covers with special properties: A, H, C, W and P.

(2) Technical specifications

① The shield cover shall be made of insulating material with small hygroscopicity, and the insulating material shall be able to meet the requirements of electrical and mechanical properties at certain high and low temperatures.

② According to the electrical properties of the shield cover, it can be divided into five grades: 0, 1, 2, 3 and 4.

③ The main surface of the insulating shield cover shall be smooth, and there shall be no surface defects such as apertures, joint cracks, floating bubbles, crevasses, unknown sundries, abrasion and scratch, and obvious machining marks on its inner and outer surfaces.

④ The shield cover shall have lifting ring, barrel hole, hook and other parts in the structure.

⑤ One or more locking parts shall be arranged on the shield cover to prevent it from slipping suddenly in use or under the action of external force.

3. Inspection Contents of Insulating Shield Cover

In order to ensure the integrity of the electrical and mechanical properties of the shield cover, a careful appearance inspection and trial assembly shall be carried out before each use of the tool, as follows.

(1) The shield cover shall be free from damage after storage and transportation, and the insulating surface of the tool shall be free from holes, bruises, scratches and cracks.

(2) The surface of shield cover shall be clean and dry.

(3) Removable parts or components of the shield cover shall be complete and intact after being assembled.

(4) The shield cover shall be operated correctly, the tool shall rotate flexibly without jamming, and the locking function shall be correct.

4. Usage of Insulating Shield Cover

(1) It is only for live working of power equipment 10kV or below, such as cross arm shield cover and insulator shield cover shown in Fig. 5-17.

Fig. 5-17　Use of Insulating Shield Cover

(2) The shield cover does not play the role in main insulation, but allows occasional short-term contact, mainly limiting the motion range of human body.

(3) The shield cover shall be used in conjunction with the human body safety protection appliances.

(4) The overlaps between the insulating shield covers or with the edges of other shielding objects during the shielding operation shall not be less than 150mm.

(5) The shielding scope of each shield cover shall not exceed the protection of the shield cover protection area.

5. Precautions for Storage of Insulating Shield Cover

(1) The insulating shield cover can not be folded and squeezed, because crease will cause the rubber to be oxidized, reducing the insulation performance.

(2) The shield cover shall be stored in a properly sized storage bag, packaged in parts and placed flat.

(3) It is prohibited to store the shield cover near the steam pipe, heat dissipation pipe or other artificial heat source and in the environment with direct sunlight, light or other light source. In particular, it shall be avoided to directly touch sharp objects and cause puncture or scratch when being stored or moved.

(4) It is prohibited to contact the shield cover with oil, acid, alkali or other harmful substances, and it shall be kept more than 1m away from the heat source. The storage environment temperature should be 10 - 21°C.

(5) The damp shield cover shall be thoroughly dried, but the drying temperature shall not exceed 65°C.

6. Test Contents of Insulating Shield Cover

The test contents of the insulating shield cover are as follows.

(1) Mechanical test. The mechanical test of the shield cover shall be done at low temperature. The mechanical property of the shield cover shall include simulated assembly test, low-temperature folding test of the soft insulating shield cover and low-temperature impulse resistance test of the hard insulating shield cover.

(2) Electrical test. It includes power frequency withstand voltage and leakage current test, certification test, ozone resistance test and special property test.

(3) Preventive tests. The preventive test of the insulating shield cover shall be conducted once every six months.

5.2.2 Insulating Barrier

1. Basic Structure And Function of Insulating Barrier

(1) Basic structure

① The insulating separator is generally made of epoxy resin FRP material, as shown in Fig. 5-18.

② The insulating separator is mainly composed of impregnated paper, cotton cloth, alkali-free glass-fiber cloth, impregnated phenolic aldehyde, epoxy resin and other materials, and which are impregnated, pressurized, dried, polished and solidified into the hard plate-like insulating material.

(2) Basic function

① The insulating barrier shall have high insulation property, corrosion resistance and high heat resistance, and play the role in temporary insulation and isolation of electrified parts, limiting the motion range of live working personnel and improving the insulation level of adjacent phases.

② In live working, the insulating barrier can also be placed between the moving and static contacts of the pull-apart disconnector, so as to prevent the disconnector from dropping by itself and delivering power by mistake.

③ Epoxy resin has good mechanical property and stable withstand voltage, and has the functions of blocking electric arc, voltage and partial current.

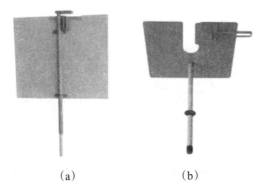

(a)　　　　　　(b)

(a) Vertical insulating barrier; (b) Transverse insulating barrier

Fig. 5-18　Sample Drawing of Insulating Barrier

2. Classification and Technical Specifications of Insulating Barrier

(1) Classification

The insulating barrier is mainly divided into the insulating barrier with handle and the insulating barrier with tether.

(2) Technical specifications

① The appearance and surface of the epoxy resin insulating board shall be smooth and flat, and the whole board shall be used for production, with uniform color. Impurities and other obvious defects are not allowed, while slight bruises are allowed, but the edges shall be cut neatly, and there are no delamination and cracks on the cross section.

② The epoxy resin insulating separator has high temperature resistance of 180~200°C, high mechanical and dielectric properties, good heat resistance and moisture resistance, and good machining property.

③ The color of epoxy resin insulating separator is beige, with good oil resistance and corrosion resistance.

3. Inspection Contents of Insulating Barrier

In order to ensure the integrity of the electrical and mechanical properties of the insulating barrier, careful inspection shall be carried out before each use. Its appearance and surface shall be flat and smooth, uniform in color, free from impurities and other obvious defects, with neat edges and no delamination or cracks on the cross section; the insulating barrier surface shall be clean and dry; the components of the insulating barrier shall be complete and intact; the insulating barrier shall be operated correctly, and the locking function shall be correct.

4. Usage of Insulating Barrier

(1) It is only for live working of power equipment 10kV or below. The insulating barrier shall be used as shown in Fig. 5-19.

Fig. 5-19 Use of Insulating Barrier

(2) The insulating barrier does not play the role in main insulation, and shall be used in conjunction with the human body safety protection appliances.

(3) The isolation scope of each insulating barrier shall not exceed the protection of the barrier.

(4) The installation shall be firm and reliable, and it is easy to disassemble it after installation.

5. Precautions for Storage of Insulating Barrier

(1) The insulating barrier cannot be squeezed and shall be placed flat.

(2) It is prohibited to contact the insulating barrier with oil, acid, alkali or other harmful substances, and it shall be kept more than 1m away from the heat source. The storage environment temperature should be 10 - 21°C.

(3) The damp insulating barrier shall be thoroughly dried, but the drying temperature shall not exceed 65°C.

6. Test Contents of Insulating Barrier

The test contents of the insulating barrier are as follows.

(1) Mechanical test. The mechanical property of the shield cover shall include simulated assembly test and low-temperature impulse resistance test.

(2) Electrical test. Including: power frequency withstand voltage and leakage current test, certification test, etc.

(3) The preventive test of the insulating barrier shall be conducted one by one once every six months.

5.2.3 Insulating Blanket

1. Basic Structure and Function of Insulating Blanket

(1) Basic structure

① The shape of the insulating blanket can be flat, slotted or specially designed to meet the needs of special use. Flat insulating blanket is shown in Fig. 5-20 and Fig. 5-21, while slotted insulating blanket is shown in Fig. 5-22.

Chapter 5
Safety Protection Devices

Fig. 5-20　Flat Resin Insulating Blanket

Fig. 5-21　Flat Rubber Insulating Blanket

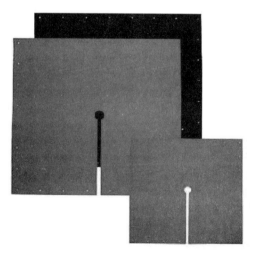

Fig. 5-22　Slotted Rubber Insulating Blanket

② The insulating blanket is generally made of epoxy resin, rubber and other insulating materials.

(2) Basic function

① The insulating blanket shall have very high insulation property, good corrosion resistance and high heat resistance, and play the role in temporary insulation and protection of electric operators from mistakenly touching the electrified body and isolating the electrified parts during live working, and improving the insulation level of adjacent phases.

② It has stable withstand voltage and the functions of blocking electric arc, voltage and partial current.

2. Classification and Technical Specifications of Insulating Blank

(1) Classification

① The insulating blanket is divided into resin insulating blanket and rubber insulating blanket.

② The insulating blanket with special properties and multiple special properties can be divided into 6 types: A, H, Z, M, S and C.

(2) Technical specifications

① The insulating blanket is divided into four grades: 0, 1, 2 and 3 according to its electrical properties.

② The insulating blanket shall be made of insulating rubber and plastic materials in seamless manufacturing technology. The edge of perforations in the insulating blanket must be reinforced with a non-metallic material, usually 8 mm in diameter.

③ There shall be no harmful irregularities on the upper and lower surfaces of the insulating blanket, and the protection area of the insulating blanket shall be clearly, obviously and firmly marked.

3. Inspection Contents of Insulating Blanket

In order to ensure the integrity of the electrical and mechanical properties of the insulating blanket, its both sides shall be carefully inspected before each use. Its appearance and surface shall be flat and smooth, uniform in color, free from impurities, pinholes, cracks, cutting and other obvious defects; the surface of the insulating blanket shall be clean and dry; the insulating blanket shall have a test certificate within the test time limit.

4. Usage of Insulating Blanket

(1) It is only for live working of power equipment 10kV or below, as shown in Fig. 5-23.

Chapter 5
Safety Protection Devices

Fig. 5-23 Use of Insulating Blanket

(2) It does not play the role in main insulation, but occasional short-term contact is allowed, and it shall be used in conjunction with the human body safety protection appliances.

(3) The isolation scope of each insulating blanket shall not exceed the protection of the protected area.

(4) After the insulating blanket is installed, it shall be securely fixed with the insulating blanket clip.

(5) The overlaps between the insulating blankets or with the edges of other shielding objects during the shielding operation shall not be less than 150mm.

5. Precautions for Storage of Insulating Blanket

(1) The insulating blankets shall be stored one by one in packing bags of sufficient strength. The insulating blanket shall be carefully placed to ensure that it is not squeezed and folded, especially to avoid direct contact with sharp objects, resulting puncture or scratching. It is prohibited to store it near steam pipe, heat dissipation pipe or other artificial heat source, and it shall be kept more than 1m away from the heat source. It is prohibited to store it in the environment with direct sunlight, light or other light source; no contact with oil, acid, alkali or other harmful substances is allowed. The storage ambient temperature should be between 10°C and 21°C.

(2) The damp insulating blanket shall be dried, but the drying temperature shall not exceed 65°C.

6. Test Contents of Insulating Blanket

The test contents of the insulating blanket are as follows.

(1) Mechanical test. The mechanical property test of the insulating blanket includes tensile strength and elongation test, mechanical puncture resistance test, tensile permanent de-

formation test and tear resistance test.

(2) Electrical test. AC voltage test includes AC voltage certification test and AC withstand voltage test.

(3) Preventive tests. The preventive test of the insulating blanket shall be conducted once every six months.

5.2.4 Insulating Mat

1. Basic Structure and Functions of Insulating Mat

(1) Basic structure

The insulating mat shall be made of rubber insulating material, and the upper surface shall be designed with anti-slip features such as wrinkle or diamond pattern to enhance the anti-slip property of the surface. The back surface may be made of cloth or other anti-slip materials, as shown in Fig. 5-24.

Fig. 5-2-8 Insulating Mat

(2) Basic function

① The insulating mat is made of rubber insulation material and laid on the ground or grounded object to protect the operators from electric shock.

② It has stable withstand voltage and the functions of blocking electric arc, voltage and partial current.

2. Classification and Technical Specifications of Insulating Mat

(1) Classification

① According to the voltage grade, it can be divided into 5kV insulating rubber mat, 10kV insulating rubber mat, 15kV insulating blanket, 20kV insulating rubber board, 25kV insulating rubber board, 30kV insulating rubber board and 35kV insulating rubber board.

② According to the color, it can be divided into red, dark black and green insulating rubber mats.

(2) Technical

① The insulating mat is divided into four grades: 0, 1, 2 and 3 according to its electrical properties.

② The insulating mat shall be made by seamless manufacturing technology.

③ There shall be no harmful irregularities on the upper and lower surfaces of the insulating mat.

④ The diameter of the insulating mat depression shall be no more than 1.6m, and the edge is smooth. When the reverse surface of the depression is spread on the thumb, there shall be no visible mark on the front side. There are less than 5 depressions, and the distance between any two depressions shall be no more than 15mm. When it is stretched, depressions or models tend to have smooth surfaces. The bumps formed by impurities on the surface do not affect the extension of the material.

3. Inspection Contents of Insulating Mat

In order to ensure the integrity of the electrical and mechanical properties of the insulating mat, its both sides shall be carefully inspected before each use. Its appearance and surface shall be flat and smooth, uniform in color, free from impurities, pinholes, cracks, cutting and other obvious defects; the surface of the insulating mat shall be clean and dry; the insulating mat shall have a test certificate within the test time limit.

4. Usage of Insulating Mat

(1) The insulating mat is only for live working of power equipment 10kV or below, as shown in Fig. 5-25.

(2) The insulating mat does not play the role in main insulation, but occasional short-term contact is allowed, and it shall be used in conjunction with the human body safety protection appliances.

Fig. 5-25 Use of Insulating Mat

5. Precautions for Storage of Insulating Mat

(1) The insulating mat shall be stored in a special box to avoid direct sunlight, rain and snow, and to prevent squeezing and sharp objects from colliding.

(2) It is prohibited to store it near steam pipe, heat dissipation pipe or other artificial heat source, and it shall be kept more than 1m away from the heat source. It is prohibited to store it in the environment with direct sunlight, light or other light source; no contact with oil, acid, alkali or other harmful substances is allowed. The storage ambient temperature should be between 10°C and 21°C.

(3) The damp insulating mat shall be dried, but the drying temperature shall not exceed 65°C.

6. Test contents of Insulating Mat

The test contents of the insulating mat are as follows.

(1) Mechanical test. Including mechanical puncture test and anti-slip test.

(2) Electrical test: Including AC voltage verification test and AC withstand voltage test which shall be done once every six months.

(3) Preventive testing item of the insulating mat include mark check.

5.3 Shield Appliance

The common shielding protection appliances used in live working of transmission line mainly include shielding clothes and anti-static clothes.

5.3.1 Shielding Clothes

1. Basic Structure and Functions of Shielding Clothes

(1) Basic structure characteristics

The complete set of shielding clothes shall include coats, pants, hats, socks, gloves, shoes and their corresponding connecting lines and connectors.

(2) Basic function

Shielding clothes for live working, also called equipotential voltage-sharing clothes, are made of uniform conductor material and fiber material. When wearing the shielding clothes, an equipotential shielding surface will be formed on all external parts of the human body in the high voltage electric field to protect the human body from damages of high voltage electric field and electromagnetic wave.

2. Classification and Technical Specifications of Shielding Clothes

(1) Classification

There are two types of shielding clothes. The shielding clothes for 110 (66)kV~500kV AC, ±500kV DC and below voltage class are Type I, and that for 750kV AC voltage class are Type II. Type II shielding clothes must be equipped with a mask, and the whole set of clothes shall be one-piece clothes, pants and hats. Type I shielding clothes are shown in Fig. 5-26, while Type II shielding clothes are shown in Fig. 5-27.

Fig. 5-26 Type I Shielding Clothes

Fig. 5-27 Type II Shielding Clothes

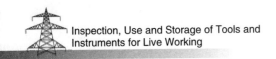
Inspection, Use and Storage of Tools and Instruments for Live Working

(2) Technical Parameters

① Shielding clothes shall be with good shielding performance, low resistance, proper flow capacity, certain flame retardancy and good wearability. Each part of shielding clothes shall be connected reliably through two detachable connectors.

② The resistance of the apparel fabric used for making the shielding clothes shall not be greater than 800m; the apparel fabric shall have certain spark resistance ability, meaning that when the high-frequency spark discharge generated by the charging capacitor, it will not be burnt away, with only carbonization and no open fire spread. After 2min of spark resistance test, the carbonized damage area of apparel fabric shall not be greater than 300mm^2.

③ The fusing current of the apparel fabric used for making the shielding clothes shall not be less than 5A.

④ The apparel fabric must be able to prevent the spread of open fire when contacting with it. The carbon length of the sample shall not be greater than 300mm; the burnt-out area shall not be greater than 100mm^2, and the burnt-out area shall not spread to the edge of the sample.

⑤ After washing for many times, the electrical and flame resistance properties of the apparel fabric do not decrease obviously.

⑥ The apparel fabric must be abrasion-resistant, so that the clothes has a certain durability value. After 500 friction tests, the resistance of apparel fabric shall not be greater than 1Ω.

⑦ For conductive fiber apparel fabric, the radial breaking strength shall not be less than 343N, the latitude breaking strength shall not be less than 294N, and the longitude and latitude breaking elongation shall not be less than 10%; for conductive coating apparel fabric, the radial breaking strength shall not be less than 245N, the latitude breaking strength shall not be less than 245N, and the longitude and latitude breaking elongation shall not be less than 10%.

⑧ The apparel fabric shall have a large permeation volume, so that wearers can feel comfortable. The air flow through the apparel fabric shall not be less than 35L/(m·2s).

⑨ After the whole set of shielding clothes are connected well, the resistance between any two farthest ends shall not be greater than 20Ω, the resistance between any two farthest ends of glove, sock, coat and pants shall not be greater than 15Ω, and the resistance of conductive shoes shall not be greater than 500Ω.

⑩ Good electrical connection between the hat and coat must be ensured. The hat must pass the shielding effect test. The shielding effect test of the hat is conducted together with the shielding property test of the whole set of clothes. For Type I shielding clothes, the protective tongue and extended edge of the hat must ensure that the exposed parts of the human body (such as the face) do not feel uncomfortable, and that the surface field strength of the exposed parts of the human body shall not be greater than 240kV/m at the highest applied voltage; Type II shielding clothes for 750kV voltage class shall be equipped with a shielding mask which shall be braided with conductive materials and flame-retardant fibers and shall have good vision, with the shielding efficiency no less than 20db; after the shielding clothes are supplied with the specified power frequency current and has been thermally stabilized for a certain period of time, the temperature rise of any part of the shielding clothes shall not exceed 50°C.

3. Inspection Contents of Shielding Clothes

The shielding clothes must be carefully inspected for voids, scratches, breakage and other defects before use; all parts are connected well, firmly and reliably. When testing the shielding clothes with special instruments, the resistance between any two farthest ends shall not be greater than 20Ω after the whole set of shielding clothes are connected well, the resistance between any two farthest ends of glove, sock, coat and pants shall not be greater than 15Ω, the resistance of conductive shoes shall not be greater than 500Ω, and the shielding efficiency of the shielding apparel fabric shall not be less than 40db.

4. Usage of Shielding Clothes

(1) When wearing, it is necessary to connect the multi-strand metal connecting wires of various parts like coat, hat, gloves, socks and shoes in accordance with the specified order. The flame-retardant underwear shall be worn inside the shielding clothes to avoid direct contact with skin, as shown in Fig. 5-28.

(2) It is strictly prohibited to short connect the grounding current, idle circuit and capacitance current of the coupling capacitor through the shielding clothes.

(3) After using, shielding clothes must be stored properly, and kept away from moisture and contaminant to avoid damages and affecting electrical performance.

Fig. 5-28 Use of Shielding Clothes

5. Precautions for Storage of Shielding Clothes

(1) The shielding clothes shall be kept in a dry and ventilated place. The shielding clothes shall be kept in a dry and well-ventilated tools and instruments warehouse as shown in Fig. 5-29, and placed on a upper shelf in a special tools and instruments box.

(2) When the shielding clothes is not worn, it shall be hung or wrapped into a cylindrical shape in a special box to prevent the break of functional fibers.

(3) The shielding clothes shall not be placed with other chemicals.

Fig. 5-29 Storage of Shielding Clothes

6. Test Contents of Shielding Clothes

The test contents of the shielding clothes are as follows.

(1) Efficiency test of apparel fabric. Including shielding efficiency test, resistance test of apparel fabric, fusing current test of apparel fabric, electrical spark resistance test, fire resistance test, wear resistance test, breaking strength and elongation test, finished product test.

(2) Test of finished shielding clothes. Including resistance test of coat and pants, resistance test of gloves and socks, resistance test of shoes, shielding effect test of hat, shielding efficiency test of shielding mask, resistance test of complete set of shielding clothes, internal

electric field strength test of complete set of clothes, test of internal current through human body of complete set of clothes, test of current capacity through human body of complete set of clothes, etc.

(3) Preventive test of shielding clothes. Including the resistance test of ready-made clothes and the shielding efficiency test of the whole set of clothes, with the test period of half a year.

5.3.2 Anti-static Clothes

1. Basic Structure and Function of Anti–static Clothes

(1) Basic structure

① The anti-static clothes for live working are made of special anti-static clean fabric. The fabric shall be special polyester filament yarn, with conductive fiber braided longitudinally or latitudinally.

② Different styles are provided according to the grade requirements, and it shall be sewed with conductive fiber to maintain the electrical continuity of each part of the clothes. The sleeve and pants have a unique double-layer structure, with conductive or anti-static threads in the inner layer to meet the requirements of a high-grade dust-free environment.

③ The whole set of high voltage anti-static clothes include coat, pants, hat, gloves and shoes, as shown in Fig. 5-30.

Fig. 5-30 Anti-static Clothes

(2) Basic function

① The anti-static clothes preventing the static agglomeration of clothing are sewn with anti-static fabric, with efficient and permanent anti-static, dust-proof performance, as well as the characteristics of thinness, slipperiness and clear texture.

② It can effectively protect the human body from high voltage electric field and electromagnetic wave.

③ It is used for tower climbing (frame) operators of transmission line and substation and ground operators who need to prevent electrostatic induction.

2. Classification and Technical Specifications of Anti-static Clothes

(1) Classification

The anti-static clothes has a split, one-piece, pullover and other styles, and socks, hats and masks can be freely combined. According to the distribution of conductive components in the fiber, the conductive fibers can be divided into 3 types: homogeneous conductive components, covered conductive components and composite conductive components.

(2) Technical specifications

① The shielding efficiency shall not be less than 28db, the resistance of apparel fabric shall not be greater than 300Ω, and the resistance of shoes shall not be greater than 500Ω.

② The apparel fabric shall have a large permeation volume, so that wearers can feel comfortable. The air flow through the apparel fabric shall not be less than 35L/(m·2s).

③ When all anti-static clothes are made of anti-static fabric, without lining. When the lining (pocket, reinforcing cloth, etc.) must be used, its exposed area accounts for less than 20% of the exposed area of all the inner surface of the anti-static clothes; mask and lining shall be detachable when it accounts for more than 20% (e.g., cold protective clothing or special clothing).

3. Inspection Contents of Anti-static Clothes

(1) Emphasis is placed on checking the anti-static clothes for any additional or worn metal items.

(2) After wearing for a period of time, the anti-static clothes shall be inspected. If the electrostatic performance does not meet the requirements, the anti-static clothes shall not be used again.

(3) The appearance shall be free from breakage, spots, dirt and other defects that affect the insulation performance.

(4) Metal accessories shall not be used on the clothes in general. Metal accessories shall not be exposed directly when they must be used (buttons, zippers, etc.).

4. Usage of Anti-static Clothes

(1) Under normal circumstances, when the explosive gas mixture continuously and fre-

quently emerges or exists for a long time, or when the explosive gas mixture is likely to appear, the minimum ignition energy of the combustible material shall be less than 0.25mJ, and the anti-static clothes shall be worn.

(2) Do not wear and take off anti-static clothes on flammable and explosive occasions.

(3) Do not attach or wear any metal items to the anti-static clothes.

(4) When the anti-static clothes are worn, anti-static shoes shall also be worn, and the ground shall be anti-static floor with grounding system at the same time.

(5) High voltage electrostatic protective clothes shall not be used as equipotential shielding clothes.

5. Precautions for Storage of Anti-static Clothes

(1) The newly sewn clean anti-static clothes can be washed directly, and the recycled clean worn anti-static clothes shall be carefully removed for oil stain (if any) before washing procedures. After washing, drying is carried out in a special clean air circulation system. After washing and drying, folding is carried out in a special clean room for washing. They shall be put into a clean polyester bag or pocket, and double-layer packaging or vacuum plastic sealing can be carried out as required.

(2) The anti-static clothes shall be stored in ventilated dry tools and instruments warehouse as far as possible. And they shall be placed on a upper shelf in a special tools and instruments box.

6. Test Contents of Anti-static Clothes

The test contents of anti-static clothes are mainly the shielding efficiency test of the whole set of protective clothes, with the test period of half a year.

Chapter 6

Common Instruments and Apparatus for Live Working

6.1 Spark Gap Detector

For many years, a great deal of research has been done on the detection of porcelain insulators in transmission and distribution lines at home and abroad. Many detection methods are put forward, such as voltage distribution measurement, infrared imaging, electric field measurement, ultrasonic measurement and so on. A large number of detection instruments are developed, such as photoelectric detection rod, self-climbing detector, ultrasonic detector, DC insulator detector and so on. But the spark gap detection device is really widely used in production practice. The spark gap detector has the advantages of simple structure, easy processing, convenient use, light weight, portability and so on, thus it is widely used in live working of transmission and distribution line.

1. Basic Structure and Function of Spark Gap Detector

(1) Basic structure

① As shown in Fig. 6-1, the spark gap detector is composed of supporting plate, contact electrode holder, spark electrode, contact electrode, connector, and the like.

② The spark electrode, the contact electrode and the holder can be made of medium carbon steel, and the contact electrode is made of round steel with the diameter of 3mm, with the chamfer in the front section.

Chapter 6
Common Instruments and Apparatus for Live Working

1-supporting plate; 2-contact electrode holder; 3-spark electrode; 4-contact electrode; 5-connector

Fig. 6-1　Sample Drawing of Spark Gap Detector

③ The supporting plate can be made of epoxy laminated glass cloth sheet.

(2) Basic function

Spark gap detector is mainly used for live detection of zero value porcelain insulator to judge the quality of porcelain insulator.

2. Classification and Technical Specifications of Spark Gap Detector

(1) Classification

① At present, spark gap detector used in China can be classified into two types: fixed type and variable type. Fixed type refers to that the gap is fixed and unchangeable in the detection process. Variable type refers to that the gap distance can be changed in the detection process and the distribution voltage of insulator can be roughly estimated.

② According to the type of the spark electrode, it can be divided into a ball-ball electrode (as shown in the right side device of Fig. 6-1) and a tip-tip electrode (as shown in the left side device of Fig. 6-1).

(2) Technical specifications

① The supporting plate shall conform to the requirements of (GB/T1303.1-2009) Industrial Rigid Laminated Sheets Based on Thermosetting Resins for Electrical Purposes—part 1: Definitions, Designations and General Requirements, and shall be 3~5mm thick and convenient for installation of holder and connector.

② After assembly of the two electrodes, the center line of the two electrodes shall be on one axis, with the maximum deviation no more than 0.5mm.

③ The telescopic insulating rod installed on corresponding voltage class when using.

④ When detecting the low and zero value insulators of insulator string in each voltage class, the corresponding electrode gap distances are shown in Table 6-1.

Table 6-1 Detecting the Corresponding Electrode Gap Distances of Insulator String in Each Voltage Class

System nominal voltage (kV)	63	110	220	330	500
Spark electrode gap distance (mm)	0.4	0.5	0.6	0.6	0.6

3. Inspection Contents of Spark Gap Detector

The spark gap detector must be inspected before use as follows.

(1) The spark gap detector shall be tested before use to ensure flexible operation and accurate measurement.

(2) The spark electrode gap distance shall be adjusted according to the voltage class of the detection equipment. The inspection of spark gap detector is shown in Fig. 6-2.

Fig. 6-2 Inspection of Spark Gap Detector

4. Usage of Spark Gap Detector

(1) Pin type insulator and suspension type insulator of less than 3 pieces shall not be tested with spark gap detector. In fact, the spark gap detection method is to determine the insulating property of the insulator by testing and short connecting 1 insulator. When the number of insulators in the insulator string is less than 3 pieces, if 1 piece is zero, it will easily cause ground short circuit fault and burn out tools and instruments during the detection, resulting in equipment accidents.

(2) When detecting insulator string with a voltage level of 35kV and above, if it is found that the number of zero insulator pieces in the same string reaches the number in Table 6-2, it shall be stopped immediately.

Table 6-2 Allowable Number of Zero Insulator Pieces in One String

Voltage level (kV)	63(66)	110	220	330	500
Number of insulators in a string	5	7	13	19	28
Number of zero insulators	2	3	5	4	6

(3) Live detection of insulator shall be carried out in dry weather.

(4) The detection shall be carried out from the conductor end to the cross arm end piece by piece.

(5) Attention shall be paid to the difference between the discharge sound when the detector is close to the conductor and the discharge sound of the spark gap to avoid misjudgment.

(6) During the measurement, the minimum safe distance between the operator and the electrified body as well as the effective insulation length of the insulating bar used with spark gap detector shall be ensured.

(7) During the detection, if there is zero value insulator, it shall be retested for 2~3 times. When spark gap detector is used, the two metal probes shall slightly contact the steel cap and steel foot of porcelain insulator. Operators shall pay attention to avoid measuring the gap deformation.

The insulator is detected using a spark gap detector as shown in Fig. 6-3 and Fig. 6-4.

Fig. 6-3 Sample Drawing 1 of Detecting the Insulator Using Spark Gap Detector

Fig. 6-4 Sample Drawing 2 of Detecting the Insulator Using Spark Gap Detector

5. Precautions for Storage of Spark Gap Detector

The spark gap detector shall be stored in the special warehouse for tools and instruments for live working, and shall be placed on the upper shelf for storage. The tool cabinet for storing the spark gap detector is shown in Fig. 6-5.

Fig. 6-5 Tool Cabinet for Storing Spark Gap Detector

6. Preventive Test of Spark Gap Detector

The preventive test period of the spark gap detector is one year, with the test types as follows.

(1) Gap adjustment and discharge test.

(2) Power frequency withstand voltage test.

(3) Switching impulse voltage withstand test.

Chapter 6
Common Instruments and Apparatus for Live Working

6.2 Insulation Resistance Tester

Insulation diagnosis is an important measure to detect insulation defects or faults of electrical equipment. Insulation resistance tester (tramegger) is a special instrument for measuring insulation resistance. At present, insulation resistance tester can be divided into two kinds: digital type and pointer type. Digital insulation resistance tester has been developed to the intelligent level of digital display. Pointer-type insulation resistance tester can be divided into electronic type and hand-operated type. At present, pointer-type insulation resistance tester is mostly used in the production site.

6.2.1 Hand-operated Insulation Resistance Tester

Insulation resistance tester, commonly known as tramegger, is a portable instrument used to test the insulation resistance of electrical equipment. Its unit of measurement is megohm (MΩ), thus it is called a megameter.

1. Basic structure and Function of Hand-operated Insulation Resistance Tester

(1) As shown in Fig. 6-6, a general megameter is mainly composed of hand generator, ratio type magnetoelectric system measuring mechanism, measuring circuit, and the like.

(2) The hand-operated insulation resistance tester has three terminals, the two larger terminals at the upper end are marked with "Earthing" (E) and "Line" (L), respectively, and the smaller terminal at the lower end is marked with "Guard Ring" (or "Shield") (G).

1-Line terminal; 2-Earthing terminal; 3-Shield end button; 4-Meter cover; 5-Dial; 6-Handle;
7-Generator crankshaft; 8-Testing electrode

Fig. 6-6 Hand-operated Insulation Resistance Tester

2. Technical Parameters

(1) When measuring the insulation resistance of equipment or line with rated voltage below 500V, 500V or 1,000V hand-operated insulation resistance tester shall be used.

(2) When measuring the insulation resistance of equipment or line with rated voltage above 500V, 1,000~2,500V hand-operated insulation resistance tester shall be used.

(3) A 2,500-5,000V megameter shall be used for measuring insulators. In general, the hand-operated insulation resistance tester with the range of 0~200M can be used to measure the insulation resistance of low-voltage electrical equipment.

(4) The hand-operated insulation resistance tester is divided into 5 grades according to accuracy: 1.0, 2.0, 3.0, 5.0, 10.0, 20.0.

3. Inspection Contents of Hand-operated Insulation Resistance Tester

The megameter must be inspected before use, the contents are as follows.

(1) Check whether the hand-operated insulation resistance tester is with good appearance and clean surface.

(2) Check whether the connecting wire of the hand-operated insulation resistance tester is in good condition and whether the connection is firm and reliable.

(3) Check the factory certificate and calibration certificate of the hand-operated insulation resistance tester and confirm that the insulation resistance tester is within the inspection period.

(4) Carry out an open circuit test on the hand-operated insulation resistance tester. Open the two connecting wires and shift the handle pointer to infinite.

(5) Carry out short circuit test on hand-operated insulation resistance tester. Slowly shake the megohmmeter handle and short connect the two connecting wires. The pointer shall point to zero. It should be noted that the shaking of megohmmeter handle can be stopped only after the two connecting wires are disconnected.

Fig. 6-7 shows the inspection of the hand-operated insulation resistance tester, and Fig. 6-8 shows the open circuit test of the hand-operated insulation resistance tester.

Fig. 6-7 Inspection of Hand-Operated Insulation Resistance Tester

Chapter 6
Common Instruments and Apparatus for Live Working

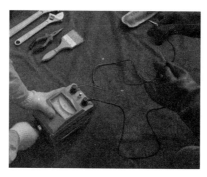

Fig. 6-8　Open Circuit Test of Hand-Operated Insulation Resistance Tester

4. Usage of Hand-operated Insulation Resistance Tester

(1) The "grounding" terminal (i.e. E terminal) of the hand-operated insulation resistance tester is reliably grounded (generally connected to a grounding body) through the test line, and the "line" terminal (i.e. L terminal) is connected to the object to be tested through the test line.

(2) After connection, shake the hand-operated insulation resistance tester clockwise, at an gradually increasing rotation speed, and then shake at a constant speed after keeping at about 120 rpm. After the rotation speed is stable and the pointer of the meter is stable, the value indicated by the pointer is the insulation resistance value of the measured object.

(3) In actual use, the two terminals E and L can also be connected at will, i.e. E can be connected with the object to be tested, and L can be connected with the grounding body (i.e. grounding), but the G terminal shall not be connected incorrectly.

(4) Precautions for use.

① Open circuit and short circuit tests shall be conducted before use. Disconnect the L and E terminals, and shake the hand-operated insulation resistance tester, then the pointer shall point to "∞"; Short connect the L and E terminals and turn them slowly. The pointer shall point to "0". Both of them meet the requirements, which indicates that the hand-operated insulation resistance tester is normal.

② When the insulation resistance of electrical equipment is measured, the power supply shall be cut off firstly, and then the equipment shall be discharged to ensure personal safety and accurate measurement.

③ When it is measured, the hand-operated insulation resistance tester shall be placed in a horizontal position, and the hand-operated insulation resistance tester shall be pressed firmly to prevent waggle during shaking. The shaking speed is 120rpm.

④ The test line shall be multi-strand flexible wire with good insulation performance. The two test lines shall not be twisted together to avoid inaccurate measurement data.

⑤ The tested object shall be discharged immediately after the measurement. Before the crankshaft of the tramegger stops rotating and the tested object is not discharged, it is not allowed to touch the measured part of the tested object by hand or remove the conductor to prevent electric shock.

Fig. 6-9 shows the use of the hand-operated insulation resistance tester.

Fig. 6-9　Use of Hand-Operated Insulation Resistance Tester

5. Precautions for Storage of Hand-operated Insulation Resistance Tester

(1) The insulation resistance tester shall be stored in a well-ventilated, dry and clean environment, and put on shelves.

(2) Properly maintain instruments and apparatus.

(3) Properly keep technical data and instructions of instruments and apparatus.

As shown in Fig. 6-10, a tool cabinet for storing a hand-operated insulation resistance tester is provided.

Fig. 6-10　Tool Cabinet for Storing Hand-operated Insulation Resistance Tester

6. Preventive Test of Hand-operated Insulation Resistance Tester

The preventive test of the hand-operated insulation resistance tester is mainly to check the hand-operated insulation resistance tester regularly. The verification period of the hand-operated insulation resistance tester shall not exceed two years.

6.2.2 Electronic Insulation Resistance Tester

1. Basic Structure And Function of Electronic Insulation Resistance Tester

(1) Basic structure

The electronic insulation resistance tester is mainly composed of a ground terminal, a shielding terminal, a wiring terminal, an instrument dial, a luminous display tube, a zero adjusting device, a test key, a band switch and a test electrode, and the functions of each part are as follows.

① Earth terminal. Connect to the shell or ground of the tested equipment.

② Shielding end. Connect to the high-voltage guard ring of the tested equipment to eliminate the influence of surface leakage current.

③ Line end. The high voltage output port is connected to the high voltage conductor of the tested equipment.

④ Double-row tick marks. The upper gear is green - 5000V/2GΩ-200GΩ, 10000V/4gΩ-400GΩ. The lower gear is red - 5000V/0-4000MΩ, 10000V/0-8000MΩ.

⑤ Green LED. Read the green gear (upper gear) when emitting light.

⑥ Red LED. Read the red gear (lower gear) when emitting light.

⑦ Mechanical zero adjustment. Adjust the position of the mechanical pointer to align with the ∞ tick mark.

⑧ Test key. Press to start the test, and then rotate clockwise to lock this key.

⑨ Band switch. It can realize output voltage selection, battery detection, power switch and other functions.

⑩ Test electrode. One end is connected to the meter, and the other end is in contact with the measured object.

(2) Basic function

Insulation resistance tester can be used to measure insulation resistance of transformers, mutual inductors, generators, high voltage motors, power capacitors, power cables, arresters, etc.

The appearance of the electronic insulation resistance tester is shown in Fig. 6-11.

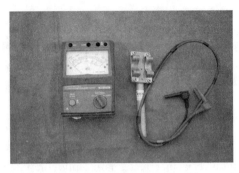

1-Earth terminal; 2-Line terminal; 3-Shielding terminal; 4- Double-row tick marks; 5-Green LED;
6-Red LED; 7-Mechanical zero adjustment; 8-Test key; 9-Band switch; 10-Test electrode

Fig. 6-11   Electronic Insulation Resistance Detector

2. Technical Parameters of Electronic Insulation Resistance Tester

(1) Output voltage: 500V, 1000V, 2500V, 5000V.

(2) Insulation resistance: 200GΩ.

(3) Accuracy: 0.05.

3. Inspection contents of electronic insulation resistance tester

The electronic insulation resistance tester must be checked before use (as shown in Fig. 6-12), as follows.

(1) Check whether the appearance of electronic insulation resistance tester is good.

(2) Check the factory certificate and calibration certificate of the electronic insulation resistance tester and confirm that the electronic insulation resistance tester is within the inspection period.

(3) Check whether the pointer of the instrument is on infinite, otherwise, adjust the mechanical zero adjustment screw.

(4) Check whether the instrument battery is in good condition.

(5) Check whether the joints are in good condition.

Fig. 6-12   Inspection of Insulation Resistance Tester

4. Usage of Electronic Insulation Resistance Tester

(1) Insert one end (red) of the high voltage test line into the LINE terminal, connect the other end to or use a hook to hang it on the high-voltage conductor of the tested equipment, insert one end of the green test line into the GUARD terminal, and connect the other end to the high-voltage guard ring of the tested equipment to eliminate the influence of surface leakage current. Insert another black test wire into the EARTH terminal and connect the other end to the shell or ground of the tested equipment. When wiring, pay special attention to the connection between LINE (red) and GUARD (green), and do not short circuit them.

(2) Turn the band switch BATT.CHECK, press the test key, and the meter starts to detect the battery capacity.

(3) Turn the band switch to select the required test voltage.

(4) Press or lock the test key to start the test. At this time, the high-voltage output indicator light above the test key will illuminate and the buzzer built in the meter will sound every 1 second, which means that there is high-voltage output at the LINE end. Note that during the test, it is strictly prohibited to touch the exposed part at the front end of the probe to avoid electric shock hazard.

(5) When the green LED is on, read the insulation resistance value of the outer ring (high range); when the red LED is on, read the insulation resistance value of the inner ring. After the test, release the test key, stop the test, wait for a few seconds, and do not remove the test electrode immediately. At this time, the meter will automatically release the remaining charge in the test circuit. Note: when the test is completed or repeated, the test object must be fully discharged to ground after short connection (the meter also has built-in automatic discharge function, but for a long time).

(6) If the second measurement is required continuously, the steps (3) to (5) can be performed. Note: If the test is not carried out for a long time, the battery in the battery compartment shall be taken out so as not to damage the instrument due to leakage of battery liquid.

(7) For two or more test pieces, such as lightning arresters and coupling capacitors, the wiring shown in Fig. 6-13 can be used for measurement. In the figure, the shielding end is connected to a flange on the lightning arrester under test. In this way, the interference current caused by the high voltage line above is shielded by the shielding terminal without passing through the main circuit under test, thus avoiding the influence of interference current. For the arrester at the top section, the upper flange of the arrester can be connected to the

EARTH of the instrument and then grounded, so that the interference current is directly connected to the earth. However, the latter cannot completely eliminate the interference.

As shown in Fig. 6-14, the insulation resistance tester is used.

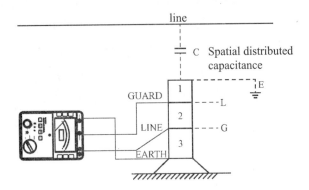

Fig. 6-13 Wiring of More Than Two Tested Items

Fig. 6-14 Use of Electronic Insulation Resistance Detector

5. Precautions for Storage of Electronic Insulation Resistance Tester

(1) The electronic insulation resistance tester shall be stored in a well-ventilated, dry and clean environment, and put on shelves.

(2) Properly maintain instruments and apparatus.

(3) Properly keep technical data and instructions of instruments and apparatus.

Fig. 6-15 shows the storage tool cabinet of insulation resistance tester.

Fig. 6-15 Storage Tool Cabinet for Insulation Resistance Detector

Chapter 6
Common Instruments and Apparatus for Live Working

6. Preventive Test of Electronic Insulation Resistance Tester

The preventive test of the electronic insulation resistance tester is mainly to check the electronic insulation resistance tester regularly. The verification period of the electronic insulation resistance tester shall not exceed 2 years.

6.2.3 Other Commonly Used Insulation Resistance Testers

There are many types of insulation resistance testers commonly used on the production site, as shown in the two types shown in Fig 6-16 and 6-17. See the relevant technical manuals for their structure, function, inspection and usage.

Fig. 6-16　Rapid Insulation Resistance Tester

Fig. 6-17　Digital Insulation Resistance Tester

6.3 Clamp Ammeter

1. Basic Structure and Function of Tong-type Ammeter

(1) Basic structure

Tong-type ammeter, is mainly composed of an electromagnetic ammeter and a through-type current transformer. The core of the through-type current transformer is made into a movable opening and is tong-shaped, hence it is named as tong-type ammeter. Fig. 6-18 shows the structure diagram of tong-type ammeter.

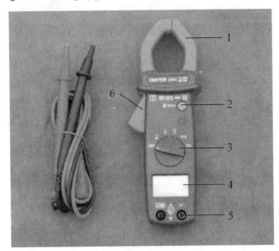

1-Bayonet; 2-Data lock button; 3-Gear selection; 4-Digital display screen; 5-Terminal block; 6-Wrench

Fig. 6-18 Structure Diagram of Tong-type Ammeter

(2) Basic function

Tong-type ammeter is a kind of portable instrument that can directly measures AC current of circuit without disconnecting the circuit, which is very convenient to use in electrical maintenance.

2. Classification and Technical Specifications of Tong-type Ammeter

(1) Classification

① Classified as pointer type and digital type according to the display mode.

② Classified as AC clamp ammeter, multi-purpose clamp ammeter, harmonic digital clamp ammeter, leakage clamp ammeter and AC-DC clamp ammeter according to the function.

③ Classified as high-voltage clamp meter and low-voltage clamp meter according to the measured voltage.

④ Classified as transformer type and electromagnetic system type according to the structure and application. Commonly used is transformer type tong-type ammeter, which consists of current transformer and rectifier instrument It can only measure alternating current. The deflection of the movable part of the electromagnetic instrument is unrelated to the polarity of current, so it can be used in both AC and DC.

(2) Technical specifications

The accuracy of tong-type ammeter mainly includes 2.5, 3.0 and 5.0 grades.

3. Check Contents of Tong-type Ammeter

The tong-type ammeter must be inspected before use as follows.

(1) Check whether the tong-type ammeter has the factory certificate and calibration certificate, and whether it is within the validity period of the test.

(2) Before use, check whether the tong-type ammeter is damaged and whether the pointer points to zero. If it is found that the pointer does not point to the zero position, use a small screwdriver to gently turn the mechanical zero adjustment knob to return the pointer to the zero position.

(3) Check whether the insulation material on the jaw has any damage such as falling off and cracking, and whether there is any obvious gap after closing. If so, it must be repaired before use.

(4) Check the whole shell of tong-type ammeter, including gauge glass, which shall be free of cracking and breakage, because the insulation of the jaw and the integrity of the meter shell are directly related to the safety of measurement and the performance of the meter.

(5) For multi-purpose tong-type ammeter, check whether the test line and meter bar are damaged, and they shall have good conductivity and be free of damage.

(6) For the digital tong-type ammeter, it is also necessary to check whether the battery in the checklist is sufficient or not, and it must be updated if insufficient.

Fig. 6-19 shows the inspection of tong-type ammeter.

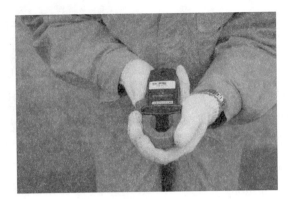

Fig. 6-19　Inspection of Tong-type Ammeter

4. Usage of Tong–type Ammeter

(1) Before measurement, the current to be measured shall be estimated first to select appropriate range. If it is not possible to estimate, a large range shall be selected first, and then reduced gradually to the appropriate gear. The shifting of range gear must be done in the case of no electricity or under the opening of the jaw so as not to damage the instrument.

(2) During measurement, the measured conductor shall be placed in the middle of the jaw as far as possible. If there is noise on the joint surface of the jaw, it shall be opened and closed again. If there is still noise, the joint surface shall be treated to make the reading accurate. In addition, the two conductors cannot be clamped at the same time.

(3) The measurement of high-voltage clamp ammeter shall be operated by two persons wearing insulating gloves and standing on insulating pad without touching other equipment to prevent short circuit or grounding.

(4) When measuring the current below 5A, in order to get a more accurate reading, if possible, the conductor should be wound several times before put into the jaw for measurement. The actual current value shall be the instrument reading divided by the number of conductors put into the jaw.

(5) During the measurement, a safe distance shall be kept between all parts of the body and the electrified body. The safety distance of the low-voltage system is 0.1-0.3m. When measuring current of each phase of high-voltage cable, the distance between the cable head and conductor shall be more than 300mm, with good insulation and measuring operation conditions. When observing the meter, a safe distance between the head and the electrified part shall be kept. The distance between any part of the human body and the electrified body shall not be less than the entire length of the clamp meter.

Chapter 6
Common Instruments and Apparatus for Live Working

(6) When measuring low-voltage fusible fuse or horizontal arranged low-voltage bus current, each phase fuse and busbar shall be protected and isolated by insulating materials, so as not to cause inter-phase short-circuit. If one phase of cable is grounded, measurement is not allowed, so as to prevent breakdown of the ground due to the low insulation level of the cable head, which shall endanger personal safety.

(7) Before and after each measurement, the switch for regulating current range shall be adjusted to the most advanced position, so as to avoid instrument damage by measuring the instrument without selecting the range.

Fig. 6-20 shows the use of tong-type ammeter.

Fig. 6-20 Use of Tong-type Ammeter

5. Precautions for Storage of Tong-type Ammeter

(1) The tong-type ammeter shall be stored in a well-ventilated, dry and clean environment, and put on shelves.

(2) Properly maintain instruments and apparatus.

(3) Properly keep technical data and instructions of instruments and apparatus.

(4) The tong-type ammeter shall be maintained regularly. During maintenance, first check whether there is any damage to all components and parts of the meter appearance, wipe off the floating dust with a soft brush or cotton cloth, then carefully wipe and dry it with a little non-corrosive solvent or clean water. During maintenance, the battery box can be opened for inspection or scrubbing, but the meter shell cannot be opened.

(5) If the tong-type ammeter is internally equipped with a battery, it must be taken out to prevent the battery from leaking and damaging the ammeter.

(6) The indoor light for the tong-type ammeter shall not be too strong, and direct sunlight is strictly prohibited on the ammeter, because it will accelerate the aging of its plastic shell under the action of sunlight, while direct sunlight shall be prevented for the liquid crystal display screen. Fig. 6-21 shows a tool cabinet for storing tong-type ammeter.

Fig. 6-21 Tool Cabinet for Storing Tong-type Ammeter

6. Preventive Test of Tong–type Ammeter

The preventive test type of tong-type ammeter is mainly to check tong-type ammeter regularly. The checking period is usually 3 months, or it shall be checked after the meter is continuously used for multiple times.

Chapter 6
Common Instruments and Apparatus for Live Working

6.4 High Voltage Phase Detector

In the construction or maintenance of the power system loop network and dual power supply power network, phase inspection of the power supplies on both sides of the closed-loop point circuit breaker is a very important test item, otherwise interphase short circuit may occur, causing terrible consequences.

1. Basic Structure and Function of High Voltage Phasing Tester

(1) Basic structure

① The high-voltage phasing tester consists of three parts: two voltage detection and emission combined devices (including insulating rod) and a wireless receiving device. Fig. 6-22 shows the structure of the HV phasing tester.

② The two transmitting devices respectively send the sampled voltage data signals to the receiving end through the high-frequency transmitting device, and the receiving end carries out corresponding processing according to the signals transmitted by the transmitter, and displays the phase relation and the frequency of the two lines on the display screen.

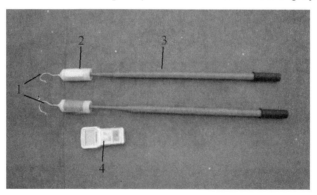

1-Contact electrode; 2-Emitter; 3-Insulating rod; 4-Receiver

Fig. 6-22 Structure of HV Phasing Tester

(2) Basic function

Perform phasing operation on power lines.

2. Classification and Technical Specifications of Hv Phasing Tester

(1) Classification

① According to the working principle of phasing meter, HV phasing meter can be divided into capacitive phasing meter and resistive phasing meter. Capacitive phasing detector is used to detect and indicate the equipment based on the phase relation of current grounding through stray capacitance. It has two types: double-pole unconnected lead and single-pole memory system. Resistive phasing meter is used to detect and indicate equipment based on the phase relation of current passing through resistive elements, such as double-pole phasing meter.

② According to the type of phasing meter, it can be divided into single phasing meter and split phasing meter.

(2) Technical specifications

① Use range: 6-35kV.

② Maximum distance between transmitters: 20m.

③ Insulating rod: 2 pieces with 4 sections in total. When in use, the two sections shall be jointed. The insulating rod is made of epoxy resin.

3. Inspection Contents of Hv Phasing Tester

The HV phasing tester must be inspected before use as follows.

(1) Check whether the appearance of the HV phasing tester is good.

(2) Check the factory certificate and calibration certificate of the HV phasing tester and confirm that the HV phasing tester is within the inspection period.

(3) Check the insulation rod withstand voltage test report. If there is no qualified insulation rod withstand voltage test report, it cannot be used. It is forbidden to use the HV phasing tester in rainy days.

(4) Check whether the meter battery is in good condition and charge it before use.

(5) Before phase test, check whether the phasing tester is normal on the same power grid, with one person for operation and one person for monitoring. Fig. 6-23 shows the inspection of HV phasing tester, and Fig. 6-24 shows the inspection of insulation rod of HV phasing tester.

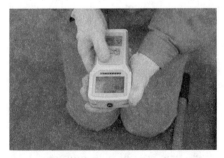

Fig. 6-23 Inspection of HV Phasing Tester

Chapter 6
Common Instruments and Apparatus for Live Working

Fig. 6-24 Inspection of Insulating Rod of HV Phasing Tester

4. Usage of HV Phasing Tester

(1) Connect the transmitter and insulating rod before use.

(2) Check the phase. Hook the metal of the phasing tester main pole (with meter pole) to any phase of the live A-side line (the meter is with indication in this case), and connect the other pole (auxiliary pole) to any phase of the B-side line. If the meter indicates zero or close to zero at this time, the phases of the two lines of the tested line are in phase. When the pointer of the watch indicates the potential difference (line voltage) between the two outputs, the two phases are out of phase.

(3) Check the electricity. According to the phasing tester assembly, the main pole is connected to the high-voltage line, the auxiliary pole is grounded or connected to the iron cross arm, and another wire can be connected. If the indicator value is large, it indicates that the line is charged, otherwise, it is not.

(4) When checking on grounding fault. In the power system with neutral point not directly grounded, it can be used to locate the grounding point, connect the main pole to the line of grounding accident, and connect the auxiliary pole to the grounding wire and metal frame or cross arm. At this time, the meter will indicate zero. When a device with accident is disconnected, the indicating value of the meter will be large, and the grounding will be eliminated.

5. Precautions for Storage of Hv Phasing Tester

(1) The HV phasing tester shall be stored in a well-ventilated, dry and clean environment, and put on shelves.

(2) Properly maintain instruments and apparatus.

(3) Properly keep technical data and instructions of instruments and apparatus. Fig.

6-25 shows a tool cabinet for storing HV phasing tester.

Fig. 6-25　Tool Cabinet for Storage of HV Phasing Tester

6. Preventive Test of HV Phasing Tester

The types of preventive tests for HV phasing tester are as follows.

(1) The HV phasing tester shall be sent for inspection on a regular basis.

(2) The HV phasing tester shall be regularly checked.

(3) The insulation rod of the HV phasing tester shall be subjected to pressure test regularly, and the period of pressure test of the HV phasing tester shall be half a year.

Chapter 6
Common Instruments and Apparatus for Live Working

6.5 Anemometer and Thermohygrometer

Due to the strict requirements on wind speed, temperature and humidity for live working of power transmission and distribution lines, it is necessary to accurately detect the wind speed and temperature and humidity on site before each live working.

6.5.1 Anemometer

1. Basic Structure and Function of Anemometer

(1) Basic structure

The anemometer consists of two parts: a wind speed measuring sensor and a display meter.

(2) Basic function

An anemometer that converts velocity signals into electrical signals. Fig. 6-26 shows the structure of anemometer.

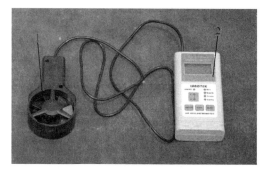

1-Sensor; 2-Display meter

Fig. 6-26 Structure of Anemometer

2. Classification of Anemometers

Anemometer can be classified as wind cup type anemometer, propeller type anemometer, hot wire anemometer, digital anemometer and acoustic anemometer according to the principle. At present, digital anemometer is most commonly used in live working.

3. Inspection Contents of Anemometer

The anemometer must be inspected before use as follows.

(1) Check whether the appearance of anemometer is good.

(2) Check the factory certificate and calibration certificate of the anemometer and confirm that the anemometer is within the inspection period.

(3) Check whether the battery of anemometer is good.

(4) Check whether the joints are in good condition.

4. Usage of Anemometer

(1) Insert the sensor into the display meter, then raise the sensor with one hand, and operate the reading of the display meter with the other hand.

(2) Keep the posture unchanged, rotate a certain angle, and measure the wind speed again.

(3) Repeat the above steps and take the maximum wind speed after several more points are measured.

(4) If the anemometer has abnormal smell, sound or smoke, or there is liquid flowing inwards during operation, turn off the instrument and take out the battery immediately. Otherwise, there will be danger of electric shock, fire or damage to anemometer.

(5) Do not expose the probe and anemometer body to rain. Otherwise, there will be danger of electric shock, fire or personal injury.

(6) Do not touch the sensor inside the probe.

(7) It is forbidden to use the anemometer in combustible gas environment.

(8) Do not drop or press the anemometer. Otherwise, malfunction or damage may be caused to the anemometer.

(9) Do not touch the sensor of the probe when the anemometer is charged. Otherwise, the measurement result may be affected or damage to the internal circuit of the anemometer may be caused.

Fig. 6-27 shows the use of anemometer.

Fig. 6-27　Use of Anemometer

5. Precautions for Storage of Anemometer

(1) The anemometer shall be stored in a well-ventilated, dry and clean environment, and put on shelves.

(2) Properly keep technical data and instructions of instruments and apparatus.

(3) Take out battery when the anemometer is not in use for a long time. Otherwise, battery may leak, causing damages to anemometer;

(4) Do not place anemometer in a place with high temperature, high humidity, dust and direct sunlight. Otherwise, air will cause damage to the internal device or deterioration of anemometer performance.

(5) Do not wipe the anemometer with volatile liquid. Otherwise, air may cause deformation and discoloration of the anemometer shell. When there are stains on the surface of anemometer, it can be wiped with soft fabric and neutral detergent.

Fig. 6-28 shows the tool cabinet for storing anemometers.

Fig. 6-28　Tool Cabinet for Storing Anemometer

6. Preventive Test of Anemometer

The preventive test type of anemometer is mainly to check the anemometer regularly, and the test cycle shall be half a year.

6.5.2　Temperature and Humidity Meter

1. Basic structure and Function of Temperature and Thermohygrometer

(1) Basic structure

The thermohygrometer consists of temperature and humidity probe and display meter.

(2) Basic function

Thermohygrometer is an instrument used to measure instantaneous temperature and humidity as well as average temperature and humidity, with functions of measuring temperature and humidity, display, record, real-time clock, data communication and over limit alarm. Fig. 6-29 shows the structure of thermohygrometer.

2. Technical Parameters of Thermohygrometer

(1) Range: humidity 0%-100%R.H., temperature $-20°C - +50°C$.

1-Temperature and humidity probe; 2-Display meter
Fig. 6-29 Structure of Thermohygrometer

(2) Resolution: 0.1%R.H, 0.1°C, 0.1°F.

(3) Accuracy: humidity is ±3%, temperature is ±1°C.

(4) Display: Double display of temperature and humidity.

3. Inspection Contents of Thermohygrometer

The thermohygrometer must be inspected before use as follows.

(1) Check whether the appearance of thermohygrometer is good.

(2) Check the factory certificate and calibration certificate of the thermohygrometer and confirm that the thermohygrometer is within the inspection period.

(3) Check whether the battery of thermohygrometer is good.

(4) Check whether the joints are in good condition.

4. Usage of Thermohygrometer

(1) Connect the temperature and humidity probe to the display meter, then raise the temperature and humidity probe with one hand, and operate the reading of the display meter with the other hand.

(2) If the thermohygrometer is with abnormal smell, sound or smoke, or there is liquid flowing inwards during operation, turn off the instrument and take out the battery immediately. Otherwise, there will be danger of electric shock, fire or damage to thermohygrometer.

(3) Do not expose the probe and thermohygrometer body to rain. Otherwise, there will be danger of electric shock, fire or personal injury.

(4) Do not touch the sensor inside the probe.

(5) Do not drop or press the thermohygrometer. Otherwise, malfunction or damage may be caused to the thermohygrometer. Fig. 6-30 shows the use of thermohygrometer.

Fig. 6-30　Use of Thermohygrometer

5. Precautions for Storage of Thermohygrometer

(1) The thermohygrometer shall be stored in a well-ventilated, dry and clean environment, and put on shelves.

(2) Properly maintain instruments and apparatus.

(3) Properly keep technical data and instructions of instruments and apparatus.

(4) Take out battery when the thermohygrometer is not in use for a long time. Otherwise, battery may leak, causing damages to thermohygrometer.

(5) Do not place thermohygrometer in a place with high temperature, high humidity, dust and direct sunlight. Otherwise, damage to the internal device or deterioration of thermohygrometer performance will be caused.

Fig. 6-31 shows the tool cabinet for storing thermohygrometer.

Fig. 6-31　Tool Cabinet for Storing Thermohygrometer

6. Preventive Test of Thermohygrometer

The preventive test type of thermohygrometer is mainly to check the thermohygrometer every year, and the test cycle shall not exceed one year.

Chapter 7

Safety and Insulating Hand Tools

7.1 Security Tools

Safety tools commonly used for live working of transmission and distribution lines include portable short circuit ground wires, safety belts, climbers, lifting boards, insulated wrenches, pin-pulling pliers, strippers, bolt clippers, etc. The inspection, use and storage of safety tools are important contents of live working of power transmission and distribution lines.

7.1.1 Portable Short Circuit Ground Wire

1. Basic Structure and Function of Portable Short Circuit Grounding Wire

(1) Basic structure

① As shown in Fig. 7-1, portable short circuit ground wire is composed of conductor wire clamp, short-circuit wire, bus clamp, ground wire, wiring clip, ground terminal clamp, grounding operating rod, etc.

② wire clamps for conductor terminal and ground terminal are die-cast of high-quality aluminum alloy, and the metal of fasteners matched are nickel plated.

③ The grounding operating rod is made of high-grade epoxy resin and glass fiber with excellent insulation performance and mechanical strength. Meanwhile, a silicone rubber sheath is provided at the operating handle to ensure safer and more reliable insulation.

④ The short-circuit wire and grounding wire are twisted with multiple strands of high-quality soft copper wires, and are covered with a soft and high-temperature-resistant transparent insulating protective layer, which can prevent the abrasion of the grounding copper wires in use and ensure the safety of operators in operation.

⑤ A new crimping process shall be adopted for the connection nose, bus clamp, grounding copper wire and outer sheath. Soft connection can effectively prevent the copper wire at the connection from breaking during use, thus improving the reliability and service life of the grounding wire.

(2) Basic function

The portable short circuit ground wire can prevent the harm to human body caused by induced voltage generated by adjacent HV electrified equipment, and can also be used for discharging the charge on the conductor in the intense electric field and releasing the residual charge of the power-off equipment.

1-Grounding operating rod; 2-Conductor wire clamp; 3-Bus clamp; 4-Short-circuit wire;
5-Wiring clip; 6-Ground wire; 7-Ground terminal wire clamp

Fig. 7-1　Portable Short Circuit Ground Wire

2. Classification and Performance Parameters of Portable Short Circuit Ground Wire

(1) Classification

① According to the use environment, it can be divided into indoor bus-type ground wire (JDX-NL) and outdoor line-type ground wire (JDX-WS).

② According to the voltage grade, it can be divided into 0.4kV, 10kV, 35kV, 110kV, 220kV, 500kV grounding wires, etc.

③ The nominal cross-sectional areas of portable short-circuit grounding wires are: $16mm^2$, $25\ mm^2$, $35mm^2$, $50mm^2$, $70mm^2$, $95mm^2$, which can be selected according to voltage grade and rated short-circuit current.

(2) Performance data

See Table 7-1 for performance parameters of 10-110kV portable short circuit ground wire.

Table 7-1 Performance Parameters of 110-110 kV Portable Short Circuit Ground Wire

Parameter \ Voltage kV	Length of insulation rod part (mm)	Length of handshake part (mm)	Length of metal joint (mm)	Joints	Rod diameter (mm)	Nominal section (mm2)	Total length (mm)
10kV	700	300	50	1	30	25	1050
35kV	900	600	50	1	30	25	1550
110kV	1300	700	140	2	30	35.67	2140

3. Inspection Contents of Portable Short Circuit Ground Wire

The portable short circuit ground wire must be inspected before use. The inspection contents are as follows.

(1) First of all, check whether the specification of ground wire is consistent with the voltage level of the working line. It is strictly prohibited to use ground wire that does not conform to the specification.

(2) Check whether the portable short circuit ground wire test certificate is within the validity period.

(3) Check the connector of the portable short circuit ground wire (cable clip or wire clamp), which shall be in good contact after installation, and have sufficient clamping force to prevent the wire from being melted due to poor contact in the event of large short-circuit current, or falling off due to electrodynamic force.

(4) Check the connection between the portable short circuit grounding copper wire and the three short-circuited copper wires to see if it is firm. Generally, it shall be tightened by screws and then soldered with tin to prevent fusing due to poor contact.

(5) Check whether the plastic sheath of the portable short circuit ground wire soft copper wire is damaged, and whether the soft drawn copper wire is worn and broken.

Fig. 7-2 shows the inspection of portable short circuit ground wire.

Chapter 7
Safety and Insulating Hand Tools

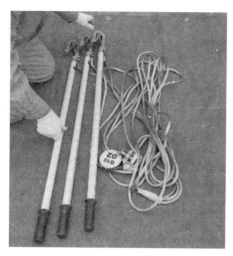

Fig. 7-2 Inspection of Portable Short Circuit Ground Wire

4. Usage of Ground Wire

(1) When in use, the connector of the portable short circuit ground wire (cable clip or wire clamp) shall be in good contact after installation, and have sufficient clamping force to prevent the wire from being melted due to poor contact in the event of large short-circuit current, or falling off due to electrodynamic force.

(2) When the portable short circuit ground wires are used, attention must be paid to the assembling and disassembling sequence. When the ground wires are used, the grounding clip must be connected first, and then the wire clip must be connected, and the contact must be good. The disassembling sequence of portable short circuit ground wires shall be opposite to this.

(3) Installation of portable short circuit ground wire must be carried out by two persons. Insulating gloves shall be worn When the portable short circuit ground wire is installed and removed.

(4) The portable short circuit ground wire shall be inspected thoroughly before each installation. The damaged portable short circuit ground wire shall be repaired or replaced in time. It is forbidden to use the non-conforming conductor as the ground wire or short-circuit wire.

(5) When it is used to release electric charge, the ground wire of the ground terminal must be fixed to the grounding body with a special wire clamp. It is forbidden to perform grounding or short circuit operations by winding.

(6) It is not allowed to connect a disconnector or fuse between the portable short circuit ground wire and the working equipment to prevent the equipment from being not grounded when it is disconnected, causing an electric shock accident to the maintenance personnel.

(7) The maximum swing range of the installed portable short circuit ground wire shall keep a sufficient safe distance from the live part.

5. Precautions for Storage of Portable Short Circuit Ground Wire

(1) Portable short circuit ground wire shall be stored in a well-ventilated and dry environment, and shall be put on the shelf for storage.

(2) Each group of portable short circuit ground wires shall be numbered, and the storage location shall also be numbered. The number of portable short circuit ground wire must be the same as the storage location number, so as to avoid accidents caused by accidental disassembly or missing disassembly of the ground wire during partial interruption maintenance in a complicated system.

Fig. 7-3 shows a tool cabinet for storing ground wire.

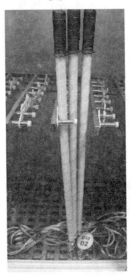

Fig. 7-3　Tool Cabinet for Storage of Ground Wire

6. Preventive Test of Portable Short Circuit Ground Wire

The preventive test of portable short circuit ground wire shall be conducted once a year. The test contents are as follows.

(1) Voltage drop test of grouped portable short circuit ground wire.

(2) Fatigue test of multi-strand flexible conductor.

(3) Grip strength test of wire clip and multi-strand flexible conductor.

(4) Test on over-tightening force of ground wire clamp.

(5) Mechanical strength test of grounding operating rod.

(6) Short circuit current test.

7.1.2 Safety Belt

1. Basic Structure and Function of Safety Belt

(1) Basic structure

As shown in Fig. 7-4, the full-body safety belt consists of a belt, a rope and metal accessories. According to the nature of the work, its structural form is also different, mainly including two kinds of safety belts: pole work safety belts and hanging work safety belts, etc.

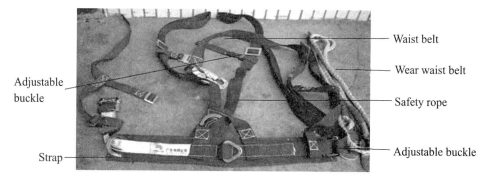

Fig. 7-4 Parts of Full-Body Safety Belt

Safety belts are made of nylon, vinylon, silk and other materials. However, due to the small amount of raw materials and high cost of silk, chinlon is currently the main material. Ordinary carbon steel or aluminum alloy steel for metal accessories.

(2) Basic function

Safety belts are protective articles for high-altitude workers to prevent falling casualties. It is widely used in the construction, installation and maintenance of overhead transmission and distribution line towers. In order to prevent workers from falling from high places, safety belts must be used. The safety belt for pole-enclosing operation is mainly used for line maintenance and patrol operation; Hanging safety belts are mainly used in construction, installation and other operations. Fig. 7-5 shows the structures of two kinds of safety belts, i.e. pole work safety belts and hanging work safety belts, and Fig. 7-6 shows the picture of three-point and five-point safety belts.

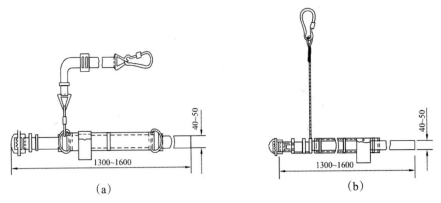

(a) Pole work safety belt; (b) Hanging work safety belt

Fig. 7-5　Structure of Safety Belt

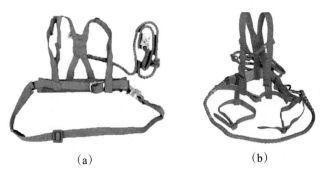

(a) Five-point safety belt; (b) Three-point safety belt

Fig. 7-6　Safety Belts

2. Classification and Technical Specifications of Safety Belt

(1) Classification

① Safety belts can be divided into three categories according to the use conditions, i.e. pole work safety belts, zone limiting safety belts and falling hanging safety belts, as shown in Table 7-2.

Table 7-2　Classification of Safety Belts According to Use Conditions

Classification	Component composition	Hanging point device
Hanging work safety belt	Belts, connectors, adjusters (adjusting buckles), safety ropes, and pole belts (pole ropes)	Main column
Zone restraint safety belt	Belts, connectors (optional), safety ropes, adjusters, connectors, pole belts (pole ropes)	Hanging point
	Belts, connectors (optional), safety ropes, adjusters, connectors, pole belts (pole ropes), pulleys	Guide rail

Chapter 7
Safety and Insulating Hand Tools

Table (Cont'd)

Classification	Component composition	Hanging point device
Falling hanging safety belt	Belts, connectors (optional), bumpers, safety ropes, connectors, pole belts (pole ropes), speed difference automatic control device	Hanging point
	Belts, connectors (optional), bumpers (optional), safety ropes, connectors, pole belts (pole ropes), self-locking devices	Guide rail

② Safety belts can be divided into three categories according to different structural forms, as shown in Table 7-3.

Table 7-3 Classification of Safety Belts by Structure

Classification	Scope of application
Diagonal safety belt	Substation maintenance test, 110kV and below substation frame work, 35kV and below distribution line aerial work
Half-body safety belt	The 220kV substation framework work and the 110kV line aerial work as well as other work where areas must be used to restrict safety belts
Full-covered safety belt	Transmission lines of 220kV and above or aerial work with a working height of more than 30m, substation framework work of 500kV and above and other work where falling hanging safety belts must be used

(2) Technical specifications

① The main belt must be a whole piece with a width of not less than 40mm and a length of 1300-1600 mm.

② The waistband is with a width of not less than 80mm, a length of 600-700 mm, and an auxiliary belt width of not less than 20mm .

③ The effective length of safety rope (including unexpanded buffer) shall not be greater than 2m, and buffer or falling protector shall be additionally installed for special needs exceeding 2m. When working at high altitude with a range of activities exceeding the protection range of safety rope, it must be used in conjunction with speed differential automatic controller. For safety belts with two safety belts (including unexpanded bumper), the single effective length shall not be more than 1.2m.

3. Safety Belt Inspection Contents

(1) Before use, the safety belt must be visually inspected. If it is found to be damaged,

deteriorated or metal accessories are broken, it shall be forbidden for use. When not in use at ordinary times, it shall also be visually inspected once a month.

(2) Before using the safety belt, it is necessary to check whether the test certificate is within the valid period. Safety belts exceeding the valid test cycle cannot be used.

(3) Before using the safety belt, a force impulse test must be carried out to check the bearing capacity of the safety belt, as shown in Fig. 7-7.

Fig. 7-7 Safety Belt Impulse Test

4. Usage of Safety Belt

(1) The safety belt shall be hung at a high place and used at a low place or fastened horizontally. The high hanging and low use is to hang the rope of the safety belt at a high place, and the operator works below the hanging point; the horizontal fastening is to tie the safety belt at the waist when the single belt is used, and the hook of the rope is hung at the same level as the safety belt. The worker shall keep a distance from the hook that approximately equals to the length of the rope. Low hanging and high use is prohibited, and the live beam clamp shall be fastened tightly.

(2) When using the safety belt in low temperature environment, care shall be taken to prevent the safety belt from hardening and splitting.

(3) The frequently used safety rope shall be subject to regular appearance inspection. In case of abnormality, new rope shall be replaced in time, and attention shall be paid to the problem of adding rope slings.

(4) The safety belt cannot be knotted to prevent the safety rope from being disconnected from the knotted place in case of impact. The safety hook shall be hung on the connecting ring instead of directly hanging on the safety rope to prevent the safety rope from being cut

off in case of falling.

(5) The various parts on the safety belt shall not be arbitrarily removed. When replacing the new rope, pay attention to the use of a noose. The belt can be used for 3 to 5 years. If the abnormality is found, it shall be scrapped in advance.

(6) When using and storing the safety belt, avoid contact with high temperature, open flames and acids, as well as hard objects with acute angles and chemicals.

Fig. 7-8 shows the wearing method of the safety belt, and Fig. 7-9 shows the using method of the safety belt when working on the rod.

Fig. 7-8　Wearing Method of Safety Belt

Fig. 7-9　Use of Work Safety Belt on Rod

5. Storage of Safety Belts

(1) Safety belts shall be stored in a dry and ventilated environment, numbered and fixed on shelves. They shall not be exposed to high temperature, open flames, strong acids, strong alkalis and sharp hard objects, and shall not be exposed to the sunlight. Tools with hooks and thorns shall not be used when moving, and shall be protected from sunlight and rain during transportation.

(2) Safety belts shall be kept clean regularly. They can be put into warm water and gently rubbed with soapy water, then rinsed with clean water and dried.

Fig. 7-10 shows the storage of safety belts on upper shelf.

Fig. 7-10　Storage of Safety Belts on Upper Shelf

6. Preventive Test of Safety Belts

The preventive test of safety belt is mainly static load test, with a test cycle of 1 year (in which the test cycle of cowhide belt is half a year).

7.1.3　Climbers

1. Basic Structure and Function of Climbers

The climbers, also known as the iron stands, are a tool for climbing the electric pole. Climbers are generally made of high-strength seamless tubes. After heat treatment, they have the advantages of light weight, high strength, good toughness, good adjustability, portability, flexibility, safety, reliability, and portability. They are ideal tools for electricians to climb concrete electric poles or wooden electric poles of different specifications.

2. Basic Types and Performance Parameters

(1) Basic type

Climbers are divided into two types: One is the climbers with iron teeth on the buckle for climbing the wooden electric pole, as shown in Fig. 7-11 (a); the other is the climbers with the buckle wrapped with rubber for climbing concrete electric pole, as shown in Fig. 7-11(b), and Fig. 7-11(c) shows the picture of the climbers. The speed of the climbers climbing the electric pole is high.

Chapter 7
Safety and Insulating Hand Tools

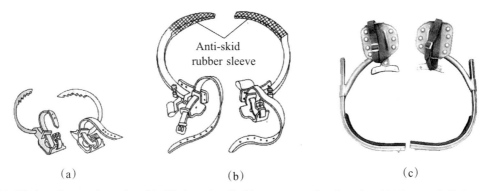

(a) Climbers for wooden poles; (b) Climbers for climbing concrete electric poles; (c) Picture of climbers

Fig. 7-11　Climbers

(2) Performance data

Refer to Table 7-4 for performance parameters of climbing pole climbers.

Table 7-4　Performance Parameter Table of Climbing Climbers

Product name	Specification	Height (m)	Material	Optional accessories
Climbers for climbing pole	JK-T-250	8	High-carbon steel	Climber strap
Climbers for climbing pole	JK-T-300	10	High-carbon steel	Climber strap
Climbers for climbing pole	JK-T-350	12	High-carbon steel	Climber strap
Climbers for climbing pole	JK-T-400	15	High-carbon steel	Climber strap
Climbers for climbing pole	JK-T-450	18	High-carbon steel	Climber strap
Climbers for climbing pole	JK-T-500	21	High-carbon steel	Climber strap

3. Inspection Before Use of Climbers

(1) Before use, the climbers must check whether its metal parts are worn, rusted and cracked, whether the movable buckle slides flexibly in the buckle body, whether the instep strap is in good condition, etc.

(2) Before using the climbers, it is necessary to carefully check whether the test certificate is within the validity period, and the climbers exceeding the validity period of the test is not allowed to be used.

(3) Before the climbers are used, a human body load impulse test must be carried out on the climbers to check whether each part of the climbers is firm and reliable.

4. Usage of Climbers

Climbers are mainly used for climbing electric poles, of which wooden pole climbers are mainly used for climbing concrete electric poles or steel pipe towers of electric power, postal telecommunications lines. The climbers of the concrete electric pole is suitable for climbing the pole of the concrete electric pole for electric power and postal telecommunication lines.

As shown in Fig. 7-12, the climbers are used to climb the pole, and as shown in Fig. 7-13, the climbers are used to lower the pole.

Fig. 7-12 Use Climbers to Climb the Pole (Upper Pole)

Fig. 7-13 Use Climbers to Lower the Pole (Lower Pole)

5. Precautions for Storage of Climbers

Climbers shall be kept in a dry and ventilated warehouse and must be numbered and stored on shelves. Fig. 7-14 shows the storage of pole-climbing tools (climbers, lifting board).

Chapter 7
Safety and Insulating Hand Tools

Fig. 7-14   Storage of Pole-climbing Tools

6. Preventive Test of Climbers

Preventive test of climbers is static load test, and the test cycle is one year.

7.1.4 Lifting Plate

1. Basic Structure and Function of Lifting Board

Lifting board, also known as triangle board, pedal board, pedal or tread. It consists of iron hook, hemp rope and wood board, as shown in Fig. 7-15. Lifting board is a tool for electrician to climb pole and work on pole. The vertical length of the rope hook to the plank is adapted to the height of the operator, and generally the length of the operator's arm is suitable.

Fig. 7-15   Picture of Lifting Board

2. Performance Parameters of Lifting Board

The pedals are made of tough wood, such as white ash and oak, into rectangular pedals with a thickness of 30-50mm, and the two ends of the manila rope are tied to the two ends of the pedals. White brown rope shall be three-strand manila rope with a diameter of 16mm. The board of the lifting board and the manila rope shall be able to bear 300kg.

3. Inspection Before Use of Lifting Board

(1) Before the lifting board is used, the pedal must be carefully checked for cracking or decay, and the rope for corrosion or breakage. If found, it must be handled in time. Otherwise, it is easy to slip and hurt people during pole climbing.

(2) Before use, it is necessary to carefully check whether the test certificate is within the validity period, and the lifting board exceeding the validity period of the test is not allowed to be used.

(3) Before use, the lifting board must be subjected to a human body load impact test to check whether each part of the lifting board is firm and reliable. Fig. 7-16 shows the load impulse test on the lifting board.

Fig. 7-16   Load Impulse Test of Lifting Board

4. Usage of Lifting Board

The usage of lifting board shall be properly mastered, otherwise it will cause personal injury or death if it is unhooked or slipped. The lifting board shall be flexible and comfortable when working on the rod, which is suitable for long-term working on the rod and can reduce the fatigue degree.

Fig. 7-17 shows the use of a lifting board for pole climbing.

Fig. 7-17 Use of Lifting Board for Pole Climbing

5. Precautions for Storage of Lifting Board

Lifting board shall be kept in a dry and ventilated warehouse and must be stored on shelves.

6. Preventive Test of Lifting Board

Preventive test of lifting board is static load test, and the test cycle is half a year.

7.2 Insulating Hand Tools

Insulating hand tools for transmission and distribution lines mainly include insulated bolt knives, insulated wrenches, insulated hand tongs, insulated wire strippers and manual wire cutters, etc. They are the most commonly used hand tools for live operators.

7.2.1 Insulating Screwdrivers

1. Structure and Function of Insulating Screwdrivers

Insulating screwdrivers are also called screwdrivers. Insulating sleeves are sleeved on the metal parts of the screw rods of common screwdrivers, and the heads of the insulating screwdrivers have two types, namely a straight type and a cross type; Insulated screwdrivers are tools used to tighten or loosen grooved screws on the head. Fig. 7-18 shows the picture of the insulating screwdrivers.

Fig. 7-18 Picture of Insulating Screwdrivers

2. Types and Performance Parameters of Insulating Screwdrivers

(1) Basic type

There are several types of insulating minus screwdrivers, including insulating square screwdrivers, insulated cross screwdrivers, insulating quick screwdrivers, insulated elbow screw screwdrivers, etc.

(2) Performance data

Insulating screwdrivers generally take the diameter of the screwdrivers x the length of the screwdrivers as specification parameters. Commonly used screwdrivers have diameters of 3mm, 5mm, 7mm, 9mm, etc., and commonly used screwdrivers have lengths of 50mm, 100mm, 150mm, 300mm, etc.

3. Inspection of Insulating Screwdrivers Before Use

(1) Check whether the handle part of the insulating screwdriver is cracked and damaged, and check whether the insulating sleeve is loose and cracked.

(2) Check whether the knife edge of the insulating screwdriver is blunt and damaged. Those with serious damage, deformation, split handle or damage to the knife edge shall be discarded.

4. Usage of Screwdrivers

(1) Appropriate insulating screwdrivers shall be selected according to the groove width and groove shape of the tightened or loosened screw head.

(2) It is not allowed to use a smaller screwdriver to unscrew a larger screw.

(3) Insulating cross screwdrivers shall be used to tighten or loosen screws with cross grooves in the head.

(4) Insulating elbow screwdrivers shall be used for screw heads with limited space.

(5) Do not tighten or loosen the screws on the workpiece held in your hand with the insulating screwdriver. The workpiece shall be fixed in the clamp to prevent injury.

(6) It is not allowed to pry open gaps or remove metal burrs and other objects by hammering the end of the insulating screwdriver handle.

(7) Insulating screwdrivers shall not be used as crowbars, and shall not be hammered on the handle by hand, nor shall wrenches or pliers be used at the handle and knife edge of the screwdrivers to increase torque force to prevent bending damage of the screwdrivers.

(8) Do not use insulating screwdriver tools to play or fight, so as to prevent personal injury.

(9) When using insulating screwdriver to screw the tool, the posture shall be correct and the force shall be appropriate. The correct usage of insulating screwdrivers is shown in Fig. 7-19.

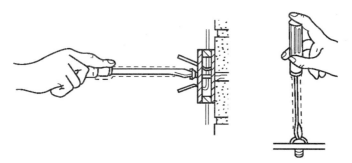

(a) Usage of larger screwdrivers; (b) Usage of smaller screwdrivers

Fig. 7-19 Usage of Screwdrivers

5. Precautions for Storage of Insulating Screwdrivers

The insulating screwdrivers shall be kept in a dry and well-ventilated tools and instruments warehouse, stored in a special tools and instruments box or placed at the fixed point of the shelf.

6. Preventive Test of Insulating Screwdrivers

The withstand voltage of the insulation part of the common insulating screwdriver is 500V, and the withstand voltage of the insulation part of the high-voltage insulating screwdriver can reach 1000V or 1500V. Preventive test for insulating screwdrivers includes appearance inspection and high-voltage power frequency withstand voltage test for insulation part. The test cycle shall be 1 year.

7.2.2 Insulating Wrench

1. Structure and Function of Insulating Wrench

Insulating wrenches are tools used to tighten hexagonal, square screws and various nuts. Insulating wrenches are made of tool steel, alloy steel or malleable cast iron. Insulating adjustable wrench is mainly composed of fixed wrench lip, adjustable wrench lip, worm gear, shaft pin, handle, etc., as shown in Fig. 7-20. Insulating adjustable wrench is a tool for tightening or loosening angular screws or nuts. The size of the wrench opening can be adjusted by rotating the worm wheel of the adjustable wrench. The fixing and forming of the wrench opening of the insulating solid \wrench cannot be adjusted, and is only applicable to screws of the same specification, as shown in Fig. 7-21.

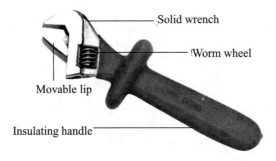

Fig. 7-20 Insulating Adjustable Wrench

Fig. 7-21 Insulating Solid Wrench

2. Basic Types and Technical Specifications of Insulating Wrenches

Insulating wrenches commonly used in live working of power transmission and distribution lines generally include adjustable wrenches, fixed wrenches and socket wrenches.

There are three types of insulating adjustable wrenches: 200mm, 250mm and 300mm (English 8in, 10in and 12in). According to the size of the nut, choose the insulating wrench of appropriate specifications, in order to avoid damage of too large wrench to the nut; Or the wrench is damaged due to too large nut.

3. Inspection of Insulating Wrench Before Use

(1) Check whether the handle part of the insulating wrench is cracked and damaged, and whether the insulating part is aged.

(2) Check whether the opening of insulating adjustable wrench can be flexibly adjusted.

4. Usage of Insulating Wrench

(1) When using, insulating adjustable wrench with appropriate specifications shall be selected according to the shapes, specifications and working conditions of screws and nuts.

(2) The opening size of the insulating adjustable wrench can be adjusted within a certain range, so screws and nuts within the opening size range can generally be used.

(3) For insulating movable wrenches, it is not allowed to use large-sized wrench to tighten screws of smaller size, which will break the screws due to excessive torque; the opening shall be adjusted according to the size of the opposite side of the hexagonal head of the screw or the hexagonal head of the nut, and the clearance shall not be too large, otherwise the head of the screw or the nut will be damaged, and it will easily slip off, causing injury accidents; the fixed jaw shall be subjected to the main acting force, the wrench handle shall be tightened towards the operator, not pushed forward, and the wrench handle shall not be lengthened arbitrarily; wrenches shall not be used as hammering tools.

5. Precautions for Storage of Insulating Wrench

Insulating wrenches shall be stored in a dry and well-ventilated tool warehouse, and shall be numbered and placed on shelves at fixed points. The storage of insulated wrenches is shown in Fig. 7-22.

Fig. 7-22 Storage of Insulating Wrench

6. Preventive Test of Insulating Wrench

The insulation part of the insulating wrench shall be subjected to high voltage power frequency withstand voltage test on a regular basis with a test cycle of 1 year.

7.2.3 Wire pliers

1. Structure and Function of Wire Pliers

Wire pliers, commonly known as calipers and hand pliers, also known as electrical pliers, are one of the safety appliances used in live working of power transmission and distribution lines. It is a tool for clamp and shearing, the structure of which is shown in Fig. 7-23 and consists of a clamp head and a clamp handle. The clamp head has four openings: jaw, tooth edge, knife edge and chopping edge; the clamp handle is provided with an insulating sleeve. Fig. 7-24 shows the picture of wire pliers.

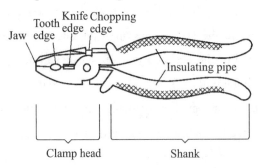

Fig. 7-23 Structure of Wire Plier

Chapter 7
Safety and Insulating Hand Tools

Fig. 7-24 Picture of Wire Clamp

2. Technical Specifications of Wire Pliers

There are 150mm, 175mm and 200mm specifications for common wire pliers. The insulated handle of common wire pliers has a withstand voltage of 500V, and the working voltage of high-voltage insulated wire pliers can reach 1000V. Fig. 7-25 shows a picture of 1000V high voltage insulated wire pliers.

Fig. 7-25 1000V HV Insulated Wire Pliers

3. Inspection Before Use of Wire Pliers

(1) Before using the wire pliers, check whether the insulating sleeve of the insulating handle is worn and cracked, so as to prevent the pliers head from contacting the charged part during work and causing accidents due to electrification of the pliers handle.

(2) Before use, pull the handle to check whether the jaws of the wire pliers are flexible and reliable.

4. Usage of Wire Pliers

(1) When using the wire pliers, the knife edge of the clamp head shall face inward, that is to say, towards itself, so as to control the position of the jaw conveniently; extend your little finger between the shanks to hold the shanks and open the clamp head.

(2) During use, pay attention not to using the knife edge to cut the wire, so as to avoid knife edge damage. The cutting edge of insulated handle wire pliers can be used to cut electric wires and iron wires. When cutting No.8 galvanized iron wire, the blade shall be used to cut back and forth around the surface a few times, and then the iron wire shall be broken with only a slight pull.

When cutting live wires with electrical wire pliers, the jaws shall not be used to cut the phase line and the zero line at the same time, or cut both phase lines at the same time, which will cause short circuit of the lines.

(3) The plier head cannot replace the hammer as a beating tool.

(4) Do not throw, smash or straighten the steel wire with the handle during the use of insulated handle wire cutters, so as not to damage the insulated plastic pipe.

(5) When winding the anchor cable with insulated handle wire pliers, clamp the wire with the teeth of the pliers and wind it clockwise.

5. Precautions for Storage of Wire Pliers

The wire pliers shall be kept in a dry and well-ventilated tools and instruments warehouse, stored in a special tools and instruments box or placed at the fixed point of the shelf.

6. Preventive Test of Wire Pliers

Preventive test for HV insulating wire pliers includes appearance inspection and high-voltage power frequency withstand voltage test for insulation part. The test cycle shall be 1 year.

7.2.4 Pin Puller

1. Structure and Function of Pin Puller

The pin puller is also called pin-pulling plier. Its jaw is hollow, and other parts are similar to wire pullers. Its function is mainly to remove the spring pin of insulator in line work.

2. Basic Types of Pin Puller

The pin puller can be divided into two types: non-live and live. Most of the power outages are hand-held, with two types, ordinary and "Z". For live working, there are two main types: hand-held and insulated operating rod. Hand-held type is the same as reimbursement pliers used during power failure. Insulated operating rod includes two types: linear pin puller and tension pin puller. Fig. 7-26 shows several common pin pullers.

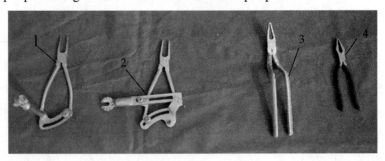

1-Tension insulator pin puller; 2-Pin puller for linear insulator; 3-220 kV hand-held pin puller;
4-10kV hand-held pin puller

Fig. 7-26 Pin-pulling Puller

3. Inspection of Pin Puller Before Use

(1) Before using the tension-resistant and linear insulator pin puller, check whether the jaw is deformed, worn, closed completely and whether the adjusting screw of the plier handle is loose.

(2) For hand-held pin puller, check whether the jaw is deformed, worn, and whether the insulation sheath is cracked and aged.

4. Usage of Pin Puller

(1) Strain and linear insulator pin-pulling plier must be matched with insulating operating rods when in use. Insulating bar shall be matched with insulating operating rods of the same voltage grade.

(2) Strain and linear insulator pin-pulling plier is suitable for the ground potential operation method for live working of power transmission lines. When strain and linear insulator pin-pulling plier is used at ground potential, the operation of insulating rod must be smooth and fast. Fig. 7-27 shows the operation of replacing the single-sheet strain insulator with strain insulator pin-pulling plier on live line.

Fig. 7-27 Operation of Replacing Single-sheet Strain Insulator with Strain Insulator
Pin-pulling Pilier on Live Line

5. Precautions for Storage of Pin Puller

Pin pullers shall be stored in a dry and well-ventilated tool warehouse, and placed on shelves at fixed points. As shown in Fig. 7-28 for pin puller storage.

Fig. 7-28 Storage of Pin Puller on Upper Shelf

6. Preventive Test of Pin Puller

Preventive test of strain and linear insulator pin puller is mainly appearance inspection insulation rod preventive test, and preventive test of insulation rod is mainly power frequency withstand voltage test; the preventive test of hand-held pin puller mainly includes appearance inspection and power frequency withstand voltage test of insulated handle. The test cycle is 1 year.

7.2.5 Insulated Conductor Stripper

1. Structure and Function of Insulated Conductor Stripper

Insulated conductor stripper, also known as insulated conductor stripping plier or conductor stripping plier, is mainly used to strip the insulation layer of insulated conductors, and is currently mainly used for stripping the insulating layer of 10kV insulated wires and high-voltage cables. Its manufacturers are from United States, Japan, Italy, China and other countries, and it is with various shapes and structures. The structure of domestic BXQ-Z-40A rotary stripper consists of a fixed knob, a feed knob, a fixed head, a cylindrical cutter, a sliding head and a handle, as shown in Fig. 7-29. The blade is made of alloy steel, and is sharp and durable. The cutter head can be disassembled and ground. The peeling thickness has feed scale. The operation is intuitive and convenient. As shown in Fig. 7-30, it is a BXQ-Z-40A rotary stripper.

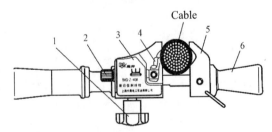

1-Fixed knob; 2-Feed knob; 3-Cylinder cutter; 4-Sliding head; 5-Handle

Fig. 7-29 Structure of BXQ-Z-40A Rotary Cutting Wire Stripper

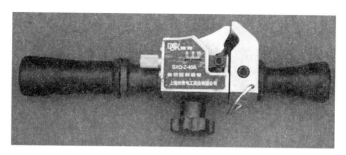

Fig. 7-30 BXQ-Z-40A Rotary Stripper

2. Basic Types and Performance Parameters of Insulated Wire Strippers

Insulated wire strippers can be divided into manual and electric types according to the operation mode, and can be divided into American, Japanese, Italian, joint venture and domestic made types according to the manufacturer's standard.

The following describes the specifications and performance parameters of several common insulated conductor strippers, as shown in Table 7-5.

Table 7-5 Specifications and Performance Parameters of Insulated Conductor Strippers

S/N	Model	Standard category	Performance data	Use
1	NP400	Japan	It is suitable for insulated wires with a diameter ranging from 12 to 32mm and an insulation layer thickness ranging from 1.4 to 4 mm.	Stripping the insulating layer of overhead insulating conductors
2	WS-50	American	Stripping line diameter is 12.7-57.1mm, stripping thickness is 8.5mm, positioning stripping length is 150 mm.	Suitable for 10kV main insulating layer and overhead insulating conductor
3	TYX-300	Joint venture	The area of the strippable insulated conductor is 25-300mm^2	Suitable for overhead insulated conductor
4	AV6220	Italy	Suitable for cables with a diameter of more than 25mm, and a cutting thickness of 5mm.	Suitable for stripping the sheath of three-core cable
5	BXQ-Z-40A	Domestic	The stripping diameter is 12-40mm, and the stripping thickness: no less than 6mm for the terminal stripping, and less than 4mm for middle section stripping.	It is mainly used for quick stripping and cutting of insulated conductor, overhead conductor, cable terminal and middle section.

As shown in Fig. 7-31, there are NP400, WS-50, TYX-300 and AV6220 strippers.

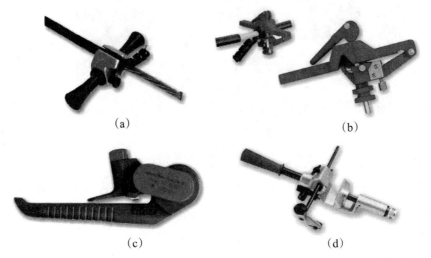

(a) NP400 stripper; (b) AV6220 stripper; (c) TYX-300 stripper; (d) WS50 stripper

Fig. 7-31　Stripper

3. Inspection of Insulating Conductor Stripper Before Use

(1) Before use, select the corresponding stripper according to the diameter of the conductor. Never use it beyond the scope to avoid damage to the blade.

(2) The cylinder cutter will inevitably be blunt if it is used for a long time. It shall be polished with oilstone in time to ensure the quality of stripping and cutting.

(3) Before use, check whether the structure of each part of the stripper is good, whether the knob is flexible and reliable, and whether there is any slip failure.

4. Usage of Insulating Conductor Stripper

The application method of domestic BXQ-Z-40A stripper is as follows.

(1) According to the structural schematic diagram of BXQ-Z-40A, loosen the fixing knob, pull open the slide plier head, clamp the stripping part of the wire, and then tighten and fix the knob.

(2) Tighten the knob counterclockwise while observing the scale feed (1mm per grid), and stop the feed until an appropriate depth is reached.

(3) Tighten the stripper. The feeding direction is from right to left, and when it is fast, push slightly to the right to make the stripper tilt to the left.

(4) When turning to a predetermined length, cancel the leftward push to turn the stripper in-situ, and the chippings will break off on their own.

(5) Turn the knob clockwise to retract the cylinder cutter (scale returns to 0), then loosen the fixing knob, pull open the slide head, take out the conductor and finish stripping.

(6) When the stripper is used, the blade feeding depth shall be appropriate. Under normal circumstances, the 0.5 ~ 1mm thick insulating layer shall be reserved along the feeding depth, so as to prevent from damaging the cutting edge of the conductor and the cylinder notch resulted from the excessive feeding depth.

For other strippers, please refer to the use manual.

5. Precautions for Storage of Insulated Conductor Stripper

The insulated conductor stripping plier shall be kept in a dry and well-ventilated tools and instruments warehouse, stored in a special tools and instruments box and placed at the fixed point of the shelf. The storage of BXQ-Z-40A stripper is as shown in Fig. 7-32.

Fig. 7-32 Storage of Stripper

6. Preventive Test of Insulated Conductor Stripper

The preventive test of insulated conductor stripper includes the visual examination and the power frequency withstand voltage test of insulation part, and the test cycle is 1 year.

7.2.6 Gear Type Bolt Clipper

There are variable cable bolt clipper used for live working of transmission and distribution line. Here will only introduce the portable gear type bolt clipper for personal use.

1. Structure and Function of Gear Type Bolt Clipper

The gear type bolt clipper is mainly used to cut the cables, including the steel-cored aluminum stranded conductor, steel strand and insulated conductor. The LK-300B gear type bolt clipper manufactured in Taiwan consists of the insulated handle, moving blade, fixed blade, cover plate, retaining knob and other parts, as shown in Fig. 7-33.

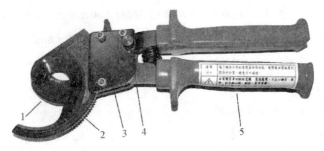

1-Fixed blade; 2-Moving blade; 3-Cover plate; 4-Retaining knob; 5-Insulated handle

Fig. 7-33　LK-300B Structure of Gear Type Bolt Clipper

2. Model and Technical Parameters of Gear Type Bolt Clipper

The gear type bolt clipper may be manufactured in USA, Japan and China, while the type and technical parameters for two types of gear type bolt clippers manufactured in USA is introduced in the Table 7-6. The RCC-325 type and RCC-500 type gear type bolt clippers are as shown in Table 7-34.

Table 7-6　The Type and Technical Parameters for Two Types of Gear Type Bolt Clippers Manufactured in USA

Brand	KUDOS (USA)	KUDOS (USA)
Model	RCC-325	RCC-500
Scope of application	325mm² and below copper aluminum cable	500mm² and below copper aluminum cable
Dimensions (L)	250mm	700mm
Weight	0.65kg	1.67kg
Characteristics	Single hand operation Immersion type anti-slip handle It may not be used to cut the hard copper wire or any steel-cored cable The cutting blade is much wide and not easily deformed, while the tooth pitch of the ratchet wheel is less and the cutting is labor-saving	

(a)　　　　(b)

(a) RCC-325 type; (b) RCC-500 type

Fig. 7-34　Gear Type Bolt Clipper

Chapter 7
Safety and Insulating Hand Tools

3. Inspection Before the Use of Gear Type Bolt Clipper

(1) Before use, please check whether the gear cutting edge of the bolt clipper is damaged and whether the bolts at each part becomes loose.

(2) Before use, please check whether the two blades are misplaced or cannot be closed, causing the cutting difficulty. If such situation occurs, it may be solved by adjust the corresponding bolts.

(3) Please check whether the rotation part for the gear of the bolt clipper is flexible. If the rotation difficulty occurs, please timely add the oil at the gear part for lubrication purpose.

(4) Please check whether the insulated handle of the bolt clipper is broken and whether there is any crack and aging. If such situation occurs to the insulated handle, it's better not to use it.

4. Use Method of Gear Type Bolt Clipper

(1) The over-range and over-load use is strictly prohibited. The gear type bolt clipper has its special rated strength. Upon use, please properly select the type and specification as per the actual demand. The substitution of the small articles to the large articles, and the cutting of articles of which the hardness is greater than that of the blade-edge of the bolt clipper, are strictly prohibited, so as to prevent from causing the tipping or turning of the blade.

(2) Please don't consider it as the ordinary steel tools and instruments, and replace other tools and instruments.

(3) Before cutting, please place the retaining knob at the use position, adjust the cutting edge as per diameter of conductor to be cut, and turn the gear by pressing and releasing the insulated handle, so as to gradually close the fixed blade and moving blade.

(4) After use, please adjust the closed moving blade and fixed blade to the open status, and timely remove the mud and other foreign matters contained in the spring and tooth space.

5. Precautions for Storage of Gear Type Bolt Clipper

The gear type bolt clipper shall be kept in a dry and well-ventilated tools and instruments warehouse, stored in a special tools and instruments box and placed at the fixed point of the shelf.

6. Preventive Test of Gear Type Bolt Clipper

The preventive test of gear type bolt clipper includes the visual examination and the power frequency withstand voltage test of insulation part, and the test cycle is 1 year.

Chapter 8

Storage and Transportation if Tools and Instruments for Live Working

The live working has been carried out in China for decades, and the live working industry has accumulated a great number of practice experiences. However, in 1990s, many accidents resulted from the non-standard use and management of tools still occur in China's live working industry, which will make it difficult for the conduct of the live working. The main reason is that the live working tools become damp during the storage and transportation process, causing the insulation performance to be greatly reduced, and even causing the occurrence of breakdown and flashover accidents. Especially in southern China, the air humidity is much greater and it is much easier for the live working tools to become damp, causing the accident hazards.

This chapter emphasizes on the basic contents for the storage and transportation of the tools and instruments for live working.

8.1 Basic Requirements of Live Working Tools and Instruments Warehouse

The live working tools and instruments warehouse is the special room where the live working tools and instruments warehouse will be kept for a long period. In order to ensure the use safety of tools and instruments, the environmental conditions, technical conditions, equipment & measurement and control device of the tools and instruments warehouse, and the warehouse information management system must conform to the requirements of relevant specifications and regulations.

Chapter 8
Storage and Transportation of Tools and Instruments for Live Working

8.1.1 General Requirements of Live Working Tools and Instruments Warehouse

The live working tools and instruments warehouse is located at the moisture-proof and well-ventilated place. If it is possible, it is better to place it at the first floor of the building, and the bottom layer shall be made into the 0.8 ~ 1.2m high overhead layer. The warehouse area shall be either specially designed as per the quantity and size of the tools, or designed as per the reference area provided in Table 8-1. In addition, please also pay attention to store by zones and provide the descriptions of partition mark, such as the voltage class, name of tools, specifications, and etc. The proportion of tools storage space and activity space shall be about 2:1. The internal space height of the warehouse shall be more than 3.0m. If the building height requirements cannot be satisfied, it shall be no less than 2.7m in general. The doors and windows shall be anti-theft, anti-dust and properly sealed. Meanwhile, they shall also be fireproof and damp-proof. The distance from the observation window to the ground shall be 1.0 ~ 1.2m. The window glass shall adopt the double-glass and the thickness for each layer of glass shall be no less than 8mm in general, so as to ensure the damp-proof and fireproof functions of the warehouse.

Table 8-1 Warehouse Area Design

Voltage class for tools storage kV	10 ~ 66	66 ~ 220	220 ~ 500
Warehouse area m^2	20 ~ 60	50 ~ 150	60 ~ 200

When the warehouse is decorated, the materials must adopt the dust-free, flame retardant, thermal insulation, damp-proof and non-toxic materials. The warehouse ground located on the first floor shall adopt the damp-proof wet insulation materials. Generally, the tools and instruments storage shelf shall be made from the stainless steel and other anti-rust materials.

The warehouse shall be provided with adequate firefighting devices. For the warehouse equipped with the automatic management system, the firefighting automatic alarming system is required. It is better to provide the independent power line for the warehouse, and the selection of line diameter shall consider the requirements of variable systems of the warehouse. The fire prevention requirements shall be satisfied.

Generally, the storage volume of the garage for the aerial devices with insulating arm shall be 1.5 ~ 2.0 times of the vehicle body. The top space shall be 0.5 ~ 1.0m, and the garage gate may adopt the thermal insulation and fireproof special garage gate which may adopt either the electric remote control method or the manual control method.

8.1.2 Technical Conditions and Facilities of Live Working Tools and Instruments Warehouse

The technical conditions of live working tools and instruments warehouse mainly include the temperature, humidity and ventilation conditions. The corresponding heating, dehumidification and ventilation equipment shall be installed within the warehouse, so as to ensure that the storage conditions of the tools and instruments for live working will be satisfied.

1. Humidity Requirements

The relative humidity of the air in the warehouse shall be no more than 60%. In order to ensure the reliability of the humidity measurement, two humidity transducers are required to be installed in each room. The design of tools warehouse shall conform to the requirements of GB/T18037-2008 Technical Requirements and Design Guide for Live Working Tools.

2. Temperature Requirement

The live working tools and protection tools shall be stored by zones as per the type of tools, and each storage zone may have the variable temperature requirements. For the storage zone for the hard insulating tools, soft insulating tools, testing tools, and shielding tools, the temperature shall be controlled within the range of 5 ~ 40°C. For the storage zone of the distribution and live working insulating shielding tools and insulating protection tools, the temperature shall be controlled within the range of 10 ~ 21°C.

In addition, considering the great indoor and outdoor temperature difference in winter in Northern China, the condensation problems will easily occur upon the warehouse-in. In such region, the warehouse temperature shall be adjusted and controlled within a certain scope as per the variation of environmental temperature. If it is hard to adjust the overall temperature of the warehouse, the tools may be firstly and temporarily stored in the temperature-adjustable preparation room before their warehouse-in. Then, after the condensation does not occur, please deliver it to the warehouse and store there.

In order to ensure the reliability of temperature measurement, two temperature transducers are required to be installed in each room of the warehouse. In order to compare the indoor and outdoor temperature difference, a temperature transducer shall be installed outdoor for the control system of the whole warehouse.

Chapter 8
Storage and Transportation of Tools and Instruments for Live Working

3. Dehumidification Facility

The dehumidification facility shall be equipped in the warehouse. The dehumidification amount shall be selected as per the volume of warehouse, which is generally selected and allocated as per the proportion of $0.05 \sim 0.2 L/ (d \cdot m^3)$. In Northern China, it may be selected and allocated as per the proportion of $0.05 \sim 0.15 L/ (d \cdot m^3)$. In Southern China, it may be selected and allocated as per the proportion of $0.13 \sim 0.2 L/ (d \cdot m^3)$. Among the above regions, for the region with higher relative humidity, the dehumidifier shall be selected and allocated as per the upper limit.

4. Drying & Heating Facility

The drying & heating facility shall be equipped in the warehouse. It is suggested to adopt the hot air circulation heating equipment. Under the condition that the even heating may be assured, the infrared heating equipment, non-luminance heating tube, new-type low-temperature radiant tube, and etc., may also be adopted. The heating power may be selected either as per the volume of warehouse, or as per the $15 \sim 30 W/ m^3$ based on the local temperature environment.

The heating equipment may be installed in an even and dispersed manner inside the warehouse, while the distance from the heating equipment or hot air port to the surface of tools and instruments shall be no less than $30 \sim 50 cm$. The distance between the installation height of hot air type drying and heating equipment and the ground shall be about 1.5m, while the low-temperature lightless heater may be placed at the height in parallel and level with the ground. The heater of garage shall be installed on top or the height of arm part. The fan inside the heating equipment shall have the delayed stop device.

5. Ventilation Facilities

The exhaust equipment shall be equipped in the warehouse. Each room must be provided with two exhaust fans, one for air intake and the other for exhaust, which will circulate the air. For the air exhaust amount, the exhaust fan may be selected as per the parameter of $1 \sim 2 m^3/ h/ m^2$. The top suction type exhaust fan shall be installed on the drop ceiling, while the axial exhaust fan shall be installed on the wall surface with the clear height of $2/3 \sim 4/5$ of the wall height inside the warehouse. The air outlet shall set the window shades or woven wire window, while the air inlet shall sets the filter screen to prevent the birds, snakes, rats and other small animals from entering into the warehouse.

6. Alarming Facility

The temperature over-limit protection facility, smoke alarm, outdoor alarm, anti-theft alarm and other alarming facilities. When the warehouse temperature exceeds 50°C, the temperature over-limit protection facility shall be able to automatically cut off the heating power source and start the outdoor alarm. Meanwhile, the temperature over-limit protection facility is also required to be normally started in case of the failure of control system. When the smoke occurs to the warehouse, the smoke alarm and outdoor alarm shall be able to automatically give out the alarm. When the non-registered staff open the gate of warehouse, the anti-theft alarm will be started.

7. Comprehensive Allocation and Selection of Warehouse Facility

Upon the comprehensive allocation and selection of dehumidification, drying & heating and ventilation facilities, it is mainly determined based on whether the temperature, humidity and control requirements are satisfied.

8. Garage for Aerial Devices with Insulating Arm

With regard to the garage for aerial devices with insulating booms, the requirements for ventilation, dehumidification and drying facilities shall be the same as the requirements for live working tools garage. The garage heater is generally installed at the location where it is easier to dry the arm or the top, and the heater is not required at the lower part.

As shown in Fig. 8-1, it is the garage for aerial devices with insulating arm.

Fig. 8-1 Garage for Aerial Devices with Insulating Arm

8.1.3 Monitoring Function and Facility Requirements

For the live working warehouse with much higher automation conditions, the monitoring system consisting of the sensing device, monitoring instrument, control device, computer, switchgear and etc., shall implement the realtime monitoring upon each parameter of the warehouse as per the setting requirements. The humidity and temperature regulation and control system of the tools warehouse, shall automatically start the heating, dehumidification and ventilation facilities as per the monitoring parameters, so as to realize the regulation and control of warehouse humidity and temperature. When the regulation and control fails and exceeds the specified value, the alarming and display shall be provided.

1. Functional Requirements

In order to ensure that the temperature and humidity environment of the tools warehouse may conform to the use requirements, it is required to set the special temperature and humidity measurement and control system. The temperature and humidity measurement and control system shall have the humidity measurement and control, temperature measurement and control, warehouse temperature and humidity setting, over-limit alarm and automatic record of warehouse temperature and humidity, display, search, report printing and other functions, and support the remote regulation and control and the search of historic records.

2. Monitoring Requirements

The monitoring system consisting of the transducer, measuring equipment, control panel cabinet and the attachments, and etc., shall implement the real-time monitoring upon the warehouse temperature and humidity, and further record and store them.

3. Regulation and Control Requirements

The humidity and temperature regulation and control system of the tools warehouse, shall automatically start the heating, dehumidification and ventilation facilities as per the monitoring parameters, so as to realize the regulation and control upon the humidity and temperature of the warehouse. When the regulation and control becomes invalid, and exceeds the specified value, the alarm and display may be provided. When the warehouse temperature exceeds the limit, the temperature over-limit protection device shall automatically cut off the heating power source.

In order to effectively ensure the safe and effective operation of the measurement and control system, the control system shall set the automatic reset device. This will ensure that the measurement and control system will immediately carry out the automatic reset due to the failure resulted from the external interference, so as to further restore the normal operation.

In order to ensure that the dehumidification device and heating device may still be put into use under the condition that the measurement and control system completely becomes invalid or is under maintenance, it is required to set the manual/ automatic shift switch and the corresponding manual switch on the control panel cabinet.

4. Element and Equipment Requirements

The technical performance and indicators of the equipment, devices and components inside the warehouse shall conform to the requirements of relevant equipment and element standards, so as to ensure the stable, reliable and safe operation of measurement and control system.

5. Display and Printing

The measurement and control system shall be able to store the one-year temperature and humidity data of the warehouse, has the report display, curve display, report printing and other functions of the warehouse temperature and humidity data at any time around the clock, and carry out the real-time monitoring and recording of the warehouse working status.

6. Remote Monitoring Requirements

The warehouse and garage shall have the remote video surveillance function. The measurement and control system shall have the Web release function, which is required to be connected with the enterprise LAN, so as to realize the remote information sharing and hierarchical management, warehouse temperature and humidity parameter setting and other functions, and realize the remote monitoring upon the warehouse environmental temperature and humidity variation conditions, and the remote management of tools and instruments, and etc.

7. Technical Performance Requirements of the Main Measurement and Control Elements

(1) Temperature measurement and control indicators: scope: $-10°C \sim 80°C$, precision: $\pm 2°C$.

(2) Humidity measurement and control indicators: scope: $30\% \sim 95\%$ R.H., precision: $\pm 5\%$.

(3) Temperature sensor indicator: range: $-40°C \sim 80°C$, with the precision of $\pm 0.5°C$ under the scope of $-10°C \sim 85°C$.

(4) Humidity sensor indicator: range: $0\% \sim 100\%$R.H., with the precision of $\pm 3\%$ under the scope of $10\% \sim 95\%$R.H.

Chapter 8
Storage and Transportation of Tools and Instruments for Live Working

As shown in Fig. 8-2, it is the monitoring system of live working tools and instruments warehouse.

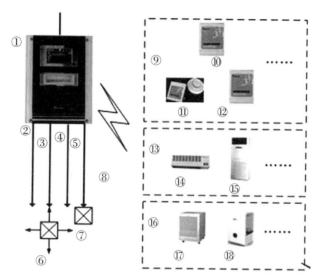

①Main intelligent control box; ②Main power supply of power load; ③Main power supply of ventilation; ④Main power supply of control system; ⑤Main power supply of lighting; ⑥Venting; ⑦Lighting; ⑧Main warehouse 1; ⑨Sensing terminal; ⑩Temperature and humidity sensor; ⑪Smoke alarm; ⑫Temperature over-limit alarm; ⑬Temperature control terminal; ⑭Heater; ⑮Air conditioner; ⑯Humidity control terminal; ⑰Dehumidifier; ⑱Humidifier

Fig. 8-2　Monitoring System of Live Working Tools and Instruments Warehouse

8.2 Management System of Live Working Tools and Instruments Warehouse

The computer management system of tools warehouse shall make the real-time record upon the tools storage conditions, warehouse-in and warehouse-out information, receipt procedure, test conditions and other information. As required, the computer management system of tools and instruments warehouse shall also have the function of WEB release implemented on the enterprise LAN and remote maintenance function. The basic function block diagram of the management system is as shown in Fig. 8-3.

8.2.1 Management System of Live Working Tools and Instruments Warehouse

The management system composition framework block diagram of live working tools and instruments warehouse is as shown in Fig. 8-3. This system block diagram may be referred to and used by each unit. Each unit may also make the modification and supplement as per the specific requirements of its own management.

8.2.2 Tools Management System of Live Working

1. System Composition Framework Block Diagram

The management system composition framework block diagram of live working tools and instruments is as shown in Fig. 8-4.

2. Tools Warehouse-in and Warehouse-out Information Management

(1) Please establish the warehouse-in record of the live working tools and instruments, including the tools and instruments manufacturer, ex-factory date, warehouse-in time, test conditions, tools code, and etc.

(2) Please establish the live working tools and instruments abandonment record, mainly including the name, code, warehouse-in time, abandonment time and abandonment reason of the tools, and etc.

Chapter 8
Storage and Transportation of Tools and Instruments for Live Working

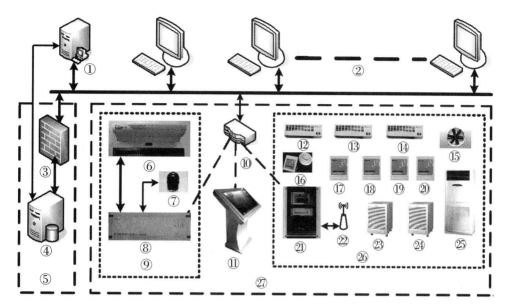

①Production of MIS; ②Enterprise LAN and office system; ③Firewall; ④System server; ⑤Tools management data center; ⑥RFID tag identification unit; ⑦Picture identification unit; ⑧Main control unit of tools management; ⑨RFID tag identification and tools management sub-system; ⑩Router; ⑪Background display; ⑫, ⑬, ⑭Heaters; ⑮Exhaust fan; ⑯Smoke alarm; ⑰Temperature over-limit alarm; ⑱, ⑲, ⑳Temperature and humidity sensor; ㉑Warehouse controller; ㉒Digital wireless communication unit; ㉓, ㉔Dehumidifier; ㉕Warehouse air conditioner; ㉖Warehouse environmental control subsystem; ㉗ Warehouse environmental control and tools management system

Fig. 8-3 Management System Structure of Live Working Tools and Instruments Warehouse

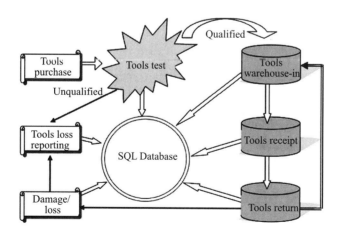

Fig. 8-4 Management System Structure of Live Working Tools and Instruments Warehouse

(3) The accidental dirt, damage, loss or other conditions of the live working tools and instruments occurring in work, must be filled in the relevant statements, and the description must be given. Meanwhile, it shall also be signed and confirmed by the relevant staffs.

3. Tools Borrowing Management

(1) For the warehouse-in and warehouse out of the live working tools and instruments, the ledger must be established, mainly recording the borrowing, return and visual examination conditions.

(2) When the working staffs receive variable tools from the warehouse, there are many methods regarding the tools selection. The tools may be arbitrarily selected through the classification, specification, tool name, bar code or brief code.

(3) After the working staffs complete the use of tools, this operation is required, and it is required to complete the return of tools. In addition, the return also has three statuses, namely, return, loss and damage. After the return of tools and instruments, they must be placed on the specified position.

(4) For the insulating tools and instruments that cannot be returned on the current day, please store them at the place where the conditions are satisfied.

4. Tools Test Management

The tools and instruments shall be regularly tested, which will greatly strengthen the safety and reliability of the tools and instruments. The test management link will be a most important link of this platform, while the live working tools and instruments test may be classified as follows: checking test, preventive test and type test. The tools may be further used after they pass through the test. Otherwise, the abandonment of tools must be applied. Please set the test type, test cycle, test date and test code as per the variable types of tools, establish the test ledger and remind the next test time of each tool and instrument. In addition, for the over-due tools and instruments, the alarm shall be given.

5. Comprehensive Query

(1) The fully intelligent integrated tool management platform has an advanced query system, which can query the inventory status, tool distribution, tool stock-in, tool collecting, tool loss reporting, electrical test and mechanical test respectively. All data are stored in SQL database and can be retrieved safely and reliably at high speed.

(2) The tools and instruments inventory system is different from the query system. The tool inventory system is mainly to let the warehouse keeper know clearly the existing tools in a warehouse. Inventory for each warehouse can be performed separately. The comprehensive query system for the warehouse of tools and instruments for live working is shown in Fig. 8-5.

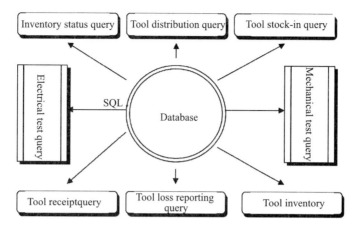

Fig. 8-5   Comprehensive Query System for the Warehouse of Tools and Instruments for Live Working

8.2.3　RFID System

Radio frequency identification (RFID) is a non-contact automatic identification technology which can automatically identify target objects and obtain relevant data through radio frequency signals. The identification can work in various harsh environments without manual intervention. At the same time, it can also identify fast moving objects and multiple tags, with quick and convenient operation. Fig. 8-6 shows an RFID system.

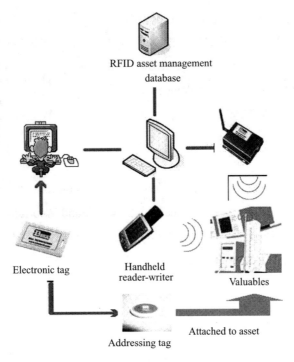

Fig. 8-3-4　RFID System

RFID system is a simple wireless system with only two basic devices, which are mainly used to control, detect and track objects. The system consists of a reader (or interrogator) and many tags (or transponders).

It is an automatic identification system consisting of radio frequency tag, reader and computer network. Generally, the reader emits energy in an area to form an electromagnetic field. When the radio frequency tag passes through this area, it detects the reader's signal and sends the stored data. The reader receives the signal sent by the radio frequency tag, decodes and verifies the accuracy of the data to achieve the purpose of identification.

Application of RFID in warehouse management of tools and instruments for live working: Tools stored in the warehouse of tools and instruments for live working can be identified by readers in each key link of the workflow after RFID tags are affixed on the tools.

8.2.4　Automatic Measurement and Control System for Warehouse Environment

1. Basic Composition of the System

The system fully adopts the latest industrial measurement and control technologies such as intelligent distribution control, intelligent neural sensing network, plug and play (PnP), etc. The hardware platform uses embedded industrial computer as the basic monitoring center of the system, the vertical LCD screen with touch as the man-machine dialogue platform, PnP humidity sensor, temperature sensor, electric quantity collector, over-limit alarm and various state quantity collectors as input devices, and the ad hoc network wireless signal exchange module as the system information exchange channel. The control output module and contactor are used as output devices, and the alarm output system is composed of dial-up language alarm, buzzer and indicator lamp,etc. With the support of Linux operating system (front control computer) and environment measurement and control software, Windows operating system (background computer), Office suite, background Web release, video monitoring, telephone alarm and other professional software, the basic functions of real-time measurement, data acquisition, data processing, equipment control, man-machine dialogue and the like are realized.

The communication between it and sensors and control equipment adopts Zigbee wireless networking technology with international free ISM band 2.4G frequency, and the reliable point-to-point communication distance is about 30m without occlusion. Due to the networking technology, each terminal has routing (cascade) function, so as long as the distance between any two terminals does not exceed 30m, all devices in the network can communicate.

Network communication is adopted between the background client system and the intelligent control box, which can remotely view the environmental parameters and equipment operation conditions in the warehouse, remotely configure the warehouse parameters and remotely control the warehouse equipment. Multiple intelligent control boxes in the same place are connected to only one background client system through routers or switches.

The telephone alarm module is generally installed on the background client system. When an abnormality occurs in a warehouse, the background client system activates the telephone alarm function. There can be a maximum of 6 preset telephone numbers.

2. System Structure

The structure of the warehouse environment automatic detection system is shown in Fig. 8-7.

3. Unique Functional Features of the System

The front-end processor uses an industrial embedded host computer (Linux platform) to realize system data acquisition and system control. It can perform the most basic functions of the system, such as protection, measurement, control, etc. It switches the appropriate heater, dehumidifier and ventilator according to the set control rules. The control rules should be flexible and diverse.

The background workstation uses Windows platform for complex data processing and man-machine dialogue, can be networked with internal LAN, and supports querying warehouse temperature and humidity, varieties, quantity, specifications, collecting and testing conditions of tools and instruments through LAN. The system has excellent anti-virus features, supports Watchdogtimer and system mapping technology, and can recover itself quickly after program running-out or system breakdown, thus preventing the system from crashing and ensuring the stability and reliability of the system.

The manual/automatic controller has manual and automatic switch, which can realize manual or automatic control. In case of smoke and high temperature alarm, it can automatically cut off the control power supply of heater, dehumidifier, air conditioner and other equipment, and force open the ventilation device.

8.2.5 Video Monitoring System

Video monitoring system is installed in the warehouse of tools and instruments for live working, so that the staff can monitor the situation in the warehouse without entering the warehouse and can also rotate the angle as required. The video monitoring system can record the warehouse in real time.

This system adopts infrared camera with pan-tilt head, which can rotate 360° in all directions. In the night vision state, the infrared camera used in this system can capture images in dark environment.

Digital Video Recorder (DVR) is a computer system for image storage and processing. It has the functions of recording, recording, remote monitoring and controlling images/voices for a long time. DVR integrates five functions, including video recorder, picture divider, pan-tilt lens control, alarm control and network transmission. This device can replace the functions of a large number of other devices in analog monitoring system, and has an advantage in price.

The storage of video data is divided into video with personnel entering and leaving the warehouse and video without personnel. The video data can be stored for 3 months, and overdue video can be automatically cleared.

Chapter 8
Storage and Transportation of Tools and Instruments for Live Working

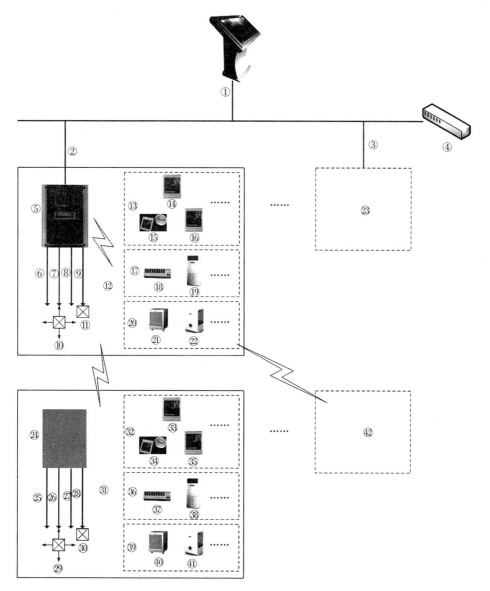

①Client monitoring terminal; ②③Network; ④Network switch; ⑤Main intelligent control box; ⑥Main power supply of power load; ⑦Main power supply of ventilation; ⑧Main power supply of control system; ⑨Main power supply of lighting; ⑩Venting; ⑪Lighting; ⑫Main warehouse 1; ⑬Sensing terminal; ⑭Temperature and humidity sensor; ⑮Smoke alarm; ⑯Temperature over-limit alarm; ⑰Temperature control terminal; ⑱Heater ⑲Air conditioner; ⑳Humidity control terminal; ㉑Dehumidifier; ㉒Humidifier; ㉓Main warehouse n; ㉔Sub-control box; ㉕Main power supply of power load; ㉖Main power supply of ventilation; ㉗Main power supply of control system; ㉘Main power supply of lighting; ㉙Venting; ㉚Lighting; ㉛Sub-warehouse 1; ㉜Sensing terminal; ㉝Temperature and humidity sensor; ㉞Smoke alarm; ㉟Temperature over-limit alarm; ㊱Temperature control terminal; ㊲Heater; ㊳Air conditioner; ㊴Humidity control terminal; ㊵Dehumidifier; ㊶Humidifier; ㊷Sub-warehouse n

Fig. 8-7 Structure of Warehouse Environment Automatic Detection System

8.2.6 Remote Monitoring System

The remote management system for the warehouse of tools and instruments for live working enables the warehouse keeper or company leader to monitor the warehouse operation in real time without entering the warehouse. The most popular control software worldwide, SymantecpcAnywhere, is taken as an example for introduction.

1. Symantecpcanywhere User Interface

The SymantecpcAnywhere user interface of the warehouse remote monitoring system is shown in Fig. 8-8.

2. Schematic Flowchart of SymantecpcAnywhere Operation

The schematic flowchart of pcAnywhere operation is shown in Fig. 8-9.

3. SymantecpcAnywhere Features at a Glance

(1) Connect to the remote device in a simple and secure manner.

(2) Easy cross-platform operation.

Fig. 8-8 Interface of Warehouse Remote Monitoring System

(3) Manage computers and quickly solve service desk problems.

(4) With simple and safe remote connection capability, it can easily operate the main console of the warehouse across regions and platforms, and temperature and humidity monitoring and tool stock-in management can be easily realized. With SymantecpcAnywhere, remote connections are simple and secure. The new gateway function enables users at the master terminal to find the required controlled terminal, which can be easily found even if the controlled terminal is located behind a firewall or router or does not have a fixed or public IP address. The basic window interface is friendly for new users and provides simple graphical

options for each task. Built-in FIPS140-2 verifies AES256-bit encryption to ensure the security of phased work. SymantecpcAnywhere can now support the connection and management of computers running MicrosoftWindows, Linux, MacOSXUniversal or MicrosoftPocketPC.

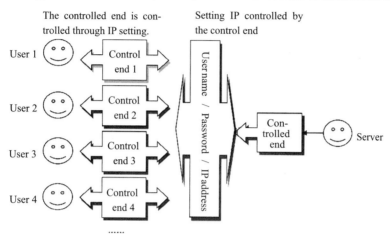

Fig. 8-9　Schematic Flowchart of Symantecpc Anywhere Operation

4. Main Features of Symantecpc Anywhere

(1) Simplified GUI and improved connectivity options.

(2) Providing powerful functions of protecting remote control, file transfer, command queue, remote management, rapid deployment and connection.

(3) Providing a safe, efficient and flexible remote management method, which can eliminate faults and repair them quickly, correct user errors and reduce the time required to solve problems.

(4) Providing the ability to manage local and remote devices, increasing the control and scope of the entire IT infrastructure.

(5) The "Connection Wizard" software in the system will guide the new user to make the initial connection between the client and the controlled terminal.

(6) Powerful file transfer function allows users to upload and download files across platforms.

(7) Mandatory password protection and login encryption can ensure that only authorized users can access the Symantec pcAnywhere controlled end.

(8) Verify encryption with FIPS140-2 up to AES256 bits.

(9) Automatic bandwidth detection automatically optimizes pcAnywhere performance for each connection type.

8.3 Storage Method of Tools and Instruments for Live Working

Tools and instruments for live working must be kept in a qualified special warehouse of tools and instruments for live working for long-term storage. Tools and instruments shall be wiped clean before being stored in the warehouse after each use. Tools and instruments for live working shall be stored in different partitions according to voltage grade and tool category, mainly classified as: metal tools and instruments, hard insulating tools, soft insulating tools, shielding protection tools, insulation shielding tools, insulation protection tools, testing tools, etc. Each tool and instrument must have a unique number, indicating the relevant mechanical and electrical performance parameters. Non-qualified tools and instruments are strictly prohibited from being stored in the special warehouse of tools and instruments for live working.

8.3.1 Metal Tools and Instruments

For the storage facilities for metal tools and instruments, the load-bearing requirements shall be considered, and the storage shall be easy to access. Multi-layer storage racks may be used. At the same time, the frequency of use shall be taken into account; clamps in metal tools shall be classified and placed with voltage class as the classification criteria. Small items such as screws are stored along with large items.

8.3.2 Hard Insulating Tools

Hard ladders, flat ladders, hanging ladders, lifting ladders, insulator cradles and the like in hard insulating tools can be stored on horizontal storage racks with intervals of over 30cm between each layer, and the height of the lowest layer to the ground shall not be less than 50cm. At the same time, bearing requirements shall be considered, and access shall be easy. The storage of insulating bars, hanging and pulling supporting rods and the like can be arranged by vertical hanging, and each rod is 10~15cm apart and each row is 30~50cm apart. When the rods are long and inconvenient for vertical hanging, the horizontal storage rack can be used for storage. When stored in layers, the rods shall be classified according to voltage class. Large-tonnage insulated lifting rods can be stored in horizontal storage racks.

8.3.3 Soft Insulating Tools

The work of vertical hanging frame can be adopted for the storage of insulating rope and rope ladder. The distance between insulating rope hooks is 20 - 25cm, and the distance between the lower end of the rope and the ground shall be no less than 30cm. Soft tools shall be coiled and stored neatly by models and specifications. The insulating rope ladder can also be stored on horizontal storage rack after being coiled.

8.3.4 Tackle

The tackle and tackle block can be stored on vertical hanging frame or on horizontal storage rack. The tackles shall be positioned and stored in groups according to their size, bearing weight and category.

8.3.5 Testing Instrument

Electroscope, phase detector, distribution voltage tester, insulator detector, psychrometer, anemometer, insulation resistance meter and other detection tools shall be placed intact to prevent collision, and can be stored with multi-layer horizontal stainless steel rack. The dry batteries in the instruments and apparatus shall be taken out and stored when not in use; the lithium batteries in the instruments and apparatus shall be charged once a month to ensure enough power; and the aged batteries shall be replaced in time.

8.3.6 Insulated Shielding Appliance

Insulating shielding appliances, such as insulating blanket, conductor shield cover, insulator shield cover, cross arm shield cover, electric pole shield cover, etc., shall be stored in bags or boxes with sufficient strength, and then placed on a multi-layer horizontal frame. It is prohibited to store them in the vicinity of steam pipes, heat dissipation pipes and other artificial heat sources, and under direct sunlight.

8.3.7 Insulation Protective Equipment

Insulating protective equipment, such as insulating clothes, insulating sleeves, insulating shawls, insulating gloves, insulating boots, etc., shall be packed by pieces. Attention shall be paid to prevent direct sunlight or being stored near artificial heat sources. It is strictly forbidden to fold them during storage. In particular, it is necessary to avoid direct contact with sharp objects, which may result in punctures or scratches.

8.3.8 Shield Appliance

Shielding appliances, such as shielding clothes, conductive gloves, conductive socks, conductive shoes, shielding masks, etc., shall be packed by pieces, stored in complete sets in bags or boxes with sufficient strength, and then placed on multi-layer horizontal frame.

The shielding clothes shall be put into a special box after using. It is strictly prohibited to knead and press. Especially, attention should be paid to prevent the round constantan alloy wire on the insulating clothes from being broken off, which may result in electric discharge during use, causing a tingling feeling to the live working personnel wearing the clothes. After use in summer, it should be soaked in 50 ~ 100 times (weight ratio of shielding clothes) of 50 ~ 60 °C hot water for 15 minutes to dissolve the sweat soaked in clothes.

8.3.9 Aerial Device with Insulating Booms

The aerial devices with insulating booms must be parked in a special garage with automatic dehumidification and drying equipment. The drying equipment must be installed at the height of the insulating boom, and the vehicle body does not need to be heated. It shall be supported by legs during long time parking. The hydraulic part of the aerial devices with insulating booms must be overhauled and renewed with oil once every two years, and the maintenance of the traveling part shall be carried out according to the requirements of the manufacturer. The storage of tools and instruments for live working is shown in Fig. 8-10.

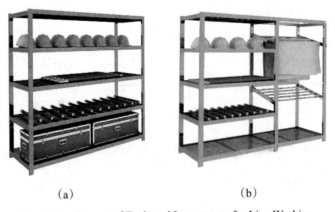

(a)   (b)

Fig. 8-10 Storage of Tools and Instruments for Live Working

Chapter 8
Storage and Transportation of Tools and Instruments for Live Working

8.4 Storage and Transportation of Tools and Instruments for Live Working

The live working of transmission line is relatively remote, the transportation duration of tools and instruments is long, and most of works on the distribution lines are in urban areas, but there are many operation points of the same type, and the time of outdoor use is also long. In order to ensure the safe use of tools and instruments for live working, strict requirements are proposed for the transportation and storage of tools and instruments for live working.

8.4.1 Storage of Tools and Instruments for Live Working

1. Tools for live working shall be stored in a well ventilated, clean and dry special tool warehouse. The relative humidity in the warehouse shall not be more than 60%. The temperature of the storage area for hard insulating tools, soft insulating tools, detection tools and shielding tools shall be controlled at 5 ~ 40 °C.

2. The warehouse for live working tools shall be equipped with thermometer, hygrometer, dehumidifier, heater with uniform radiation, and enough tool racks, hangers and fire extinguishers.

3. Tools and instruments for live working shall be kept by specially assigned person, and tools for live working shall be numbered and registered uniformly, and test, maintenance and use records shall be established. All kinds of tools and instruments shall have complete factory instructions, test cards or test reports.

4. Upon expiration of the service life of tools and instruments for live working, they shall be immediately scrapped according to the ledger. According to the usage situation, it is estimated that the purchasing plan for tools and instruments for live working will be made on an annual basis.

5. Defective live working tools shall be repaired in time, unqualified tools shall be scrapped, and shall not be used again. Unqualified tools and instruments shall not be placed in the warehouse for live working.

6. Recharge the charging equipment regularly to avoid battery damage due to power loss.

7. The warehouse for live working and live working tool car shall be sorted out every

month. Drying or visual inspection and maintenance shall be conducted for the tools and instruments, and any problems found shall be timely reported to the special responsible person.

8. Periodical preventive test and querying test shall be performed on tools and instruments for live working.

9. The overhead aerial devices with insulating booms shall be stored in a dry and ventilated garage, and the insulating part shall be provided with moisture-proof measures.

10. In humid areas or wet seasons, when live working tools are to be used outdoor for more than 24 h, special live working toolshed car must be equipped with drying and dehumidification equipment, temperature and humidity auto-control system, and shall be implemented according to the standards for live working warehouse and live working tool car. The live working toolshed car must be dedicated, the vehicle-mounted generator and the temperature and humidity control system must be in good condition and ready for the drying and dehumidification work anytime and anywhere. The interface of the temperature and humidity control system of the live working toolshed car is shown in Fig. 8-11.

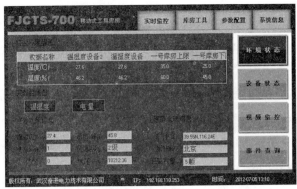

Fig. 8-11　Interface of Temperature and Humidity Control System for Live Working Toolshed Car

8.4.2　Precautions for Tools and Instruments for Live Working During Transportation

1. Tools and instruments for live working must be packed with special clean canvas bags before they are taken out of the warehouse for loading. During transportation, live working tools should be packed in special tool bags, tool boxes or special tool car to prevent dampness and damage. When conditions permit, live working mobile warehouse car can be used for transportation of tools.

2. Tools and instruments for live working shall be placed by different categories in the

Chapter 8
Storage and Transportation of Tools and Instruments for Live Working

transportation process, and necessary fixing measures shall be taken to avoid any collision and wear caused by vehicle bumping. Aluminum alloy tools, clamps and fixtures with low surface hardness and metal machines and tools that should not be collided shall be provided with special wooden and leather tool boxes during transportation. Each box shall be limited to one set of tools, and loose parts shall be fixed in the box. Metal tools shall be placed in the lower layer, insulating tools shall be placed in the upper layer and fixated, and shall not be mixed during shipment. The layout of insulating tools is shown in Fig. 8-12.

Fig. 8-12   Placement of Insulating Tools

3. Insulating hard ladder, insulating bar, insulated insulator cradle and other items shall be placed horizontally in the car, and the tool bag shall be intact. Corrosive, paint and sundries shall not be stored in the compartment.

4. When it is taken out for continuous work, drying equipment shall also be equipped. After returning to the camp every day, the insulating tools and instruments shall be dried for a period of time for use the next day.

References

[1] Hu Yi. Test Methods for Live Working Tools and Safety Tools [M]. Beijing: China Electric Power Press, 2003.

[2] Hua Jie. Special Teaching Material for Qualification Certification and Training Of Live Working Personnel. Transmission line [M]. Beijing: China Electric Power Press, 2011.

[3] Human Resources Department of State Grid Corporation of China. General Textbook for Professional Ability Training on Productive Skills Personnel · Basic Knowledge of Live Working [M]. Beijing: China Electric Power Press, 2010.

[4] uman Resources Department of State Grid Corporation of China. Special Textbook for Professional Ability Training on Production Skills Personnel · Live Working on Power Distribution Line [M]. Beijing: China Electric Power Press, 2010.

[5] Human Resources Department of State Grid Corporation of China. SGCC Special Textbook for Professional Ability Training on Production Skills Personnel - Live Working on Transmission Line [M]. Beijing: China Electric Power Press, 2010.

[6] Operation and Maintenance Department of State Grid Corporation of China. Textbook for Training of Non-interruption Working on 10kV Cable Line [M]. Beijing: China Electric Power Press, 2013.